Electro-Catalysis at Chemically Modified Solid Surfaces

CATALYTIC SCIENCE SERIES

ISSN 1793-1398 (Print)
ISSN 2399-4495 (Online)

Series Editor: Graham J. Hutchings *(Cardiff University)*

Catalysis is at the forefront of the chemical industry and is essential to many fields in the chemical sciences. This series explores all aspects of catalysis in authored and edited volumes drawing on expertise from around the globe in a focussed manner. Volumes are accessible by postgraduate students and professionals in academia and industry.

Published

Vol. 16 *Electro-Catalysis at Chemically Modified Solid Surfaces*
by Jacques Simonet

Vol. 15 *Noble Metal Noble Value: Ru-, Rh-, Pd-catalyzed Heterocycle Synthesis*
edited by Xiao-Feng Wu

Vol. 14 *Enantioselective Titanium-Catalysed Transformations*
by Hélène Pellissier

Vol. 13 *Gold Catalysis: An Homogeneous Approach*
edited by F. Dean Toste and Véronique Michelet

Vol. 12 *Catalysis by Ceria and Related Materials (Second Edition)*
edited by A. Trovarelli and P. Fornasiero

Vol. 11 *Supported Metals in Catalysis (Second Edition)*
by J. A. Anderson

Vol. 10 *Concepts in Syngas Manufacture*
by J. Rostrup-Nielsen and L. J. Christiansen

Vol. 9 *Deactivation and Regeneration of Zeolite Catalysts*
edited by M. Guisnet and F. R. Ribeiro

Vol. 8 *Petrochemical Economics: Technology Selection in a Carbon Constrained World*
by D. Seddon

Vol. 7 *Combinatorial Development of Solid Catalytic Materials: Design of High-Throughput Experiments, Data Analysis, Data Mining*
edited by M. Baerns and M. Holeňa

More information on this series can be found at http://www.worldscientific.com/series/css

(Continued at end of book)

Series Editor: Graham J. Hutchings

Electro-Catalysis at Chemically Modified Solid Surfaces

Jacques Simonet
Université de Rennes 1, France

World Scientific

NEW JERSEY • LONDON • SINGAPORE • BEIJING • SHANGHAI • HONG KONG • TAIPEI • CHENNAI • TOKYO

Published by

World Scientific Publishing Europe Ltd.
57 Shelton Street, Covent Garden, London WC2H 9HE
Head office: 5 Toh Tuck Link, Singapore 596224
USA office: 27 Warren Street, Suite 401-402, Hackensack, NJ 07601

Library of Congress Cataloging-in-Publication Data
Names: Simonet, Jacques.
Title: Electro-catalysis at chemically modified solid surfaces / by Jacques Simonet (Université de Rennes 1, France).
Other titles: Electrocatalysis at chemically modified solid surfaces
Description: New Jersey : World Scientific, 2017. | Series: Catalytic science series ; volume 16 | Includes bibliographical references.
Identifiers: LCCN 2016049882 | ISBN 9781786342430 (hc : alk. paper)
Subjects: LCSH: Electrodes--Surfaces. | Surface chemistry. | Electrocatalysis. | Electrochemistry.
Classification: LCC QD571 .S4385 2017 | DDC 541/.3724--dc23
LC record available at https://lccn.loc.gov/2016049882

British Library Cataloguing-in-Publication Data
A catalogue record for this book is available from the British Library.

Desk Editors: Chandrima Maitra/Mary Simpson/Shi Ying Koe

Typeset by Stallion Press
Email: enquiries@stallionpress.com

Printed in Singapore

Preface

Professor Jacques Simonet has made, throughout his great career, very significant and original contributions in many fields. What impresses me is the wealth of his contributions. He designed and built new electrode materials for electro-catalysis and applied them to electro-catalytic reactions. He developed new approaches in electro-organic synthesis using redox catalysis and in generating organic free radicals at specifically designed electrodes. He also developed novel conducting polymers.

The book focuses on recent research, mostly published in the course of the last 10–12 years, in the field of catalytic materials, their use in chemical transformations, and in the development of new reactions. The material presented consists essentially of the author's contributions. I mention here the ones I consider as the most original and important. The synthesis of new alloys of palladium (Ag–Pd and Cu–Pd) under very simple conditions does bring a new light in analysis and modification of electroactive molecules. In the field of carbonaceous material electro-chemistry (glassy carbon (GC), graphite, graphene, etc.), new composite electrodes such as gold–graphene and silver–graphene, which disclose major catalytic properties, are described. Novel materials have been created by an efficient modification of those types of carbon under judicious cathodic or anodic doping. Immobilizing pyridines and bipyridines on metals deposited on GC provides electrodes, very useful in chemical catalysis, such as gold–pyridines and GC–gold–bipyridines. Precious metals (Pt, Pd), which until now were considered as totally inert toward chemicals and electron transfer, are reported to give (via the reversible insertion in the metal bulk of a large variety of electrolytes) new solid phases (named iono-metallic). The latter are highly malleable and react with appropriate chemicals, via a simple exchange of inserts, to create new 2D and 3D catalytic materials. Many transition metals (Au, Fe, Ti, Cu, etc.) reversibly absorb, under cathodic polarization, large amounts of small size electro-active molecules like CO_2 and O_2. Other small molecules such as CO, SO_2, and nitrogen oxides might be absorbed also. Therefore, a useful tool for depollution! The anodic synthesis of ionophoric redox polymers deposited at solid surfaces is another interesting contribution. The book reveals many domains to explore.

In this book, Professor Simonet takes great care to provide useful keys, and simple recipes and *modus operandi,* which would allow students and scholars to easily conduct their own exploration. This comprehensive publication of his recent research provides graduate and post-graduate students, and researchers in chemistry, electrochemistry, and catalysis with an in-depth insight into key developments in the field of catalytic materials and their applications.

Jean Lessard
Emeritus Professor
Université de Sherbrooke
Québec, October 2016

About the Author

Jacques Simonet received his Ingenieur Chimiste degree from the University of Clermont-Ferrand (France) in 1962 and immediately took a position of Chercheur Scientifique at the Centre National de la Recherche Scientifique (CNRS). He received the Doctorat d'Etat from the same University in 1967, and in 1972, he spent one year as a NATO fellow at the University of Aarhus (Denmark) in the research group of Professor Henning Lund. In 1980, he became Director of Recherche and simultaneously took over the direction of the Electro-chemistry groups at both the University of Rennes (France) and the Catholic University in Angers (France). Dr. Simonet has been guest of the Monsanto Company (St. Louis, 1976) and of the Electro-synthesis Company (Buffalo, 1993) and has been a visiting faculty member at the University of Aarhus (1974 and 1975) and at the Indiana University, Bloomington (1995, 1998, 1999, 2000, and 2003).

Organic electron-transfer reactions have long been the focus of Dr. Simonet's work. In the course of his investigations, he has found new electron-labile protecting groups. In particular, he has made important contributions to the electro-chemistry of organic compounds containing sulfur (sulfones, sulfoxides, and sulfonium derivatives) and nitrogen (amines and sulfonamides). In collaboration with Professor Henning Lund, he discovered redox catalysis in organic chemistry, as well as perturbed redox catalysis in which radical coupling takes place. He was the first to anodically polymerize poly-ethers and crown ethers to generate new materials capable of extracting very low concentrations of heavy-metals and/or radioactive cations from aqueous solutions. His research concerning the insertion and /or intercalation of organic or inorganic cations into graphite and noble metals (platinum and palladium) had led to the development of new kinds of modified solid surfaces.

Dr. Simonet is the author of more than 400 research articles and has presented numerous papers and plenary lectures at national and international symposia devoted to electro-chemistry. He has written 20 review articles and book chapters, and holds 12 technical patents. He was appointed Fellow of the Electro-chemical Society (2005), won the Prix de Chimie Organique of the Société Chimique de France (1973), the Manuel Baizer Award of ECS (2000) and distinguished as Fellow in the Institute of Advanced Study in Indiana University (1995).

At the present time, Dr. Simonet serves as Director of Research Emeritus at the University de Rennes in the Electro-chemistry group.

Acknowledgments

The present monograph covers a part of my scientific activities mainly from the last 10–15 years, and is devoted to aspects of electrochemistry on fields quite different to those I was interested in (namely Organic Electro-chemistry) in the 40 years before this. On becoming an Emeritus Researcher at the Université of Rennes, I had to adapt my subjects to those of my host team more oriented to material electro-chemistry generation, surface modifications, grafting of organic radicals (MaCSE research team at Institut des Sciences Chimiques de Rennes, UMR CNRS 6226).

In France, the position of Emeritus Researcher (at least in CNRS[1]) theoretically has many benefits, however there is little opportunity to guide young scholars to learn about the principles of research, meaning he must build this up again by having direct contact with the bench. For me, another form of contact with real organic chemistry (that smells and burns the ends of fingers) was a new revelation and boosted me in the magic world of personal discovery with its joys and mistakes.

These 15 years of discovery allowed several discoveries: materials, composites, new alloys, nano-structured electrodes, catalysis,

[1] National Center for Scientific Research, or CNRS in French, is a public organization under the responsibility of the French Ministry of Education and Research.

the insertion of electrolytes within precious metals which permit the reactivity of electro-active gases and small active molecules as well as π-acceptors. These are all more or less fully presented within 19 chapters of the book. During this quite active period, I have to stress that I greatly benefited from the strategic support of two prominent scientists in the field of both solid state and organic electro-chemistry:

1. ***Professor Philippe Poizot***, now at *Université de Nantes, Institut des Matériaux Jean Rouxel (IMN-UMR CNRS 6502), who permitted the characterization of several modified metal-based electrodes including peculiar graphene/gold or silver composites.*
2. ***Professor Viatcheslav Jouikov***, *Université de Rennes 1, Institut des Sciences Chimiques de Rennes, for the synthesis in strategic organic compounds for elaboration of new modified solid phases in particular in the domain of graphene modification, and that of silicon organic compounds.*

I also remain *particularly* grateful to Professor Henning Lund (Aarhus) who invited me in 1972 for a long post-doctoral stay and who became a friend that I always admired for his advice and his kindness. I would also like to list kind colleagues of whom I still have the pleasure to meet in different places over the world: Professor Jean-Michel Savéant (Paris), Professor Albert Fry (Middleton), Professor Jean Lessard (Sherbrooke), Professor Hans Schäfer (Münster), Professor Jim Utley (London), Professor Dennis Peters (Bloomington), and many others: I apologize for not mentioning everybody.

Finally, I would like to acknowledge Professor André Tallec (now retired) for our 20-year collaboration in Rennes and fine discussions about chemistry. I show gratitude toward Dr. Marc Fourmigué, Director of the research group MACSE (centered on conducting materials and surface electro-chemistry), who provided me with a personal laboratory that enabled some of the results displayed in the present volume. I will not end this list of greatly esteemed colleagues without mentioning Professor Richard Compton (Oxford) for his strategic support, which has been fully appreciated in the course of the last 10 years.

Jacques Simonet
Rennes, August 2016
jacques.simonet@univ-rennes1.fr
jacques.simonet@dbmail.com

Contents

Preface v

About the Author vii

Acknowledgments ix

Chapter 1 How to Modify Solid Electrodes for Generating New Interfacial Processes — Another Approach in Electro-Chemical Catalysis 1

1.1. Introduction 1

1.2. About Redox Catalyses 4

1.2.1. Homogeneous redox catalysis 4

1.2.2. Heterogeneous redox catalysis 6

1.2.3. Arenization of glassy carbon 6

1.3. Building and Reduction of Electrophilic Layers onto Carbons 8

1.4. Intrinsically Catalytic Electrodes 10

1.5. Triggering of Radical Chains via Electro-Chemical Activation 11

1.6. Modified Carbons as Electrode Materials 14

1.6.1. Some unusual reactions with carbons 14

1.6.2. Composite carbonaceous materials as multi-layer reactive systems 16

12.3. Results 233
12.3.1. Reduction of alkyl iodides and bromides 233
12.3.2. Functionalization of GC 236
12.3.3. Vinylation of carbon surfaces 238
12.3.4. Immobilization of alkanoid acid chains at carbon 240
12.3.5. An efficient alkyl-ferrocenyl capping of polarized GC 240
12.3.6. About the immobilization of tetrathiafulvalenes 242
References 245

Chapter 13 Grafting at Solid Surfaces via Trimethylsilyl Derivatives 247
13.1. An Overview 247
13.2. Oxidation of Allyltrimethylsilane (ATMS) at Carbon Anodes 251
13.3. Benzyltrimethylsilane (BTMS): Oxidation at Carbon Anodes 255
13.4. Oxidation of Phenyltrimethylsilane at GC Electrodes 256
13.5. TMS as a Vector for the Immobilization of RX at Solid Surfaces 256
13.5.1. Grafting of alkyl halides at carbons 256
13.5.2. Scission of organo-trimethylsilanes at gold and platinum 260
13.5.3. The versatile use of TMS–CH_2–I under cathodic or anodic electron transfer 262
13.6. Conclusion 264
References 265

Chapter 14 Cathodic Corrosion Induced by Free Alkyl Radicals at Silvered Carbon Substrates 267
14.1. Introduction 267
14.2. Generation of Free Alkyl Radicals at GC Surfaces 269
14.2.1. Methodology 269
14.2.2. Evidence for an alkyl radical reactivity toward GC surfaces 272
14.2.3. Preliminary step: A radical grafting at GC? 280
14.3. Polarized Graphite as a Trapping Material for Alkyl Radicals 281
References 284

Chapter 15 Silver–Graphene and Gold–Graphene Electrodes: Two New Catalytic Materials 287
15.1. Introduction 287
15.2. Ag–Graphene Surfaces 288
15.2.1. Electro-chemical properties of Ag–graphene layers 289
15.2.2. Modification of the graphene layer by addition of *in situ* generated free radicals 293
15.3. Electro-Chemical Exfoliation of Natural Graphite at a Gold Substrate 296
15.3.1. How to prepare Au–graphene working electrodes 296
15.3.2. Au–graphene interface: Formation and structure 297
15.3.3. Au–graphene interface as a spin-trap 302
References 303

Chapter 16 Anodic Coupling of Aromatic Ethers: The Synthesis of Ionophoric Redox Materials Deposited at Solid Surfaces 307
16.1. Introduction 307
16.2. Anodic Trimerization of Orthodimethoxy Benzenes 309
16.2.1. Oxidation of orthodialkoxybenzenzes 309
16.2.2. A first approach toward the anodic synthesis of ionophores 312
16.3. The Concept of Ionophoric Polymers Based on Anodic Oxidation of Aromatic Ethers 313
16.3.1. Voltammetries and electro-chemical conversion of dibenzocrown ethers into ionophoric polymeric materials 315
16.3.2. About complexing properties of poly-dibenzo crown ethers 319
16.4. Generation of Poly-Electrides 321
References 323

Chapter 17 Attachment of Pyridines at Electrode Surfaces 327
17.1. Introduction 327
17.2. Experimental Conditions 329
17.3. First Evidence for the Immobilization of 4-(7-iodoheptyl)pyridine 330
17.4. Reactivity of Bromo- and Chloroalkyl Pyridines as Solid Surfaces 334
17.5. Introduction to a General Mode of Bipyridine Graftings 336
17.6. About Complexation of Transition Metal Cations by Bipyridine Layers 338
17.7. Conclusion 339
References 339

Chapter 18 The Insertion of Electrolytes into Platinum and Palladium under Cathodic Super-Dry Conditions 341
18.1. Introduction 341
18.2. Experimental Conditions Required for a Cathodic Charge of Platinum and Palladium 345
18.3. Platinum Charge in Super-Dry Conditions 346
18.3.1. Behavior in the contact of alkali metal salts 346
18.3.2. Charge process in the presence of TAAX salts 352
18.3.3. Evidencing smooth platinum surface changes by SEM 353
18.3.4. Following of the platinum charge–discharge in real time: Use of *in situ* EC–AFM 356
18.3.5. Modified platinum phases as reducing reagents 361
18.4. Cathodic Charge of Palladium in Super-Dry Media 364
18.5. Concluding Perspectives 367
References 369

Chapter 19 The Electro-Chemical Insertion of π-Acceptors in Platinum and Palladium: Reactive Organo-Metallic Phases 371
19.1. Introductory Concept 371
19.2. Reversible Electrical Charge of Palladium in the Presence of Different Salts in Super-Dry Electrolytes 374
19.3. Charge–Discharge of Palladium–Graphene Layers 380
19.4. Super-Oxide as a Donor toward Charged Palladium and Platinum 383

19.5. Additional Comments and Concluding Remarks 384
References 385

Chapter 20 Cathodic Carboxylation of Metals in Thick $\{M{-}CO_2\}_n$ Layers: Electro-Chemical Sequestration of CO_2 387
20.1. Introduction 387
20.2. Reactivity of Carbon Dioxide at a Negatively Polarized Gold Electrode 389
20.2.1. Comments on electro-chemically driven insertion/desorption of CO_2 393
20.2.2. Additional remarks on the reversible sequestration of CO_2 by gold 394
20.3. Carboxylation of Silver 395
20.4. Carboxylation of Palladium, Platinum and Rhodium 399
20.4.1. Palladium 399
20.4.2. Platinum 399
20.4.3. Rhodium 401
20.5. Carboxylation of Copper 404
20.6. Conclusive Remarks About the Reductive Insertion of Small Electro-active Molecules into Transition Metals 405
References 406

Index 409

Chapter 1

How to Modify Solid Electrodes for Generating New Interfacial Processes — Another Approach in Electro-Chemical Catalysis

1.1. Introduction

The word "catalytic" comes from Ancient Greek and can be translated as "a breaking down". The word "catalysis" was introduced in the first half of the 19th century by the Swedish Scientist J. J. Berzelius who applied it to the observation that in chemical reactions, the normal barriers can be removed by the help of external reagents. Nowadays, the term "catalysis" remains attached to the petroleum refining science introduced more than half a century ago. It mainly refers to the removal of sulfur and nitrogen from naphthas in order to produce fuel (gasoline, diesels, etc.). Refining reactions are conducted in the presence of catalysts often constituting of metals and metal oxides (mixed metal oxides, MMOs) in order to decrease cracking temperatures and improve selectivity of chemical reactions [1, 2].

In electro-chemistry, the term *catalysis* is quite recent and remains mainly linked to an easiness in interfacial reactions. Therefore, for more than 30 years, many reactions often considered as "catalytic" were published; it is not the aim of this book, especially within this introduction, to make a complete list of all the diverse reactions considered as catalytic. The purpose of the present anthology derived from research activities of the electro-chemistry group of Rennes mainly obtained in the course of the last 10–15 years is to list a rather homogeneous series of new processes devoted to original solid surfaces that permit acceleration of electrode processes, tricky surface reactions, easiness of electron transfers (homogeneous or heterogeneous), capping or grafting of different carbonaceous surfaces and metals. The creation of new alloys of palladium, generation of new ionophoric for capping solid interfaces, the highly unexpected insertion of salts in platinum, palladium, rhodium, as well as the "stocking" of small electro-active particles (such as CO_2, O_2 or SO_2) within a large series of transition metals under cathodic specific conditions are presented. An electro-chemistry of small metallic particles (Ag, Cu, Pd, etc.) and the new access to nanoflakes of graphene permits one to present a novel approach of uncommon interfacial reactions easily nicknamed as catalytic. New processes that enable the modification (insertion) of platinum and palladium by reduced forms of π-acceptors or charged graphene flakes are described.

It may then appear that most of the 19 chapters that follow this introduction possess an undeniable character of novelty both in the domain of new catalytic reactions, the chemical and electro-chemical modifications of conducting solid surfaces, especially in the field of carbons, precious metals and copper. It is obvious however that some parts of the book could present a superficial or unbelievable aspect since both the ideas and the significant results do not go sufficiently into enough details about the scientific aspect and the possible applications. It could actually be the responsibility of the reader to freely reproduce some results and use some ideas (good or judged as incomplete) presented in this research.

Among the well-established *electro-catalytic* processes, redox catalysis remains an intriguing and powerful method that specifically belongs to organic electro-chemistry. This technique corresponds to the innovative and important progress in the field of indirect electrolyses. The first evidence of the indirect reduction of an alkyl-halide RX by the anion radical of naphthalene was reported by Fry [3] while the wide scope of redox catalysis (kinetics and synthetic applications) was unambiguously disclosed by Lund and co-workers [4–6, 8, 16] in the course of 1970s by using polarographic currents of Koutecky's theory. Meanwhile, other authors were also interested in this form of catalysis [7, 9, 10]. Based on this indirect process, kinetic and thermodynamic results were successfully achieved by Savéant [11–15] by using the voltammetric technique. Thus, for the first time, thermodynamic data and standard potential values concerning reduction of $R^{\bullet}$ radicals (issued from the cleavage of alkyl and aryl halides, (RX's and ArX's)) were appreciated in this way. Thus, RIs (alkyl iodides) in general and certain RBrs (alkyl bromides) may lead to onc-electron scissions at cathodes. Therefore, selective one-electron processes and subsequently produced free radicals were considered to be of high interest also for synthesis and they prompted many authors in the course of last three decades to revisit redox catalysis in homogeneous and heterogeneous phases: one method corresponds to the electro-chemical alkylation of arenes A (used as a mediator) via the radical coupling of $A^{\bullet -}$ with electro-generated radicals $R^{\bullet}$ in solution or at the interface.

Specifically in the domain of electro-catalysis (that uses the preeminent role of the interface), different metallic electrodes chemically modified by using specific palladium alloys were employed and permitted generating not only specific reactions depending on the nature of the electrode, but also disclosing an easy way for producing free alkyl radicals [17, 18] at potentials > −1.5 V vs. SCE. Homo-coupling reactions in high yield at surfaces strongly activated by chemical or electro-chemical grafting were developed. The use of transition metals such as Ag [20–22], Pd [23], Cu [24] as well as Pt [25] were promoted. The aim of this extended survey, although not complete, is to underline some possibilities both in solution, at the interface, and inside the conductor bulk (carbons, metals) that may not always be chemically modified and may contribute to accelerate the electron transfer rate.

Most electro-chemical reactions take place at the electrode interface by means of a heterogeneous primary electron transfer. Other electron transfer reactions in solution, such as disproportionation are not excluded. However, the heterogeneous electron exchange process is often complicated and consequently slowed down by the intrinsic nature of the interface. Therefore, the scope of electro-chemistry can be widened by the use of organic or inorganic mediators (viewed as electron carriers between the conducting interface and the solution) for acting as reducing or oxidizing reagents. Ideal mediators may then be regarded as an ionic (such as the electrolyte itself) or molecular electrode of a well-defined potential where inhibiting processes are effectively excluded. However, other kinds of electro-generated species possessing not only redox but alternatively nucleophilic or electrophilic properties may react with organic substrates according to processes where single-electron transfer should not play an essential role.

Electro-generated intermediates in general react in liquid electrolytes (solvents or fuses). More sophisticated indirect electrochemical and catalytic processes have been recently developed principally by using electrodes specifically coated with conducting organic polymers. Additionally, the electron itself (possibly transferred by indirect means for higher selectivity) may contribute to initiate organic reactions according to so-called electro-catalytic processes. Lastly, let us mention that homogeneous redox reactions of electro-generated nucleophiles and electrophiles in solution toward solvent or electrolyte are not taken into consideration.

1.2. About Redox Catalyses

1.2.1. *Homogeneous redox catalysis*

First of all, voltammetries of a mixture of an arene and an alkyl halide {A + RX} was achieved both at supposed inert electrodes (like glassy carbon). Results are summarized by voltammetries exhibited in Figure 1.1. They were obtained in aprotic conditions (aprotic polar solvent, DMF) at a glassy carbon electrode. The example concerns pyrene that exhibits two reduction peaks (one-electron each,

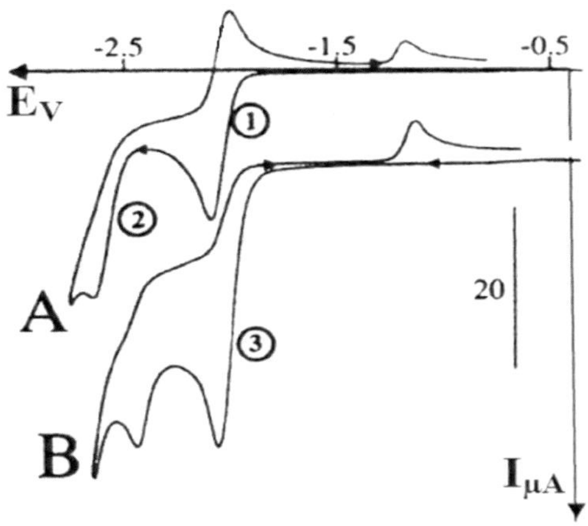

Figure 1.1. Pyrene (concentration 13 mmol L^{-1}) in a solution of $TBABF_4$ in dimethylformamide (DMF). Voltammetries at a glassy carbon electrode (area: 0.8 mm^2). Scan rate: 50 mV s^{-1}. Reference electrode: SCE (saturated calomel electrode). (A) Response of pyrene alone. (B) Response of the same system after addition of 1-bromopropane (22 mmol L^{-1}) [4].

steps (1) and (2), curve A) and the first one (formation of the anion radical) is fully reversible. In the presence of an alkyl-bromide RBr (namely 1-bromopropane), the redox catalysis process may take place and leads to the current increase of this first peak (becoming step (3), curve B) and the appearance of its total irreversibility for large concentrations of RBr. Scheme 1.1 aims to summarize the main reactions found for mixtures {A + RBr}. The homogeneous reduction of RBr via the reduction by the arene A anion radical then corresponds to Eqs. (1) and (2) of Scheme 1.1. Subsequent reactions such as the reduction of the alkyl-radical $R^{\bullet}$ were shown to occur owing to the potential range generally observed with most of arenes: at $E < -1.5$ V vs. SCE. In cases where the free alkyl radical is not reduced fast, the coupling reaction according to Eq. (4) may occur. Lastly, the organic anion issued from the coupling may afford an S_N2 reaction with RX possessing some electrophilic affinity (Eq. (5)). In other cases, the protonation reaction may take place (for example, a mono-alkylation according to (Eq. (6)). Thus, Eq. (4) might be considered as a side reaction and strongly

Redox catalysis

$$A + e^- \rightleftarrows A^{-\bullet} \quad (1)$$

$$A^{-\bullet} + RX \underset{k_b}{\overset{k_f}{\rightleftarrows}} A + R^\bullet + X^- \quad (2)$$

$$R^\bullet + A^{-\bullet} \text{ (or } e^-) \longrightarrow A + R^- \quad (3)$$

$$A^{-\bullet} + R^\bullet \longrightarrow AR^- \quad (4)$$

$$AR^- + RX \longrightarrow AR_2 + X^- \quad (5)$$

$$AR^- + \text{proton source} \longrightarrow ARH \quad (6)$$

Scheme 1.1. A general mechanism for a redox catalysis on the basis of an indirect reduction of a RX compound (here: an alkyl halide). In the scheme above, at a given electrode and the absence of a specific catalytic brought by the solid electrode, A is more easily reduced than RX. The reaction (2) is fundamental and supposes that a fast homogenous dissociative electron transfer occurs between the arene radical anion and the RX compound (see Ref. [4]).

perturbs the redox catalysis process (progressive consumption of the redox mediator). On the other hand, the reaction certainly presents some interest in organic preparative electro-chemistry since it is permitting a possible alkylation onto active sites belonging to aromatic systems.

1.2.2. *Heterogeneous redox catalysis*

In general, the redox catalysis process displayed in Scheme 1.1, implies that both **A** and **RX** are dissolved in the electrolyte. One may also propose a certain number of catalytic systems where **A** compounds (arenes, π-acceptors) or **RX** moieties can be grafted onto solid electrodes such as carbon or gold. Such systems could be considered as modified electrodes where the heterogeneous electron transfer mimics a form of redox catalysis.

1.2.3. *Arenization of glassy carbon*

Thus, the grafting and/or the physical adsorption of arenes **A** onto a solid electronic conductor allows one to considerably enlarge the reactivity of the solid conductor in terms of electrode efficiency. In Scheme 1.2, one considers a carbon electrode covered by a dense layer of **A**.

In order to gather ideas about the heterogeneous redox catalysis bound to the concrete grafting of **A** (Figure 1.2), we show below quite a simple experiment, easily achieved in a laboratory and clarified by the easy immobilization of pyrene or naphthalene (via bromomethyl substituents). Coating solid surfaces by a dense layer of arene further transformed into a coating of reduced form (Scheme 1.2) by different methods of electron transfer processes (a dense layer of radical ions that may be regarded as quite a stable structure of a macro-electron donor essentially in dry medium). In anhydrous solvents (such as dioxane or tetrahydrofuran), the layer of pyrene, anthracene,

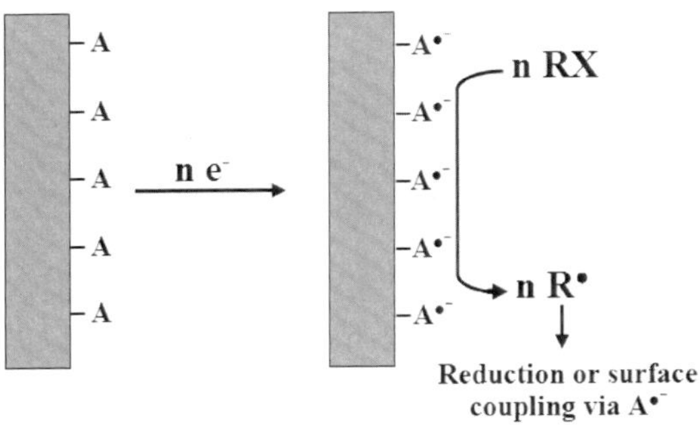

Scheme 1.2. Schematic representation of heterogeneous redox catalysis, quite similar to the system described in **1.1:** the mediator **A** is strongly immobilized (e.g., via the existence of a carbon link) onto the solid electrode. If the cathode is polarized at a potential E allowing one to partially reduce **A** into its anion radical **A**$^{\bullet-}$, the electrode starts being globally a reducer toward **RX** possibly not active at the potential E. In the case where the cleavage of the C–X bonds is observable, radicals R$^{\bullet}$ may react onto **A**, and also onto the carbonaceous material of the cathode (generally glassy carbon and also graphite, graphene, fullerenes, or conducting unsaturated organic polymers). Also, in this example, the global reaction may be very complex not only since the **A** layer is reduced, but also because the carbon electrode could be cathodically charged and therefore being a reducer by ET toward **RX**, and also possibly considered as a poly-nucleophile. This unexpected possibility that belongs to most solid materials will be developed in several chapters of this book (carboxylation, S_N2 reactivity toward a lot of aromatic **ArXs** and functionalized alkyl-halides **RXs**) [54].

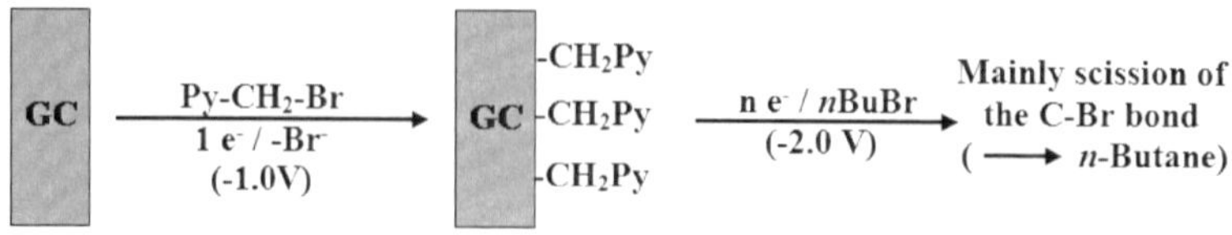

Figure 1.2. Modification of a carbon surface by means of 2(bromomethyl)pyrene which is used afterwards as a charged surface via the one-electron reduction of immobilized pyrene [52].

or naphthalene can be reduced by sodium or lithium. At the end of this book, several processes achieved in super-dry media may bring quite unexpected results and permit the reduction of palladium or platinum as well as the definitive insertion of certain π-acceptors for generating *new classes* of modified solid 3D electrodes. This could open a new gate in the modified electrode world where "solid" insertion of π-acceptors should compete with more classical methods such as radical grafting.

1.3. Building and Reduction of Electrophilic Layers onto Carbons

In the field of heterogeneous redox catalysis and electro-catalysis, the grafting of polarized carbon surfaces by alkyl halides could still be successfully achieved (see Chapter 13). As displayed by Scheme 1.3, the process is far from being conventional since C–I bond candidates for inducing a scission under electron transfer are in principal oriented toward the electrolytic bulk. Under these conditions, the mediator **A** has to get over the poly-alkyl barrier to be reduced. Such layers could be obtained (preliminary evidence in Chapter 13) by a new anodic process of grafting of iodo-alkyl chains $-(CH_2)_n$–I built with $1 < n < 8$. One may also independently use a cathode close to the modified layer and then permit electron transfer that appears to be another type of redox catalysis (Scheme 1.3). Thus, one may imagine, in the course of the bond cleavage and the concomitant formation of short life radicals, the hooking of large and poorly soluble radicalophilic molecules or particles (coronene, helicene, graphene). This proposal does not seem unrealistic just because it experimentally appeared that such grafted C–I bonds are cleaved at potentials ≫ –1 V vs. Ag/AgCl and certainly disclose interesting applications based on the electro-chemical stability of free radicals produced within –1 V < E < 0 V range.

The immobilization aryl halides moieties {–CH_2–Ph–X} (X = I or Br) could also be achieved (Figure 1.3) via substituted benzyl bromides and related compounds.

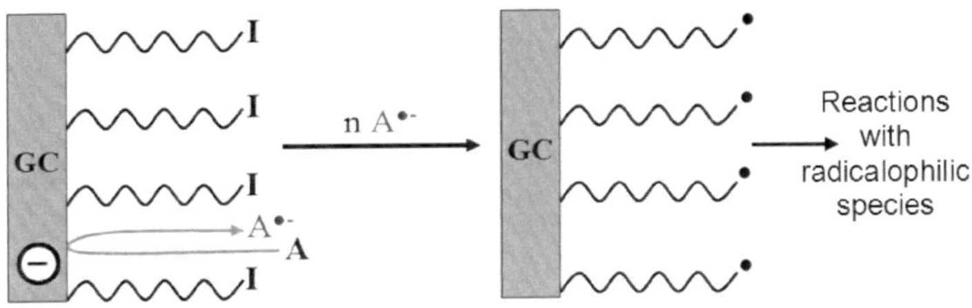

Scheme 1.3. Electro-chemical presentation for the reduction via ET of a poly-alkyl iodide layer immobilized at a glassy carbon surface. The fate of free radicals produced (homo-coupling, disproportionation, H transfers from the solvent, coupling with $A^{\bullet -}$) is not determined [52].

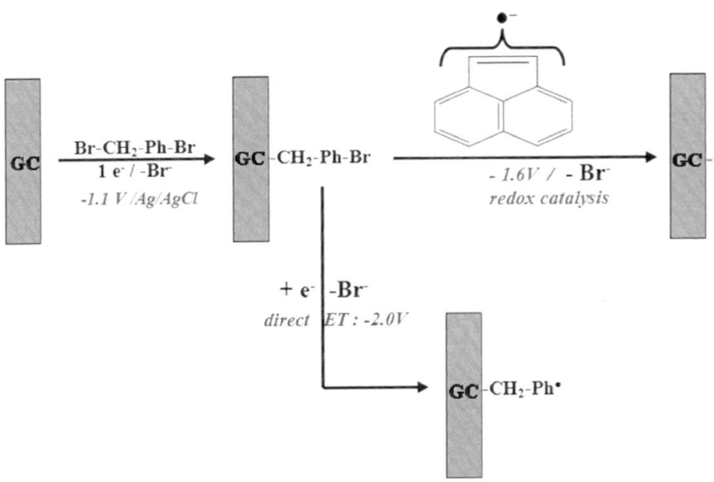

Figure 1.3. Immobilization of –CH_2–Ph–Br at a carbon electrode associated to two kinds of scissions at very different potentials: first, heterogeneous redox at –1.6 V via the anion radical of acenaphthylene, second, direct ET at –2.0 V. Interestingly, the integration of electricity is not the same between those two processes the ratio is 1e (direct according to II) vs. 2e (indirect III) and supposes two very different modes of reduction caused by the implication of the catalyst as shown in Figure 1.4 [54].

1.4. Intrinsically Catalytic Electrodes

During the last decade, new and suitable metallic electrodes with impressive catalytic capacities have been found. Among all those new catalytic electrodes, silver electrodes [19–22] appear mainly convenient for activating the catalytic scission of carbon–halogen bonds (Scheme 1.4). Within the same period, palladium electrodes also showed powerful capabilities to operate the catalytic reduction of alkyl and aromatic organic halides [23]. Additionally, syntheses at room temperature of silver–palladium and copper–palladium alloys were set up and new catalytic electrodes easy to make and convenient to use were described (see Chapters 2 and 3).

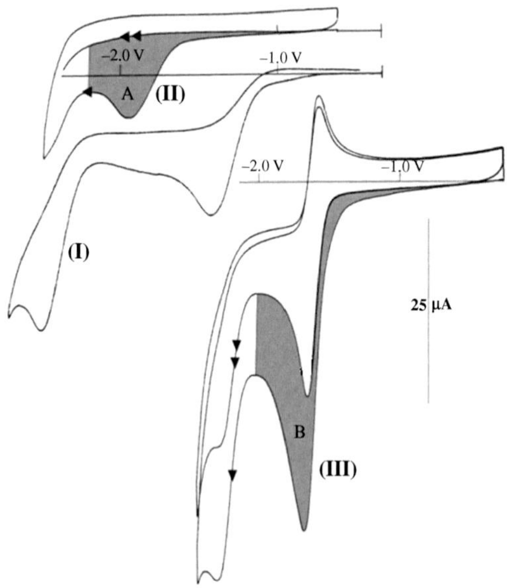

Figure 1.4. Reduction of 4-bromobenzylbromide at a GC electrode **I**. In **II**, the voltammetry of the deposit **A** at the carbon electrode, and in **III** the indirect reduction in a sole scan of the deposit according to **B** obtained in the presence of acenaphthylene (gray zones). One notes that electricity amounts between **A** and **B** are in the ratio 1/2 and would suggest that reduction processes are very different ((1e toward 2e) possibly explained by a fast coupling of aryl radicals present at the interface in the **A** process) [54].

$$RX + e^- \xrightarrow[E_R]{Ag\text{-}Pd} R^\bullet + X^- \quad (7)$$

$$2\,R^\bullet \xrightarrow{Dim} R\text{-}R \text{ or } \xrightarrow{Disp} R(-H) + R(+H) \quad (8)$$

$$R^\bullet + A \longrightarrow AR^\bullet \quad (9)$$

$$AR^\bullet + R^\bullet \longrightarrow AR_2 \quad (10)$$

$$AR^\bullet + HS \xrightarrow{solvent} ARH + S^\bullet \quad (11)$$

$$AR^\bullet + e^- \longrightarrow AR^- \quad (12)$$

$$AR^- + RX \longrightarrow AR_2 + X^- \quad (13)$$

Scheme 1.4. Use of an alloy Ag–Pd as active electrode. General mechanism for a heterogeneous one-electron catalysis ($E_7 \gg E^0_A$ and $> E^0_{R^\bullet/R^-}$) [23, 24].

The fundamental difference between a redox catalysis (homogeneous, see this chapter, Section 1. 2.) and the use of a specific catalytic electrode (see, for example, Chapter 2, case of Ag–Pd electrode) is presented below in the case of the concomitant presence of **RBr** and an arene **A**.

1.5. Triggering of Radical Chains via Electro-Chemical Activation

The evidence of the role of free radicals was clearly demonstrated by Kornblum [26], Russel [27] and Bunnett [28] in the field of nucleophilic substitution reactions via chain process initiated by a single-electron transfer (Scheme 1.5) thanks to the presence of reducing species (photochemical induction, dissolved metal, amalgams or/and solvated electron). The condition is the presence of a nucleophile that participates in the substitution with the fundamental inequality $E^0_{ArX} > E^0_{ArNu}$ essential to the development of a chain. Savéant demonstrated (Scheme 1.5) the role of the induction by means of conveniently polarized electrodes [29].

In the domain of radical chains triggered by electron transfer, let us also quote Diels–Alder reactions (types of [2 + 2] and [2 + 4] developed by Bauld through the oxidizing induction brought by

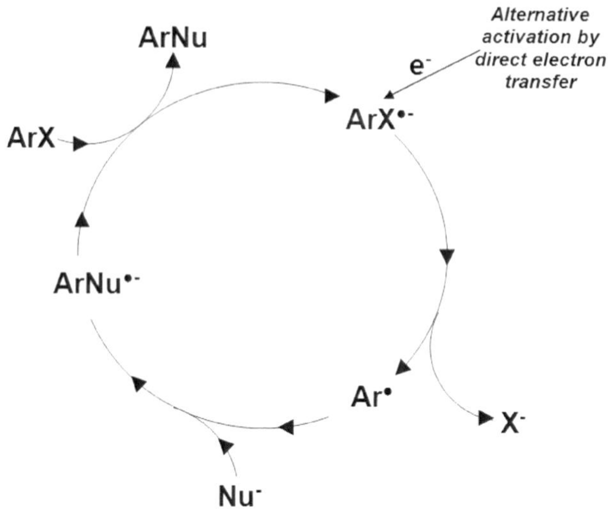

Scheme 1.5. Substitution of an activated aromatic halide **ArX** (SNR1 type reaction according to Savéant) by a nucleophile **Nu⁻** (example that corresponds to Ref. [29]: 2-chloroquinoleine in the presence of an arylthiolate as nucleophile).

cation radicals [30] as sources of four and six-membered rings and already fully developed in purely organic synthesis as symbolized by Scheme 1.6). Let us note that the reaction is formally induced by [1 + 2] or [1 + 4] additions. In electro-chemistry, the triggering of chains was first induced by electro-generated cation radicals [30, 31].

Very specifically, Diels–Alder reactions now triggered by anion radicals for cyclodimerizations of strongly activated olefins, in particular by aryl sulfonyl functions, were developed for the first time by means of electro-chemical activation [32–34]. The reaction selectively triggered by means of electron transfer could be afterwards generalized by means of very elemental experimental conditions, for example by simple galvanostatic electrolyses with two platinum electrodes and simply achieved in DMSO containing $LiClO_4$. That permitted, with very small amounts of electricity, the *quantitative* access to the corresponding cyclodimer easily transformed by means of electro-generated bases in aprotic media to the corresponding

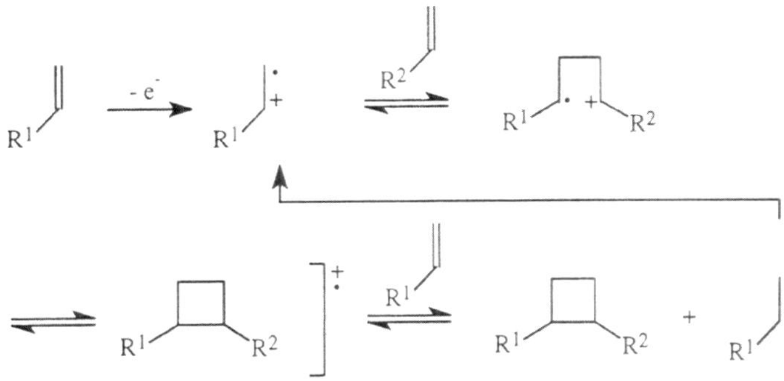

Scheme 1.6. Cyclodimerization reactions [2 + 2] induced per anodic electron transfer [31]. It should be named as [1 + 2] process.

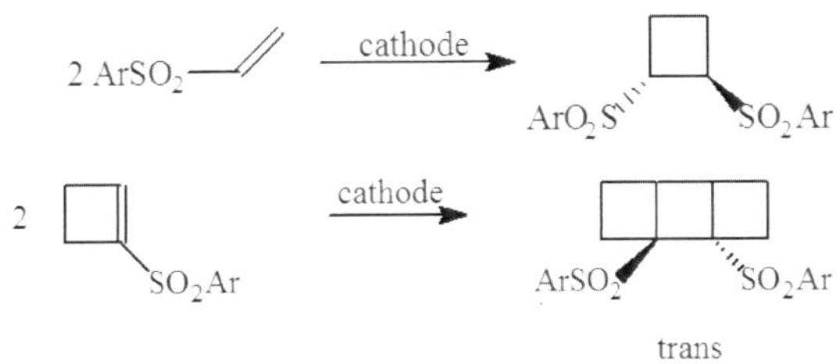

Scheme 1.7. Catalytic cyclodimerization of activated olefins via electron transfer: from cyclobutanes to ladderanes [31, 32, 35].

cyclobutenes. The reaction, quite certainly applicable to most strongly activated olefins via electron-withdrawing substituents, was revealed as being totally triggered by an electron transfer under very simple conditions. Additionally, an easy pathway to [n]ladderanes was also proposed [35] (Scheme 1.7).

Interestingly, the matching reaction type alternatively using an anodic activation was proposed by Utley [36] for a selective access to cyclodimers of vinyl pyridines and vinyl quinoleines.

With a large number of geometrical isomers (case of activated olefins) with high rotation barrier, the electro-chemical activation may present a real interest in synthesis (specific obtaining of *E* and Z isomers) especially when $E^0_Z > E^0_E$, that is the case for enol ethers, disulfones or Z 1,2-diarylsulfonylphenylethylenes [37, 38].

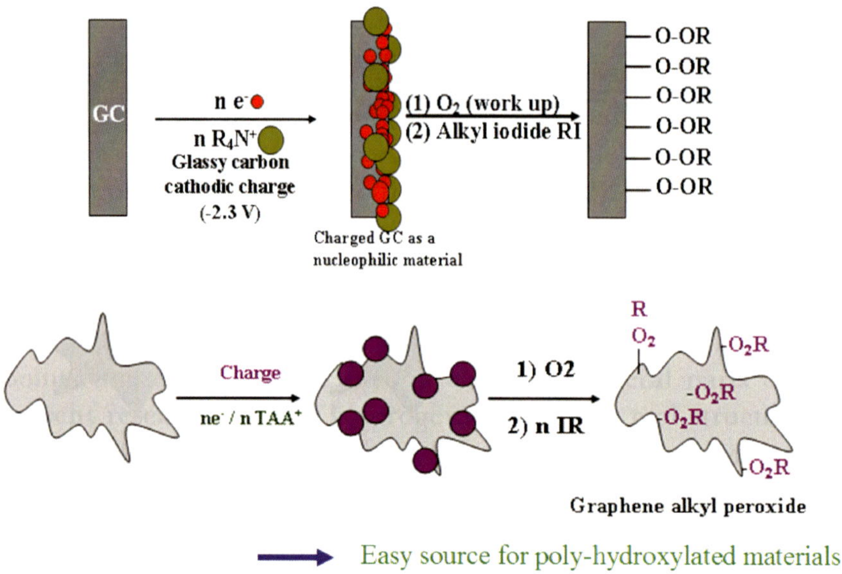

Scheme 1.10. The simple cathodic charge of many carbons in aprotic oxygen-free media may transform material into a poly-nucleophilic species that may react with di-oxygen during the workup: this process leads to a facile source of peroxidate groups easily functionalized by a large series of electrophilic species (e.g., RIs) further cleaved by electro-chemical reduction [56]. This process (taking advantage of the scission of the –O–O– bond) could represent a general source of hydroxylated materials (obtained by scission of the O–O bond) simply functionalized by varied substituents. Examples connected with the hydroxylation and carboxylation of carbons and graphene are fully described further (Chapter 11).

test and propose new solid electrodes (copper, rhodium, iron, titanium) evoked or described in this volume, the use of metals (copper, gold, palladium, silver or alloys) could help the conversion in very small size nanoparticles used in suspension in the cathodic bulk or deposited at the surface of a so-called inert conducting material like carbons.

The use of $Ni^{(0)}$ and $Pd^{(0)}$ catalysts in the presence of adequate ligands has been widely promoted in homo-coupling reactions, but interesting cross-coupling reactions (in the presence of electrophiles) were also abundantly reported [43–45]. For example, the homo-coupling of ArX type substrates passes through, in the presence of electro-generated Ni^0 catalyst by the essential formation of an activated intermediate complex of the form $ArNiXL_2$, where adequate ligands L present in solution are generally triphenylphosphine or 2,2′-dipyridine (Bipy). The high interest prompted us to conceive a process for building Bipy solid electrodes where both electron transfers and chemical reaction may simultaneously occur. These kinds of pyridine electrodes, preliminarily immobilized onto GC [46] are made increasingly precise with 2,2′-bipyridines (Scheme 1.11) in Chapter 17, especially at gold substrates.

Alternatively, the use of different metals as catalysts (under the classical forms of nanoparticles or galvanostatic deposits) is underlined, which additionally stresses their interest for generalized graftings and chemical modifications at these surfaces. Under this aspect, the innovative process (as mentioned hereafter) which introduce reducing platinum and palladium with concomitant *insertion of electrons and salts* for accessing [48] new classes of materials that are both exceptional catalysts and reducing agents (Figure 1.6). Under some aspects, this reaction resembles the cathodic charge of HOPG

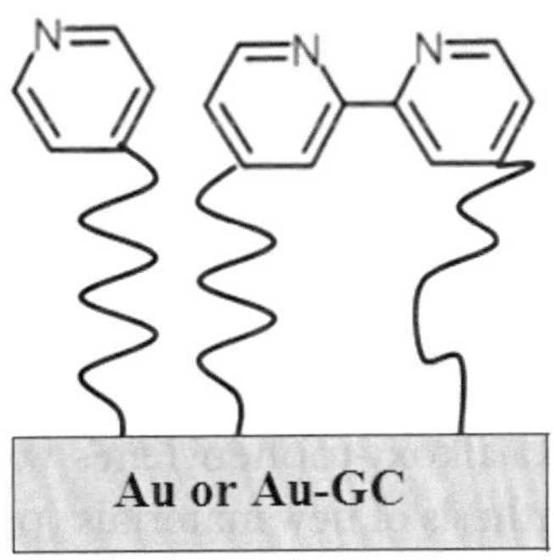

Scheme 1.11. A view of a pyridine electrode usable for catalytic purposes implying Ni, Pd and Cu ions. The length of arms (alkyl chains) is here in C_7 or C_8, aims in view of flexibility. Until now, the grafting of pyridine and BiPy [46, 47] has been successfully achieved. Any other adaptation pathways applicable to BiPy coverage is quite certainly conceivable. The immobilization of mono-dentate ligands (like triphenyl phosphine) is at present in progress [47].

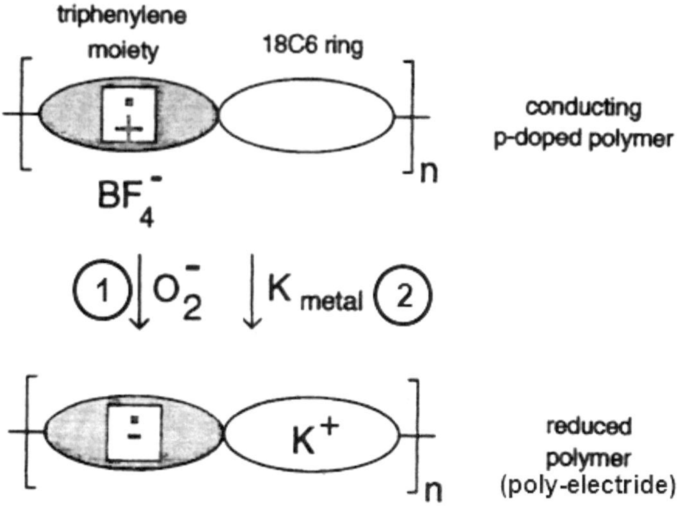

Scheme 1.13. Generation of a poly-electride by means of: (1) reduction by superoxide of the doped form of the poly-crown-ether and (2) the reduction by an alkaline metal in liquid ammonia [52].

polar solvents rendered super-dry and therefore "aprotic" by addition of activated neutral alumina. The level of moisture may reach very small values < 100 ppm: thus, at negative potential < –2 V vs. Ag/AgCl, there is no hydrogen evolution, but the obtaining of a reduction process often reversible which is assigned to the charge of the metal together with the insertion of the salt (mainly the cation) contained in the liquid electrolyte (see Chapter 18). The Rennes group has accumulated evidence on the cathodic charge of platinum (reduction of the metal with concomitantly reversible insertion of the electrolyte) that confirms this metal is electro-chemically corrodible. This opportunity then opens huge possibilities in electrochemical catalysis, energy storage and new chemically modified electrodes. Reduced or charged platinum becomes (stable under inert atmosphere) a paramagnetic species and may be used as a reducing agent. As fully developed in Chapter 19, a thick ionometallic layer (poorly conducting) is easily obtained and in the presence of an electrolyte $A^{-}M^{+}$ the stoichiometry of the layer was found to correspond to

$$\{Pt_2^{-}, M^{+}, MA\}$$

Many chemical "Lego" bricks may be obtained just by operating on the structure and the composition of this iono-metallic complex by changing both anions and cations at the condition that A^{-} or/and M^{+} are electro-inactive within a potential range up to –2.0 V for participating in the building and transformation of such intriguing materials. As shortly developed in this book (Chapter 19), the new challenge is based on the exchange of A^{-} (inorganic anion inserted first in the metal) with electro-generated π-acceptors reduced form (**π–A**$^{-\bullet}$) present at the Pt surface. Therefore, an exchange reaction takes place at the interface that permits — according to the amount of electricity concerned and the total duration of the electrolysis — the indirect insertion of **π–A** inside quite a thick (say several μm) organo-metallic layer. In the course of the workup, the layer is oxidized (presence of di-oxygen) and allowing under certain conditions the **π–A** compound is found to be ***really*** and definitively encrusted in Pt. This may represent a totally new and intriguing mode of immobilization of organic compounds in precious metals (palladium reacts quite similarly). The fate of the regenerated acceptor into the metal depends on its solubility in solvents (like acetone or tetrahydrofuran) used in the course of the workup. When the use of an efficient solvent is envisaged, one may easily obtain (without Pt loss of mass) a spongy material of high specific surface area (Figure 1.7).

On the contrary, **[π–A]**s of large molecular mass (example coronene, acenaphthenequinone), which are much less soluble in current solvents, may be efficiently trapped in Pt and Pd and then heterogeneous catalysis is activated by electro transfer. Also, organic poly-anions derived organic polymers (case of poly-fluorenes transformed by means of electro-generated bases [53]) could lead to materials of increased conductivity. Lastly, let us mention the case of graphene nanoflakes negatively doped that may produce, together

$$\{Pt_2^-, M^+_2, A^-\} + \pi\text{-}A^{\bullet -} \rightleftharpoons \{Pt_2^-, M^+_2, \pi\text{-}A^{\bullet -}\} + A^-$$

$$\{Pt_2^-, M^+_2, \pi\text{-}A^{\bullet -}\} + 2\,O_2 \xrightarrow{\text{Work up}} 2\,Pt + \pi\text{-}A + 2\,\{O_2^{\bullet -}, M^+\}$$

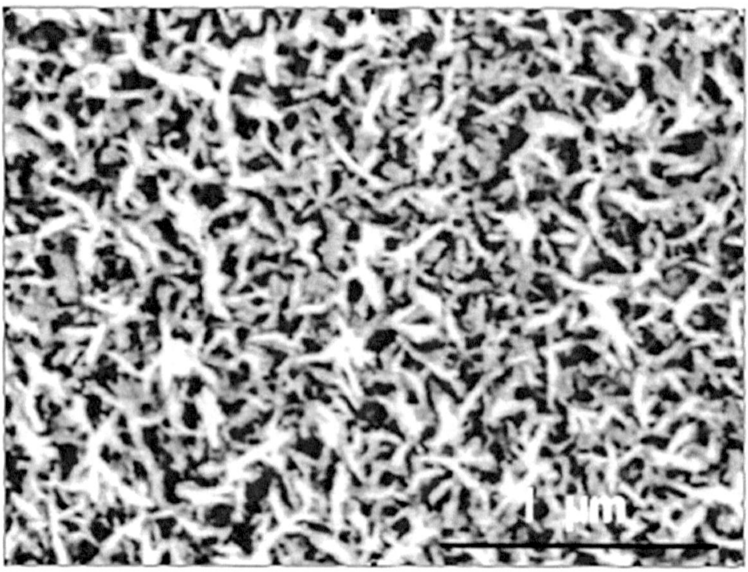

Figure 1.7. Principle of the sequestration of organic acceptors in a Pt bulk under potentiostatic reduction at < −2 V (assuming then a concomitant reduction of pristine Pt and **π–A**) [48]. The example above corresponds to the insertion of 1,4-diacetylbenzene when reduced at a Pt cathode in DMF containing CsI. The cathodic charge is 22 Ccm^{-2}. After reduction, the Pt electrode is left in contact with air and the modified iono-metallic layer is destroyed. The picture displays the aspect of the layer after rinsing with acetone (extraction of diacetylbenzene, formation a porous Pt layer) and then drying at 60°C for 10 min (more details in Chapter 19).

with Pt or Pd, versatile materials of which the conductivity of the external layer may depend on the applied potential [54].

1.10. Sequestration of Carbon Dioxide Gas into Transition Metal Bulks

Very recently, the opportunity to get specific cathodic reactions of electro-active gas such as carbon dioxide and di-oxygen at different metallic electrodes in polar solvents containing exclusively tetramethylammonium salts (TMeAX) under super-dry conditions was revealed [55]. Thus, conditions do not favor a conventional reduction of CO_2 into oxalate, methanol or carbon monoxide, but allow CO_2 to participate in a cathodic charge of the metals (Pt, Pd, Cu, Au, Ag, Ti, Rh, Fe, etc.) leading to a progressive formation of a thick inhibiting layer that blocks, more or less, the conductivity within the cathodic domain. The main interest of the process could probably lie in the anodic instability of the electro-generated layer where it implies carbon dioxide (Figure 1.8). Preliminary results support the fact that a cathodic reaction is compatible with the stocking of CO_2 anion radical in the metal.

The inset shows that carboxylation in surfaces for many metals was feasible. Modification of the carboxylate function with π-acceptors was shown as effective.

The metal carboxylation may occur only in the vicinity of the interface with very short current impulses. This procedure leads to strongly modified interfaces notably with gold, palladium, platinum,

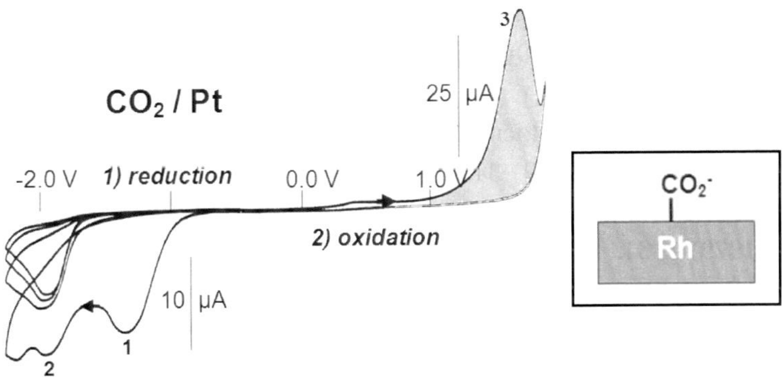

Figure 1.8. Reduction of a saturated solution of CO_2 at room temperature at a smooth platinum electrode (electrolyte DMF + TMeABF4). Four cathodic scans followed by an oxidation process: obtaining the step (in gray) that allows restoring *totally* the electrode and sweeps off a further cathodic inhibition. The cathodic step (1) is assigned to the one-electron reduction of CO_2 (catalysis at Pt) followed by the reaction onto the reduced form of platinum in (2) (as mentioned in Chapter 18). Current integration of the anodic peak (3) allows estimation of the total amount of CO_2 sequestrated in the metal: Γ_{CO_2} $\approx 10^{-7}$ mol·cm^{-2}. Similarly, carbon dioxide may be fixed onto rhodium (inset).

[50] V. Le Berre, L. Angély, N. Simonet-Guéguen, J. Simonet, *New J. Chem.* **13** (1989) 131.
[51] J.L. Dye, *Science* **301** (2003) 607.
[52] M.G. De Backer, F.X. Sauvage, E.B. Mkadmi, J. Simonet, *New J. Chem.* **17** (1993) 201.
[53] J. Rault-Berthelot, L. Angely, J. Delaunay, J. Simonet, *New J. Chem.* **11** (1987) 6.
[54] J. Simonet, Unpublished observations.
[55] J. Simonet, *Electrochem. Commun.* **58** (2015) 11.
[56] V. Jouikov, J. Simonet, *Electrochem. Commun.* **59** (2015) 40.

Chapter 2

Palladized Silver as Cathode Material: Catalytic One-Electron Scission of Carbon–Halogen Bonds

2.1. Introduction

Silver electrodes modified by "palladization" following a novel and simple process exhibit exceptional catalytic properties, especially in the scission of certain chemical bonds. In this chapter, we focus on the cleavage of primary alkyl halides used as standard systems in order to evaluate the catalytic activity of these modified electrodes. In electro-chemistry, modified solid electrodes (made by covering an Ag/Pd-based alloy formed at room temperature) exhibited large potential shifts compared to more classical electrodes such as glassy carbon (GC) as well as smooth silver. Specifically, palladium-doped silver surfaces are used for cleaving both carbon–iodide and carbon–bromide bonds. In general, experimental conditions take advantage of the use of organic polar solvents (*N,N*-dimethylformamide (DMF), acetonitrile, and propylene carbonate (PC)) containing tetraalkylammonium salts as electrolytes. While smooth silver cathodes generally facilitated two-electron cleavage reactions, palladized silver interfaces, owing to the moderate applied potential, always lead to one-electron processes with most carbon–halide bonds in which a free alkyl radical is formed as a transient.

2.2. Interest for Highly Catalytic Solid Electrodes

Studies in the field of multi-purpose electrodes usable in preparative electro-chemistry remain an important goal for achieving new reactions [1]. Thus, new electrode equipment could lead to unexpected results. Obviously, from now, the use of mercury as a cathode material is avoided due to environmental concerns [2] while weak hydrogen over-voltage renders the use of platinum, within the considered cathodic range, inconvenient. Alternatively, different carbon surfaces (may used as working electrodes, but their neutrality toward electro-chemical ion insertion and their tendency to graft electro-generated species (like highly reactive free radicals within a well-defined potential range) are considered a disadvantage [3]. Similarly, electrodes covered with functionalized conductive polymers could be used in interfacial synthesis by mimicking the principle of organic chemistry on solid support [4], especially when organo-metallic centers and/or metallic micro-aggregates are incorporated into the matrix. Quite obviously, it is expected that the future of the electro-chemical synthesis lies in the development of new solid electrodes that make use of silver [5] or those covered with (or modified by) ad-atoms [6].

Especially within the cathodic range, two-electron cleavage reactions may be of some interest for deprotection processes (such as the reduction of sulfonamides — scission of C–S bonds — in the deprotection of amines [7]). But the success of achieving such cathodic deprotections may depend on the electro-chemical potential necessary to easily cleave the carbon–heteroatom bonds. Most of the reported reactions of this type occur at potentials more negative than −2 V vs. SCE, and consequently do not allow the use of aqueous media at most polarized solid metals. As mentioned elsewhere in this book, the interest in smooth palladium and palladized surfaces [8] would be helpful to improve the ease of carbon–halide bond scission [9], in particular those of alkyl iodides [10–12]. We now present an easy way to produce a thin (but stable) layer of silver–palladium material, which is particularly useful to strongly speed up bond scission that is often described as having high activation energy for achieving the first electron transfer [13–18]. Thus, the process used for building such modified surfaces is simply chemical *palladization* of both smooth silver and silvered solid conducting substrates (the latter being produced by electro-deposition of Ag^0 possibly as nanoparticles). In this way, a wide range of bond scissions could be achieved with organic substrates such as primary alkyl-halides RX (with X = Br and I). Reactions are compared to those currently observed at GC electrodes (chosen by convention and not by catalytic materials) as well as palladized surfaces achieved by galvanostatic deposits (onto platinum or GC substrates). The advantages of these new Ag–Pd surfaces (catalytic character, stability, simplicity of use, and ease of formation) are highlighted. Results were compared to those already obtained at smooth silver cathode [19–22, 34]. The use of interfaces such as Ag–Pd is presented in terms of chemical stability, gains in energy, and also catalytic efficiency [35–37].

2.3. Chemical Doping of Silver by Palladium

The modification of silver by palladium was simply achieved by contact of smooth silver metal with a fresh aqueous solution of palladium sulfate in sulfuric acid. Thus, 1 g of $Pd(SO_4).2H_2O$ (Alfa Aesar) was dissolved in 100 mL of 0.1 N H_2SO_4 solution. After a few seconds, a shiny layer progressively covered the silver surface due to the displacement of Ag^0 by palladium cations. Palladium nitrate could be used as well. The thickness of the layer strongly depends on the time of contact with the palladium solution. It is recommended to leave rest of the silver in the palladium solution for at least 1 h before rinsing by water. Palladization resulted in a quasi-instantaneous dull deposit onto the smooth Ag surface apparently due to the displacement of silver by palladium ($E°$ (Pd^{2+}/Pd) = 0.92 V/ENH and $E°$ (Ag^+/Ag) = 0.799 V/ENH) concomitantly with an unexpected reaction leading to a specific Ag–Pd interface at room temperature.

First of all, all the used electrodes used in voltammetry had an apparent surface area of 0.78 mm^2. Silver, GC, pure palladium, and platinum disks when used as substrates were always carefully polished with silicon carbide paper (Struer) or with Norton polishing paper (types 02 and 03). Before use, the electrodes were rinsed twice with water, then alcohol, and finally acetone. Finally, they were dried with a hot air stream. Palladized platinum and palladized GC electrodes — employed for comparison — were obtained by means of a galvanostatic deposit using a solution of palladium chloride ($PdCl_2$ at the concentration 10 g L^{-1} in aqueous 0.1 N HCl). In general, such galvanostatic deposits corresponded to amounts of electricity from 10^{-3} to 20×10^{-3} C cm^{-2} with currents of the order of a few hundred micro-amps.

As will be shown, Ag–Pd electrodes could be alternatively generated from silvered surfaces. These surfaces were generally obtained by galvanostatic deposits of silver onto a large range of solid conductors from a solution of $AgNO_3$ (10 g L^{-1} in a 0.1 N HNO_3 solution). These deposits elaborated onto GC, smooth silver, or platinum were achieved thanks to fixed current electrolyses (0.5 mA mm^{-2}). Silver deposition corresponds to electricity amounts between 10^{-3} and 4×10^{-3} C mm^{-2}. Treatments by $PdSO_4$ were the same as those described above with silver plates.

2.4. Structural Characterization of the Ag–Pd Interface

When contact times were shorter than 1 min, the as-produced Ag–Pd layer appeared to be stable, and the sonication procedure had no visible effect on the adhesion of micro-crystals on the silver substrate. SEM images in Figure 2.1 show the spectacular surface evolution of the substrate while a growing "palladization" takes place. In the initial stages of the film growth, very small grains of spheroid-like shape cover the surface of the substrate. As expected, the particle size progressively increases with much longer contact times, yielding a quasi-total covering of the substrate after 5 min as shown in Figure 2.1(C). Although Pd and Ag exhibit similar energy diagrams, multi-elemental analyses of those layers performed by EDS after Pd–L_{a1} radiation at various time intervals, located at 2.8387 eV in energy, do not overlap. For comparison,

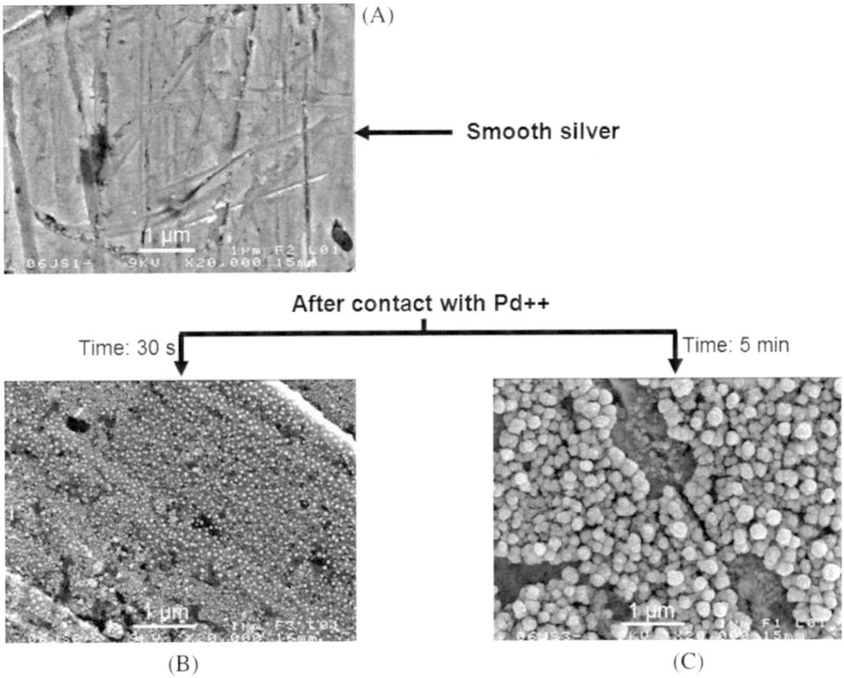

Figure 2.1. SEM images of the smooth silver surface and after immersion in a $PdSO_4$ solution for 30 s (B) and 5 min (C) [37].

Figure 2.2 shows two spectra, before and after surface modification, leading to the conclusion that those spheres do contain palladium after the chemical treatment. However, it was difficult to evaluate the Ag/Pd ratio of spheres since the thickness of the layer is not high enough to suppress the strong contribution of the silver substrate.

Therefore, a careful TEM study was performed for determining the texture, structure, and composition of the Ag–Pd layer. The bright field image (Figure 2.3(A)) reveals the presence of a dendritic-like growth from the palladium-based layer obtained with a solution of palladium sulfate (size < 20 nm). The HRTEM image (Figure 2.3(B)) of one part of the bright field image shows agglomerated nanoparticles composed of dendrite [23, 24]. These well-crystallized nanoparticles (around 5–10 nm) display a homogeneous

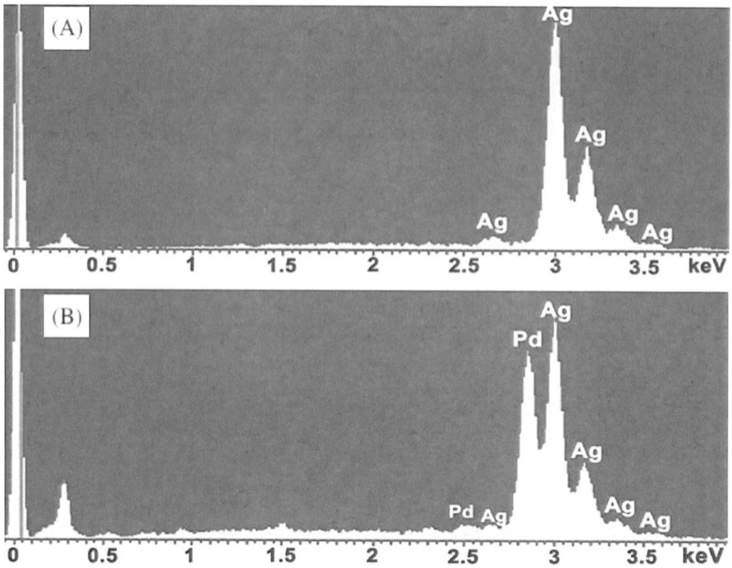

Figure 2.2. EDS spectrum of a smooth silver sample before (A) and after (B) chemical modification by palladium sulfate in contact for > 5 min (b) [37].

morphology. The SAED pattern (Figure 2.3(C)) of the corresponding picture 3(a) is made of diffraction circles owing to the fact that all nanoparticles present multi-directional orientations. By using the software Process Diffraction, a line profile of the electron diffraction pattern enables one to plot a graph similar to X-ray diffraction (Figure 2.3(D)). This permits determination of the fact that the layer is composed of an Ag/Pd alloy [25] (lattice parameter: 4.0499(8)Å) and also PdO and Ag. Indeed, as recently mentioned, Ag/Pd alloys [26] can slightly decompose in air into PdO and Ag due to oxidation. The EDS analysis (not shown) obtained on the sample that corresponds to Figure 2.3 systematically indicates the presence of a homogeneous alloy that shows a composition of 80% Ag and 20% Pd. As reported by Rao [27], the lattice parameter of Ag/Pd alloys changes with temperature, and therefore is in good agreement with an alloy composition that we have established (i.e., $Ag_{0.8}Pd_{0.2}$), when beam heating is around 200°C.

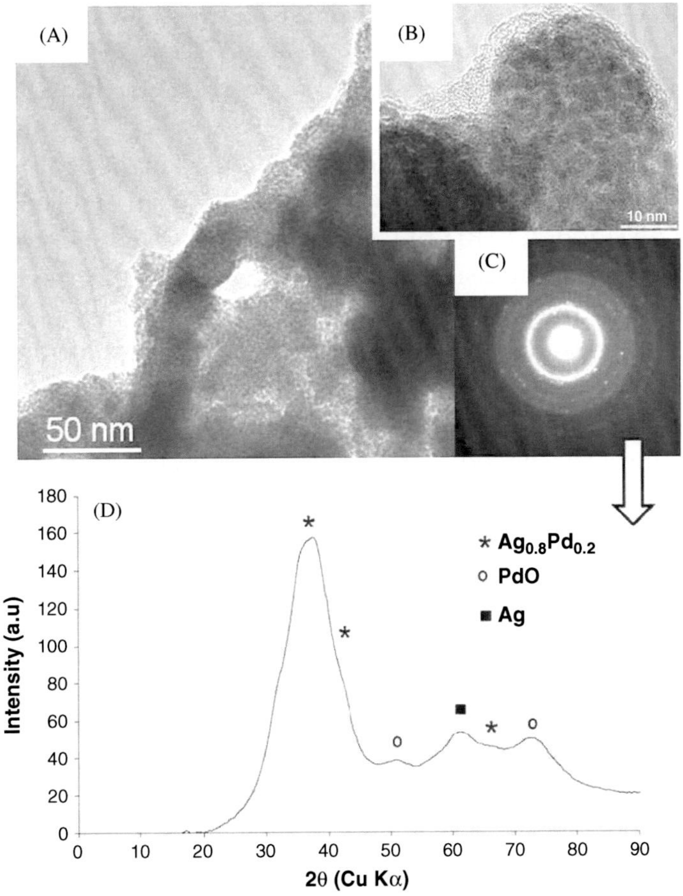

Figure 2.3. (A) Bright field image of the Ag/Pd alloy. (B) HRTEM picture showing the nanoparticles of the alloy. (C) SAED pattern of the TEM picture showing the polycrystalline nature of the layer. (D) Line profile of the SAED pattern characteristic of the $Ag_{0.8}Pd_{0.2}$ alloy, PdO, and Ag [37].

2.5. Palladized Silver Interfaces for a Catalytic Reactivity of Primary Alkyl Iodides

Electro-chemical data collected for Ag–Pd surfaces with a series of primary alkyl halides dissolved in DMF or PC are rather similar to those earlier reported for other kinds of Pd-based surfaces such as palladized platinum or palladized GC [8]. As can be seen in Table 2.1, it appears that positive potential shifts for the reduction step using an Ag–Pd surface is significant toward both GC ($\Delta E > +0.9$ V) and smooth silver ($\Delta E > +0.1$ V).

This effect does not depend much on the length of alkyl chains. Catalysis is expected to occur when induced by an antecedent reaction of alkyl iodides (RI) onto silver before the first electron transfer. The cause of catalysis could be clarified by dipping smooth silver into liquid RI (or their concentrated solutions in a polar solvent like DMF): in this way, a reaction between the metal and the organic iodide is clearly proved. Prolonged contacts (e.g., >30 min) definitively bring large discrepancies on the silver surface. Figure 2.4 depicts the surface modification of silver maintained in contact with 1-iodobutane or 1-iodooctane, leading to a huge structural change over time as clearly seen in the SEM analysis. Large black zones at the surface are assigned to the presence of organo-metallic or fully organic compounds that are subjected to reduction by silver. Moreover, on reduction, modified silver surfaces led (Figure 2.4(B)) to the growth of crystals clearly identified as homo-dimers R–R. Those results support establishing the chemical reactivity of silver toward primary alkyl iodides.

Consequently, it appears that, at pure silver electrodes, the reaction depends on the contact time between Ag and RI. Thus, experiments described above support that primary RIs are *chemically* reduced by silver. The effective reducing capacities of silver metal toward carbon–iodine bonds are expected to follow the overall mechanism given in Scheme 2.1 where the first and second electron transfers could depend on the experimental conditions. For example, experiments achieved in solvents of high polarity such as PC could help the preliminary dissociation of the C–I bond and therefore favor the occurrence of a one-electron process that permits the formation of an alkyl free radical at the level of peak 1 (Figure 2.5 at Ag electrode) as can be seen from coulometric measurements gathered in Table 2.1. This first peak is followed by a second peak that could account for the reduction of this free radical as a transient species. Interestingly, the potential range found for peak 2 in

Table 2.1. Voltammetric and coulometric data relative to some primary alkyl iodides in 0.1 $TBAPF_6$/PC. Scan rate: 50 mV s^{-1}.

		Glassy carbon electrode		Ag–Pd electrode		Coulometric measurements			
						Smooth Ag cathodes		Ag-Pd cathodes	
Entry	RX (concentration: mM)	E_{p2}(V)	$\mathbf{I}_{peak}$ (μ_A)	E_{p2} (V)	$\mathbf{I}_{peak}$ (μ_A)	E (V)	F mol-1	E(V)[a,b,c]	F mol^{-1}
1	1-Iodopropane (10)	−1.92	14.0	−1.17	11.2	−1.6	1.1	—	—
2	1-Iodobutane (8.8)	−1.93	10.6	−1.10	8.3	−1.6	1.1	−1.2	1.1
3	l-Iodohexane (7)	−1.91	8.2	−1.14	6.6	−1.5	1.2	−1.1	1.0
4	l-Iodooctane (7)	−1.91	6.1	−1.10	6.0	−1.6	1.3	−12	1.1
5	1-Iodohexadecaue (5)	−1.83	3.4	−1.12	4.5	−1.5	1.1	−1.3	1.0

Notes: Potentials are referred to aqueous SCE. Apparent area of all electrodes: 0.8 mm^2. The treatment of the Ag electrode was achieved by a dipping time of 1 min in Pd sulfate.

[a]Potentials and currents of the first peak.

[b]All coulometric measurements were achieved in DMF.

[c]At $E < -1.5$ V, electrolyses at smooth Ag yielded in all cases to two-electron processes: 2 +/−0.2 F mol^{-1} [39].

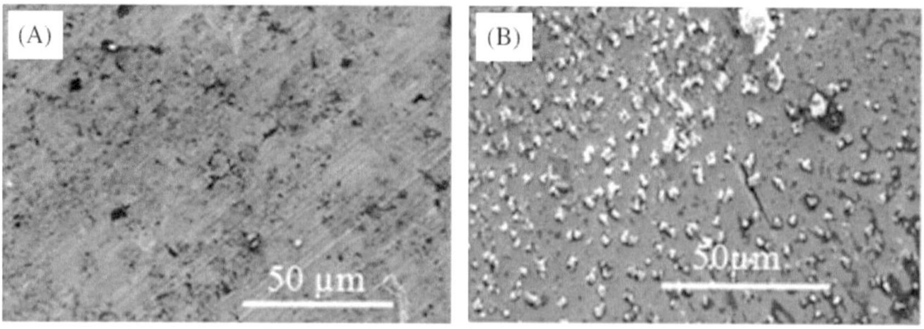

Figure 2.4. SEM images of pristine silver plates immersed (time of contact: 30 min) in alkyl iodides. (A) Silver after contact with pure 1-iodooctane. (B) Interface A after reduction at −1.4 V vs. SCE [11].

$$RI + Ag^0 \xrightarrow{\text{slow}} R\,Ag(+),\, I^- \xrightleftharpoons{\text{Salt } M^+A^-} R\,Ag(+),\, A^- \quad (1)$$

$$R\,Ag(+),\, I^-/A^- + e^- \xrightarrow{E_1} R^\bullet + I^-/A^- + Ag^0 \quad (2)$$

$$R^\bullet + e^- \xrightarrow{E_2} R^- \xrightarrow{\text{Proton source}} RH \quad (3)$$

Scheme 2.1. The catalytic electro-reduction of alkyl iodides by silver or Ag–Pd [11].

Figure 2.5 (around −1.5 V for the threshold of this peak) is in agreement with the reduction standard potentials estimated [28] for a large number of free alkyl radicals possibly generated by the present method. In general, the second step is rather short and it is expected that free alkyl radicals formed during the first reduction process are strongly adsorbed at the metal surface (formation of SAMs) or dimerize in the case of short alkyl chains. More specifically, fixed potential electrolyses at smooth silver (applied potentials < E_2) show that the global reduction may reach 2 F mol^{-1}. It is remarkable that the reduction of RI compounds can occur in two separate steps (example of dimers by coupling of the free radical $R^\bullet$ with tertiary alkyl iodides at silver cathodes under comparable but not identical conditions can be seen [29]). It is quite possible that adsorption of

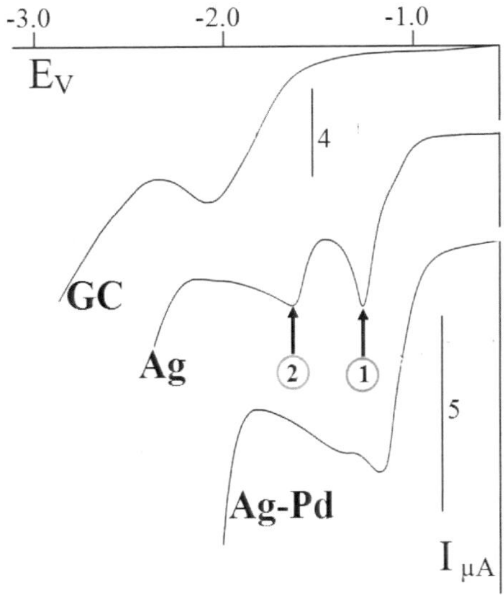

Figure 2.5. 1-Iodohexadecane. Saturated solution in 0.1 M $TBAPF_6$/PC. Scan rate: 50 mV s^{-1}. Reference electrode: SCE. Voltammetries at different working electrodes of the same surface area (0.78 mm^2). **GC**: glassy carbon, **Ag**: smooth silver with the separation of two steps (**1**) and (**2**), and **Ag–Pd** interface [11].

free radicals at the silver surface may inhibit their reduction (occurring in peak 2). The addition of free radicals onto metallic interfaces has found good evidence in recent literature [3, 27]. The effective separation of two reduction steps in the case of RIs could be applied for the modification of silver surfaces (Figure 2.6).

Lastly, the adsorption of free radicals onto silver could provoke the pre-step (to step (1)) observed with *all* the primary alkyl iodides in PC. If we refer to the classification by Laviron [30–32] on polarographic and voltammetric currents, this would be caused by the fast adsorption of the reduction product (the free alkyl radical in our case).

Electrolyses achieved at the potential of step (1) for generating modified interface by dipping of Ag in RI permitted merging of crystals identified as homo-dimers R–R generated from $R^{\bullet}$ (hexadecane coming is weakly soluble in PC). Additionally, the

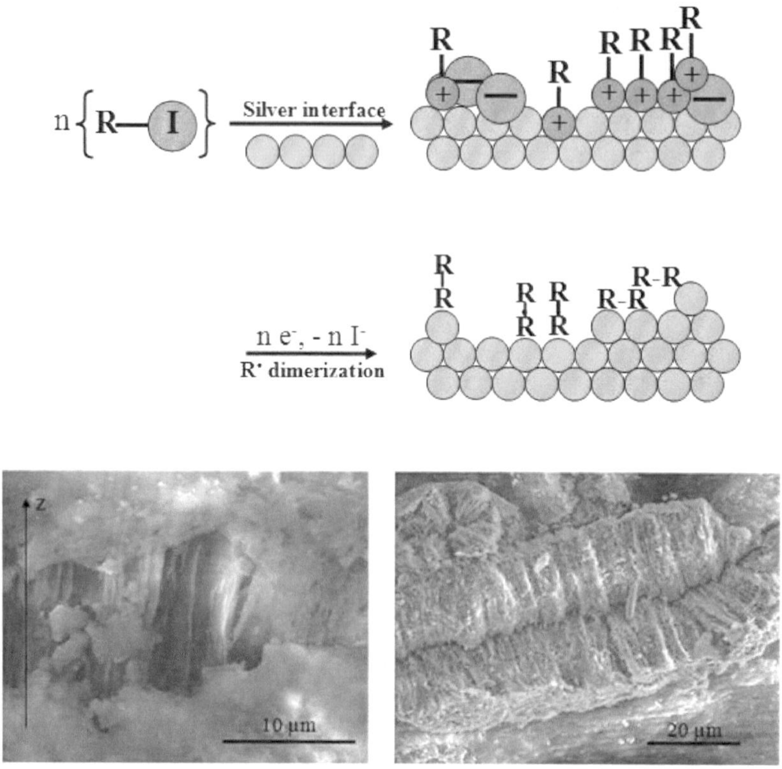

Figure 2.6. *Top*: Representation for a reductive homo-coupling of 1-iodohexadecane onto the Ag working electrode in PC + 0.1 M $TBAPF_6$ with (1) a preliminary reduction by silver followed by (2) an additional electron transfer. *Bottom*: SEM images: views of thick crystallized dimer layers (growing perpendicular to the Ag interface) [39].

reactivity of Ag toward RI compounds appears to be quite close to that already reported for copper toward alkyl halides (in particular bromides) [32]. Thus, such a substitution (fully discussed in Chapter 9) yields R–Cu(I), X^- intermediates, which are themselves electroactive in a two-electron reduction process that generates RH species.

On the contrary, palladized silver surfaces exhibit only one step when reacting with RIs (diffusion controlled for concentrations smaller than 10 mM and proportional to the organic iodide concentration) (Figure 2.5). There is no evidence for the existence of an inhibition process as described at the silver surface and the resulting peak was found to be shifted to less cathodic potential toward the one already described at a pure silver surface. The absence of inhibition of the reaction at the Ag–Pd surface in their reaction with RI compounds can be related to the lack of pure silver zones at the interface. This would mean that the use of palladized silver as cathode material specifically involves an one-electron reduction. This conclusion is supported by coulometries at Ag–Pd cathodes (Table 2.1, reductions in DMF at about –1.5 V vs. SCE, entries 2–5), which establish that RI reduction under these conditions (irrespective of the organic solvent's polarity) could obey a pure one-electron process. Also, reduction steps obtained at Ag–Pd surfaces show distinctly well the cathodic limit of –2 V observed with most of the palladized surfaces under comparable experimental conditions. These results are in full agreement with data [8] observed for palladized electrodes (via galvanostatic deposits at carbon or gold) in their reaction with a large series of RI compounds.

Thus, the use of Ag–Pd surfaces certainly brings a real improvement by the catalytic effect both in terms of positive potential shift and electrolysis specificity.

2.6. Reduction of Alkyl Bromides

Alternatively, most of the primary alkyl bromides exhibit an irreversible two-electron step at GC electrodes at quite negative potentials (< –2.4 V vs. SCE) in DMF containing TAA^+ salts [26]. Under the same conditions, the use of smooth silver as a cathode material [19] induces significant potential shifts toward less cathodic potentials. However, under the conditions given in Figure 2.7 (with 1-bromooctane — curve B) and Figure 2.8 (with 1-bromobutane — curve B), the reduction step splits into two well-divided steps (I) and (II). In DMF/TAAX salts, step (I) at a rather less negative potential (in general, half-potential $E_{p/2}$ around –1.3 V) is small and assumes a kinetic character. The sum of peak currents of (I) and (II) is roughly equal to the limit current of the two-electron step usually obtained

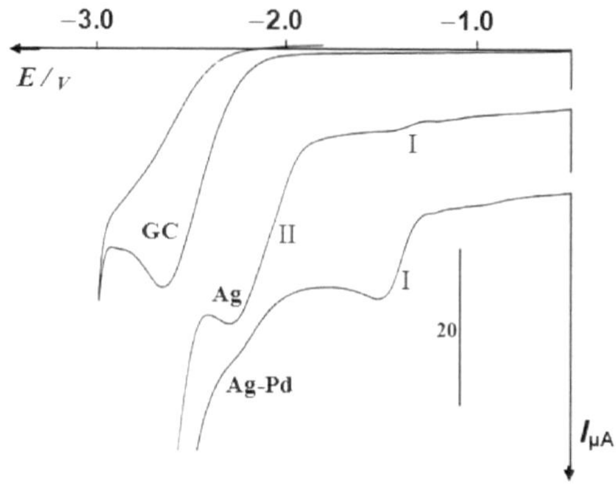

Figure 2.7. 1-Bromooctane (concentration: 10 mM). Solution in DMF containing 0.1 M $TBAPF_6$. Scan rate: 100 mV s^{-1}. Reference electrode: aqueous SCE. Voltammetries at different electrode materials (surface area: 0.78 mm^2) with GC, smooth silver with the obtaining of two steps (I) and (II), and then Ag–Pd (one step (I)) [38].

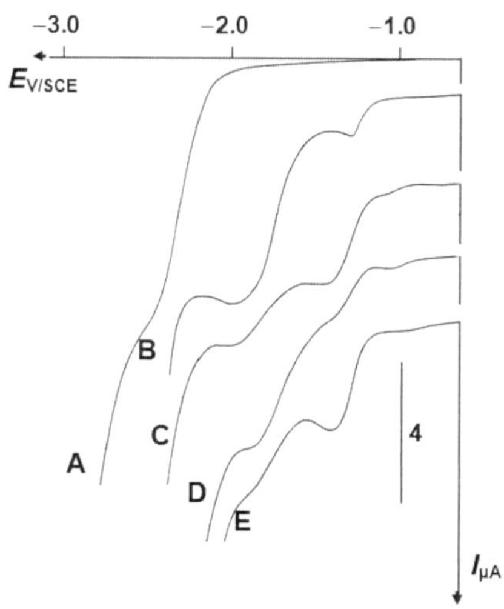

Figure 2.8. 1-Bromobutane. Concentration: 6.2 mmol L^{-1}. Electrolyte: 0.1 M $TBAPF_6$ in DMF. Scan rate: 50 mV s^{-1}. Reduction at different working electrodes (apparent surface areas: 0.78 mm^2): (A) GC, (B) smooth silver, (C) AgPd electrode, (D) platinum electrode covered with a silver galvanostatic deposit of 6×10^{-3} C mm^{-2}, and (E) the latter electrode treated by palladization (then leading to a Pt/AgPd electrode) [39].

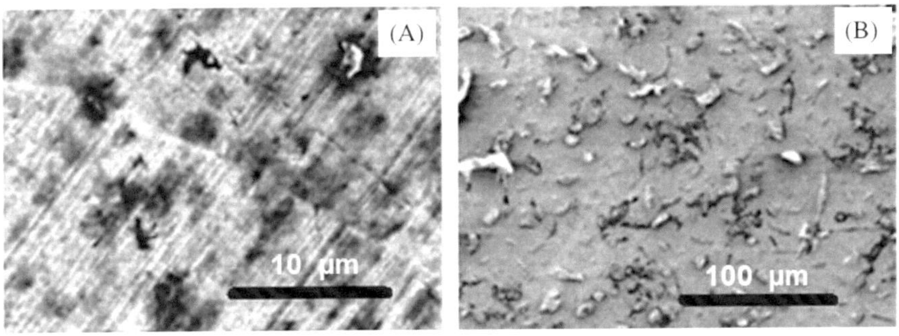

Figure 2.9. SEM images for the chemical reaction of silver pristine in contact with pure 1-bromooctane. (A) Contact time: 30 min. Rinsing with alcohol and then acetone. (B) After reduction of the former surface (without contact with air) maintained at a potential of −2.0 V vs. SCE until nil current. Electrolytic solution: 0.1 M $TBAPF_6$/DMF. Organic deposit identified as hexadecane crystals onto the surface after reduction according to mode **B** [37].

at GC. Peak (I) might point to an activation by silver explained possibly by a reaction between Ag and most RBr compounds (analogous to the previously described reactions with RI compounds). A reaction between Ag and RBr shown in Figure 2.9 using 1-bromooctane as an example produces crystals of homo-dimers after electroreduction of the pre-deposit.

Keeping in mind that the cathodic reduction of RBr compounds at solid interfaces (such as platinum or GC) is much less easier than the one observed with the corresponding iodides, the catalytic effect of silver depends on several parameters: the activation process specific to RBr compounds and its kinetics at the metal surface, the choice of the electrolyte anion, the adsorption of the electrolyte at silver as well as the preparation of the surface.

As a matter of fact, the reduction of RBr compounds follows, at least in DMF, an overall two-electron process. Definitively, reaching rather negative potentials is necessary to obtain a uniform current density (fixed $E_{Red} < -1.6$ V) due to the fact that activation by silver is too slow.

Therefore, it appears difficult to differentiate between the two processes proposed in Scheme 2.2. It is clear that free alkyl radicals

50 mA cm^{-2}) could be performed several times by using the same electrode. However, the exact nature of this electro-catalytic process was not totally clarified so far. The unexpected synergy resulting from the close association of [Ag+Pd] needs additional confirmation.

References

[1] H.J. Shäfer, in *Encyclopedia of Electrochemistry*, eds. A.J. Bard and M. Stratmann, Vol. 8, Wiley-VCH, Weinheim (2004).
[2] O.R. Brown, in *Physical Chemistry of Organic Solvent Systems*, ed. A.K. Covington, Plenum Press, New York, p. 747 (1973).
[3] E. Coulon, J. Pinson, *Langmuir* **17** (2001) 7102.
[4] E. Steckhan, in *Organic Electrochemistry*, eds. H. Lund and O. Hammerich, M.M. Dekker, New York, Basel, p. 1103 (2001).
[5] S.B. Rondinini, P.R. Mussini, F. Crippa, G. Sello, *Electrochem. Commun.* **2** (2000) 491.
[6] G. Kokkinidis, *J. Electrochem. Chem.* **201** (1986) 217.
[7] R. Kossai, J. Simonet, G. Jeminet, *Tetrahedron Lett.* **12** (1979) 1059.
[8] J. Simonet, *J. Electroanal. Chem.* **583** (2005) 34.
[9] J. Simonet, *Electrochem. Commun.* **7**(1) (2005) 74.
[10] J. Simonet, *Electrochem. Commun.* **7**(6) (2005) 619.
[11] J. Simonet, L. Laffont, P. Poizot, *J. Electroanal. Chem.* **591** (2006) 19, P. Poizot, L. Laffont, J. Simonet, *J. Electroanal. Chem.* **622** (2008) 204.
[12] P. Poizot, V. Jouikov, J. Simonet, *Tetrahedron Lett.* **50** (2009) 822.
[13] J.M. Savéant, *J. Am. Chem. Soc.* **109** (1987) 6788.
[14] C. Andrieux, I. Gallardo, J.M. Savéant, K.B. Su, *J. Am. Chem. Soc.* **108** (1986) 638.
[15] J.M. Savéant, *J. Am. Chem. Soc.* **114** (1992) 10595.
[16] C. Andrieux, I. Gallardo, J.M. Savéant, *J. Am. Chem. Soc.* **111** (1986) 1620.
[17] J.M. Savéant, in *Advances in Physical Organic*, ed. T.T. Tidwell, Vol. 35, Academic Press, New York, p. 177, (2000).
[18] J. Grimshaw, J.R. Langan, G.A. Salmon, *J. Chem. Soc. Faraday Trans.* **90** (1994) 75.
[19] P.R. Mussini, S. Ardizzone, G. Gappelletti, M. Longhi, S. Rondinini, L.M. Doubova, *J. Electroanal. Chem.* **552** (2003) 213.

[20] S. Ardizzone, G. Cappelletti, M. Lioudmila, P.R. Mussini, S.M. Passeri, S. Rondinini, *Electrochim. Acta* **48** (2003) 3789.

[21] S. Rondinini, A. Vertova, *Electrochim. Acta* **49** (2005) 4035.

[22] G. Fiori, S. Rondinini, G. Sello, C. Alberto, M. Cirja, L. Conti, *J. Appl. Chem.* **35** (2005) 363.

[23] W.B. Pearson, in *A Handbook of Lattice Spacings and Structures of Metals and Alloys,* eds. W.B. Pearson and G.V. Raynor Pergamon Press, p. 528 (1967).

[24] K.W. Wang, S.-R. Chung, T.P. Perng, *J. Alloys Compd.* **417** (2006) 60.

[25] C. NagaBhushana Rao, K. Krishna Rao, *Can. J. Phys.* **42** (1964) 1336.

[26] D.G. Peters, in *Organic Electrochemistry,* eds. H. Lund and O. Hammerich, M.M. Dekker, New York, Basel, (Chapter 8) p. 341 (2001).

[27] J. Pinson, F. Podvorica, *Chem. Soc. Rev.* **34** (2005) 429.

[28] D. Occhialini, K. Daasberg, H. Lund, *Acta Chem. Scand.* **47** (1993) 1100.

[29] C.A. Paddon, F.A. Bhatti, T.J. Donohoe, R.G. Compton, *J. Phys. Org. Chem.* **20** (2007) 115.

[30] C. Degrand, E. Laviron, *Bull. Soc. Chim. Fr.* (1968) 2238.

[31] E. Laviron, *Bull. Soc. Chim. Fr.* (1968) 2256.

[32] E. Laviron, *J. Electroanal. Chem.* **12** (1966) 516.

[33] R.B. Rieke, D.E. Stack, B.T. Dawson, T.C. Wu, *J. Am. Chem. Soc.* **58** (1993) 2483.

[34] A.A. Isse, A. De Giusti, A. Gennero, *Tetrahedron Lett.* **47** (2006) 7735.

[35] J. Simonet, *Electrochem. Comm.* **10** (2008) 647.

[36] P. Poizot, L. Laffont-Dantras, J. Simonet, *Electrochem. Comm.* **10** (2008) 864.

[37] P. Poizot, J. Simonet, *Electrochim. Acta* **56** (2010) 15.

[38] P. Poizot, L. Laffont-Dantras, J. Simonet, *J. Electroanal. Chem.* **624** (2008) 52.

[39] J. Simonet, unpublished results.

Chapter 3

A Disordered Copper–Palladium Alloy Electrode: Catalytic Scission of Carbon–Halogen Linkages

3.1. Introduction

A recurrent goal for most electro-chemists is to successfully achieve new reactions by developing an ideal and convenient multi-purpose cathode in the field of organic electro-chemistry [1]. Unfortunately, within the cathodic range, mercury as a cathode material is definitively prohibited due to for environmental concerns [2]. Platinum has a too small hydrogen over-voltage that renders its common use difficult within a wide cathodic range, especially beyond −1.5 V vs. SCE. Different carbon interfaces (such as graphite, glassy carbon, carbon felts, etc.) may prove to be useful working electrodes, but they are not always inactive and their response may depend on the electrolyte and the applied potential. Moreover, their neutrality toward insertion of electro-chemical ions as well as their tendency to graft free radicals have already been reported [3]. In another domain, modified electrodes superficially bearing functionalized conductive polymers were claimed to be useful in interfacial synthesis since they can mimic some principles of organic chemistry on solid supports [4]. Consequently, some innovative insights into electro-chemical synthesis could be linked to the development of very specific solid electrodes showing exceptional catalytic capacities. Quite certainly, following this way, smooth silver [5] as well as other solid metal electrodes modified by ad-atoms [6] could be used as alternatives when they remain associated with particular electrolytic reactions.

So, two-electron cathodic bond cleavages involving halides, sulfones, sulfonamides, or tosylates are of importance in organic synthesis since they can be elegantly applied to deprotection processes [7]. However, most of the described reactions are reported to be slow electro-chemical processes and depend on the electro-chemical potential necessary to cleave certain carbon–heteroatom bonds. Such cleavage reactions were shown to occur at quite negative potentials, basically lower than −2 V vs. SCE. Thus, the choice of the right solvent is crucial since wet organic media are often unsuitable when most of the solid metals are considered as cathodic materials. Additionally, since highly reducing potentials are needed to achieve bond cleavages, the use of super-dry solvents may lead to unexpected discrepancies depending on the nature of the electrode material (see Chapter 18). Transition metals like platinum could also react with the electrolyte to give complex iono-metallic layers of weak electronic conductivity [8]. In particular, such metallic interfaces were formed with tetraalkyl ammonium salts and alkali metal iodides to yield deposits (strongly reducing agents) that were then considered as modified electrodes [9].

On the other hand, while searching in the domain of new electrodes, one additionally could account for the tailoring of surfaces by means of a specific deposit of a catalyst. Using such alternatives, the activation may become so large that the nature of cathodic reactions totally change. Large potential shifts can be observed, which enable, in principle, to basically change the overall process. Following this way, convenient methods were preliminarily developed for producing modified electrodes based on the facile capping of a copper–palladium alloy [10]. Initial results have shown that the as-obtained Cu/Pd interface appears particularly efficient in speeding up the scission of carbon–heteroatom bonds that have a high activation energy to specifically achieve the first electron transfer

[11–14]. In the present approach, we focused our investigations on the cathodic reactivity of carbon–halogen bonds.

In this chapter, the nature of the copper–palladium layer produced onto copper substrates by displacement reactions from several Pd(II)-based precursors is studied and the efficiency of the as-produced surface toward the electro-chemical cleavage of several alkyl halides RX (with X = Br and I), commonly reported to occur with very large activation energy at regular solid metallic electrodes [11–14], is investigated. In this approach, several organic halides are used as probes to compare their cleavage modes with those conventionally observed at glassy carbon electrodes on one hand and palladized surfaces on the other, especially achieved by electro-chemical deposits (onto platinum or glassy carbon substrates under the same experimental conditions). The advantages of these new composite Cu–Pd surfaces appeared to be their huge catalytic character, stability, and ease of synthesis. The use and specificity of these interfaces are discussed, especially in terms of potential shift values related to their catalytic efficiency. It is also necessary to keep in mind that the specificity of such interfaces is to promote unexpected one-electron processes, which involve the transient formation of free alkyl radicals. With aryl halides, however, most of the reduction reactions remain typically two-electron processes.

3.2. How to Generate a Palladized Copper Alloy at Room Temperature

The copper–palladium interface is simply obtained by dipping a smooth clean copper substrate (for example, electrode, grid, or sheet), previously cleaned with acetone, into a fresh acidic bath of Pd(II) for 15 s. In the present method, three different solutions were prepared by dissolving a palladium salt $Pd(Y^{n-})_{2/n}$ (with Y = SO_4^{2-}, NO_3^- and Cl^-) into the corresponding acid. Fortunately, these salts are commercially available. More concretely, 1 g of $Pd(SO_4)\cdot 2H_2O$ (Alfa Aesar) was dissolved in 100 mL of 0.1 N H_2SO_4 solution. This procedure led to a quasi-instantaneous deposit onto the copper surface owing to the displacement of copper by palladium cations together with the formation of a palladized copper interface. The shiny layer appeared to be quite stable, and the sonication procedure did not show any effect on its adhesion to the copper substrate. To prevent any residual of acidic impurities possibly adsorbed onto the surface, a preliminary cleaning step is recommended by dipping the modified electrode into a diluted aqueous solution of tetramethylammonium hydroxide followed by rinsing with clear water, alcohol, and finally acetone before drying with hot air flow (at about 60°C). In principle, such as-made electrodes were easily reused, leading to coherent data.

3.2.1. *Texture and structure analyses*

Modified surface samples were first examined by X-ray diffraction measurements at room temperature. However, the nanometric feature sizes of the deposits made the correct identification of the as-produced material difficult. To clearly determine the structure/texture of the copper–palladium interface, investigations were conducted using high-resolution transmission electron microscopy (HRTEM). Since copper grids for microscope are commercially available, such materials were directly used as the substrate to study the Cu–Pd interface. More precisely, three samples were specifically prepared for TEM investigation by dipping a copper grid into a fresh acidic bath of $Pd(Y^{n-})_{2/n}$. Following this procedure, electron transparent specimens were obtained. The TEM and HRTEM imaging were performed using an FEI TECNAI F20 S-TWIN device. The elemental composition was also determined by energy dispersive spectroscopy (EDS) at a nanometer resolution. Diffraction patterns were performed using the selected area diffraction (SAED) mode or by Fourier transform of the HRTEM imaging.

3.2.2. *The palladized copper interface*

A TEM study was carried out on all the samples in order to precisely measure the texture, structure, and composition of the palladium-based layer. The bright field image (Figure 3.1(A)) exhibits

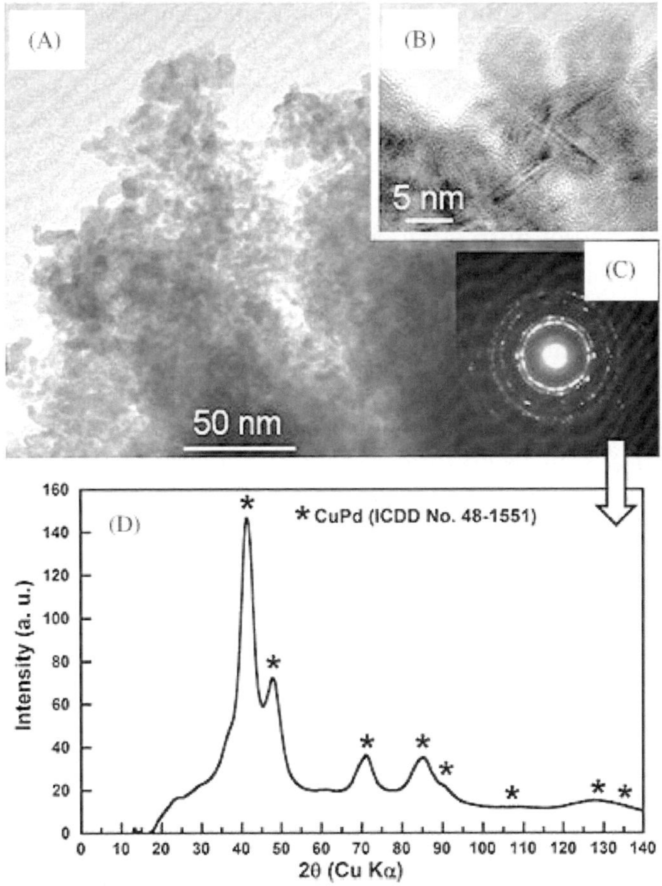

Figure 3.1. (A) Typical TEM picture showing the as-produced palladiated layer. (B) HERTEM picture showing the formation of crystallized nanoparticles. (C) SAED pattern of (A) and its corresponding line profile (D) that is characteristic of the "CuPd" phase (space group: Fm3m) [17].

a dendritic-like growth of the layer obtained with the solution of palladium sulfate (size < 50 nm). The HRTEM image (insert Figure 3.1(B)) of one part of the bright field image represents one 12 nm nanoparticle that is composed of the dendrite. Particles (dimension: around 10–15 nm) are homogeneous and possess a well-crystallized morphology. Samples prepared with palladium chloride and palladium nitrate solutions showed the growth of a non-dendritic palladium-based layer, but the layer always corresponds to the juxtaposition of nanoparticles, the size of which is closely dependent on the precursors and varies from 3 to 8 nm. All of these nanoparticles are well crystallized. SAED patterns obtained on these three samples are identical as shown in Figure 3.1(C). They are composed of diffraction circles due to randomly oriented nanoparticles. Using the "process diffraction" software, a line profile of the electron diffraction pattern enables to plot a graph similar to the X-ray diffraction pattern and determines that the layer can be related to the CuPd disordered alloy (ICDD card No. 48-1551, space group: Fm3m). The EDS analysis obtained on these layers indicates systematically a Cu/Pd ratio near to 1, which well supports the formation of a CuPd alloy. So for each sample, it is concluded that the given operating mode leads to Pd-based modified surface that corresponds to a thin layer of the stoichiometric CuPd alloy. As mentioned in a preliminary paper [16], the sparingly soluble CuCl compound could also be incorporated into the layer when using palladium chloride as a salt (the transient [Cu(I)] in the redox process is stabilized with chloride used in excess).

3.3. Cathodic Cleavages of Carbon–Halogen Bonds at Cu–Pd Electrodes

In this section, we discuss some cathodic reductions of organic halides. The conditions always correspond to the use of organic polar solvents (essentially DMF or acetonitrile) and TAA^+X^- salts.

3.3.1. *Scission of primary alkyl iodides (RIs)*

When using Cu–Pd electrodes, particularly those obtained from palladium sulfate and nitrate solutions, data regarding electrochemical reduction of primary alkyl halides appeared to differ from those already described with other conventional solid electrodes such as glassy carbon cathodes. Several alkyl iodides such as 1-iodobutane, 1-iodohexane, 1-iodooctane, and 1-iodohexadecane

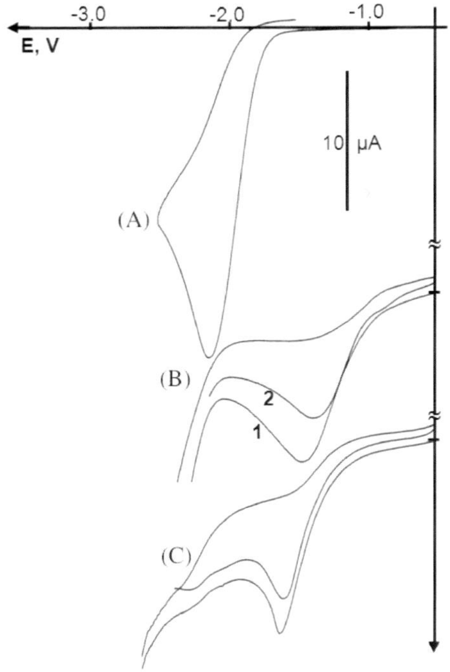

Figure 3.2. Voltammetries of 1-iodohexane [17] (concentration: 9 mM) in DMF containing 0.1 M $TBABF_4$, recorded at different micro-electrodes. Scan rate: 100 $mV \cdot s^{-1}$. Saturated calomel electrode. (A) Response at a glassy carbon electrode ($S = 0.8$ mm^2), two scans. (B) Responses at a palladized platinum electrode ($S = 0.8$ mm^2). (C) Response at a Cu–Pd modified electrode elaborated from a palladium sulfate solution ($S = 1.6$ mm^2).

were tested. Thus, 1-iodohexane (Figure 3.2) shows a strong activation process at any Cu–Pd electrode. This activation could be first quantified by the positive shift toward data obtained from a glassy carbon electrode assumed to be the ideal type of non-active electrode toward RX molecules. Thus under such "regular" conditions (use of a conventional glassy carbon electrode), the half-peak potential (first scan) of 1-iodohexane ($E_{p/2} = -1.93$V vs. SCE) is assigned to a two-electron step classically considered not being disturbed by this material without any electro-chemical specificity. On the contrary, the use of a Cu–Pd electrode dipped in a $PdSO_4$-based bath yields a cathodic step that is spectacularly shifted and exhibits a much smaller limit of current density ($E_{p/2} = -1.44$ V). The diffusion-controlled electro-chemical process was confirmed, and found to be strongly irreversible. The transfer coefficient estimate obtained from half-peak width measurements was found to be lower than 0.2. The catalytic efficiency of Cu–Pd interface was also compared to these palladized interfaces such as smooth platinum and polished palladium. As already reported [15], palladized surfaces such as Pt/Pd electrodes even led to a higher potential shift ($E_{p/2} = -1.36$ V) when compared with a classic carbon electrode. All kinds of palladized surfaces exhibit highly significant potential shifts with all tested RIs. However, when using as-prepared Cu–Pd electrodes, peak currents are twice smaller, suggesting that the overall electro-chemical reaction could turn out to be a one-electron process. Coulometric measurements confirmed this finding. Additionally, it has been observed that smooth copper also exhibits an one-electron step with butyl iodide, but located at more negative potentials, typically −1.82 V vs. SCE. Apparently, the occurrence of copper substrate with organic halides is highly complex since it corresponds to a reaction between CuI and the RX compound as discussed in Chapter 9.

Whatever be the nature of the modification process, the one-electron reductions of RIs at Cu–Pd surfaces could be confirmed using micro-coulometric measurements in millimolar amounts. At low reduction potentials ($E_r < -1.2$ V vs. SCE), the measured charge values are systematically very close to 1 $F{\cdot}mol^{-1}$ (Table 3.1). Since most of nitrones classically used as spin markers (being not electro-active within the potential range concerned) do not disturb the electro-chemical reductions of RIs, complementary investigations for establishing the presence of free radicals as transient in the presence of *n-tert*-butyl-α-phenyl nitrone (TBPN) were performed. In all cases, a sharp one-electron process was observed while genesis of paramagnetic nitroxides was detected. Thus, it may be established that 1-iodobutane leads to a six-ray ESR signal with the coupling constants $a_H = 3.16$ G and $a_N = 14.9$ G. This remains in good agreement with the already described data obtained at regular palladized surfaces [17].

Table 3.1. Voltammetric and coulometric data relative to some primary alkyl iodides.

RI	Glassy carbon electrode[a] $E_{p/2}$ (V)	Glassy carbon electrode[a] I_p (μA)	Cu–Pd cathode[b] $E_{p/2}$ (V)	Cu–Pd cathode[b] I_p (μA)	Palladized Pt cathode[c] $E_{p/2}$ (V)	Palladized Pt cathode[c] I_p (μA)	Coulometries[d] $E_{applied}$ (V)	Coulometries[d] n ($F\cdot mol^{-1}$)
1-iodobutane (9 mM)	−1.93	23	−1.44	27	−1.36	12	−1.2	1.0
1-iodohexane (9 mM)	−1.92	22	−1.37	28	−1.17	11	−1.2	1.1

Notes: Electrolyte: 0.1 M $TBABF_4$ in DMF. Scan rate: 100 $mV\cdot s^{-1}$. All potentials are referred to aqueous SCE.

[a]The surface area of the carbon electrode was 0.8 mm^2 whereas

[b]those of the Cu-Pd electrodes elaborated from palladium sulfate or nitrate solutions were 1.6 mm^2.

[c]Palladiated platinum electrodes showing an apparent surface area of 0.8 mm^2 were prepared from a galvanostatic Pd deposit onto Pt disk electrode using a solution of $PdCl_2$ in 0.1 N HCl solution (ratio: 0.1 g of salt in 10 mL of acid). The metallic deposit corresponded to an electricity amount of $5\cdot 10^{-3}$ $C\cdot mm^{-2}$.

[d]Coulometries were performed using a two-compartment cell with a Cu–Pd grid electrode obtained by dipping the corresponding copper substrate into $PdSO_4$ solution.

3.3.2. *Cu–Pd electrodes for the scission of C–Br bonds*

In general, it is well established that primary alkyl bromides with long chains of RBr compounds can be reduced at different solid electrodes, especially glassy carbon: for all the series, RBr compounds are cleaved according to a two-electron irreversible step at quite reducing potential values (i.e., $E_{p/2} < -2.5$ V vs. SCE). In well-dried polar solvents, smooth palladium electrodes also yield a main reduction step that occurs within a comparable potential range < −2.4 V (Figure 3.3). Additionally, the reduction of short alkyl chain of RBr compounds at palladized electrodes can be effective at much less negative potentials than −2 V, but currents are generally small. These reduction steps display a kinetic character, but this feature totally vanishes upon recurrent scans when the length of the alkyl chain increases. At a smooth copper interface, a reduction step was generally observed beyond −1.8 V together with the possible occurrence of an adsorption-like step possibly assigned to the reduction of copper oxide at moderate potentials. On the contrary, Cu–Pd modified electrodes yielded amazing results: much larger

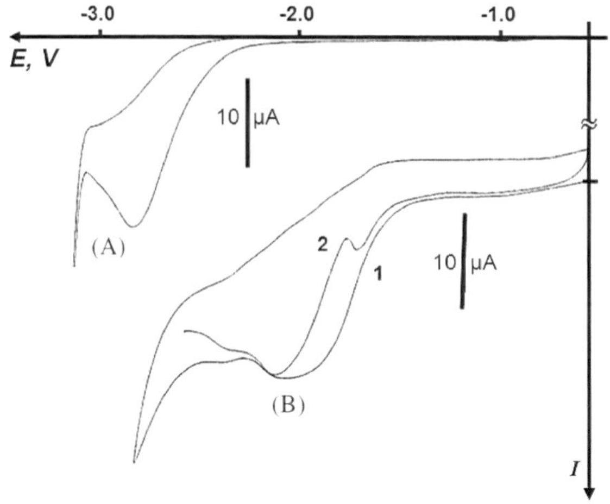

Figure 3.3. Voltammetries of 1-bromodecane (concentration: 9 mM) in 0.1 M $TBABF_4$ using DMF as solvent, recorded at different micro-electrodes. Scan rate: 100 $mV \cdot s^{-1}$. (A) Response at a glassy carbon electrode ($S = 0.8\ mm^2$). (B) Response at a Cu–Pd electrode obtained from a palladium chloride solution ($S = 1.6\ mm^2$). First two sweeps [10].

reduction steps for RBr compounds could systematically be observed much before –2.0 V (e.g., Figure 3.3 in the case of 1-bromodecane). The nature of the step obtained from the second scan is generally S-shaped (shown from the overall current to be close to an one-electron process), and this behavior could suggest a kind of self-inhibition possibly due to the irreversible adsorption — or a homo-coupling — of the transient free radical at the electrode surface. Similarly, the second scan of an impurity — "free" solution also exhibits such a potential shift, which probably suggests that catalysis is slowed down by a reduction of free active surface. The regeneration of the activated surface can only be achieved by rinsing the electrode according to the procedure described above. However, a pre-peak of a quite large current obtained with a freshly produced micro-electrode depends on the nature and the concentration of RBr compounds. The total height of the overall cathodic step is a linear function of RBr concentration. With R = *n*-propyl, *n*-butyl, *n*-pentyl, *n*-hexyl, *n*-octyl, and *n*-decyl, the total current of the reduction step (found to be globally diffusion-controlled) is

Table 3.2. Macro-reductions of some RBr compounds at Cu–Pd electrodes [15, 30].

	Glassy carbon electrode ($S = 0.8$ mm^2)		CuPd cathode ($S = 1.6$ mm^2)			Coulometry of the corresponding solution at a CuPd grid	
RBr	$E_{p/2}$ (V)	I_p (μA)	$E_{p/2}$ (V)	I_p (mA) (total current)	$E_{prepeak}$ (V)	$E_{applied}$ (V)	n (F·mol^{-1})
1-Bromopropane (15 mM)	−2.54	45	−1.83	40	−1.42	—	—
1-Bromopentane (12 mM)	−2.74	45	−2.02	40	−1.62	−1.87	1.1
1-Bromohexane (11 mM)	−2.62	32.5	−1.86	30	−1.75	−1.72	1.1

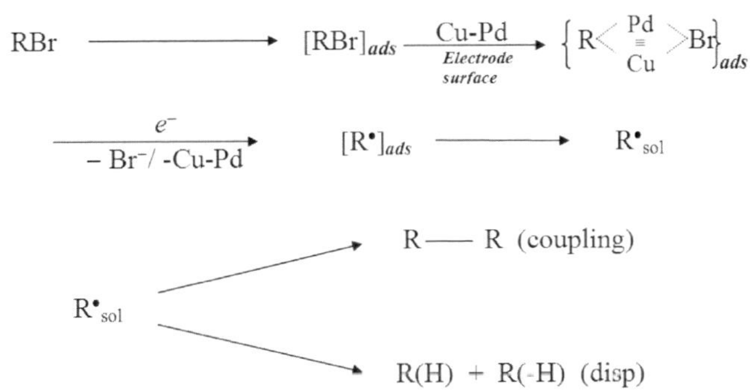

Scheme 3.1. Mechanistic proposal at Cu–Pd electrodes selectively permitting a one-electron scission of C–Br bonds applied to non-activated alkyl bromides [17].

roughly half of the current observed at a glassy carbon electrode. This experimental fact is a good argument in favor of an one-electron reduction (Table 3.2). Within the scanned potential range (i.e., $-2.5 \leq E \leq -0.5$ V vs. SCE), there is no appearance of a second step that could be assigned to the reduction of the free radical release from RBr reduction. Importantly, it is worth emphasizing that there is no evidence of a partial reduction of the alkyl halide onto a pure copper cathode at potentials above −2 V. Scheme 3.1 summarizes the global two-electron uptake of RBr compounds at Cu–Pd cathodes.

3.4. Capping Carbon Surfaces by Cu–Pd Layers

In order to produce cheap and strongly activated electrodes, Cu–Pd deposits onto glassy carbon electrodes were done. For that, a galvanostatic copper deposit was obtained from a copper nitrate solution prepared by dissolving 0.1 g of salt in 100 mL of 0.1 N HNO_3. The charge was limited to 5×10^{-3} C·mm^{-2} whereas the current value was fixed at 0.5 mA. After obtaining a copper deposit onto glassy carbon ($\approx$ 0.2 μm estimated average thickness), the electrode was shortly dipped into a palladium sulfate solution. The glassy carbon surface turned out shiny. Its use as an electrode in voltammetry showed attractive results since the activation looked extremely good. If the main reduction step declines upon repetitive sweeps, a short standby at 0 V may regenerate most of the original current. It is expected that through the copper deposit, a finely divided layer of Cu–Pd alloy could be obtained, which operates as a quite large activated surface. So far, this capping procedure was only achieved onto a carbon substrate (Figure 3.4).

The latter observation would suggest that the use of such Cu–Pd thin layer electrodes at much less negative potentials than those

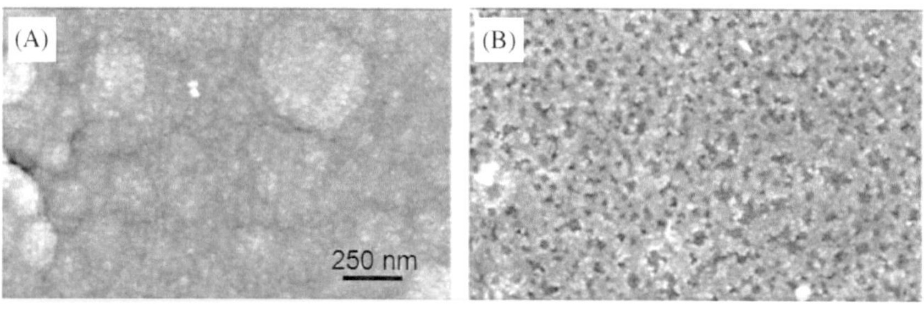

Figure 3.4. SEM images of the Cu–Pd layer deposited at GC surface (A) before and (B) after reduction of 1-bromodecane (concentration 2×10^{-2} M L^{-1}). The alloy layer (shown in (A)) has been obtained after a dipping of a copper sheet into $PdSO_4$ (see experimental section) during 2 min. The structural change of the layer (consecutively to the catalytic reduction of the RBr compound) was obtained by electrolysis at −1.9 V vs. SCE after that 2 C cm^{-2} have passed through the cell is amazing. Average current density: 0.5 mA cm^{-2} [17].

already reported with conventional electrode materials (as developed above) could be efficient, at least with RBr compounds, to extend possibilities when exclusively one-electron processes are needed. In order to confirm this observation, a complete series of coulometric experiments were performed for a large series of primary alkyl halides. Cu–Pd electrodes formed from palladium sulfate or nitrate solutions (obtaining of a bright metallic deposit) could be reused for a large number of experiments without any apparent change in efficiency. Finally, product analyses by GC/MS of the electrolysis products showed that R–R dimers and/or mixtures of R(H)/R(–H) in equal amounts were obtained with R = C_8, C_{10}, and C_{12}.

3.5. Comments and Conclusion

In order to correlate the electro-chemical efficiency of our as-prepared electrodes, the primary objective, irrespective of the palladium precursor used, was to unambiguously identify the structure of the deposited layer. For that, the electronic diffraction technique proved to be effective since the nanometric character of the metallic particles rendered a valid identification difficult using a conventional XRD analysis. With all deposits, the line profile of the electron diffraction enabled to identify the layer as a disordered CuPd alloy thanks to a perfect match with the diffraction data given by Nekrasov [18] and more specifically by Zhu *et al.* [19]. Indeed, the latter performed a thorough study of clusters of disordered CuPd (i.e., nanoparticles) via a theoretical approach using the bond order metal simulation (BOS) model and the corrected effect medium (CEM) theory. Their simulation model predicts diffraction patterns and relative peak intensity, which are in good agreement with the reported experimental data.

The TEM investigations had demonstrated the elaboration of the disordered CuPd phase, hence the data regarding the Cu–Pd system must be considered (Table 3.3) since this structure is not expected at room temperature.

A noteworthy result of this study clearly showed that the developed work succeeded in synthesizing very simply nanoparticles of the disordered CuPd alloy by plunging a copper substrate (grid,

Table 3.3. Crystal structure information for the *fcc* and *bcc* CuPd alloys.

Phase-Formula	CuPd	"CuPd(α)"	CuPd(β)
Pearson symbol	cF4	tP4	cP2
Space group	$Fm\bar{3}m$	$P4/mmm$	$Pm\bar{3}m$
Strukturbericht designation	$A1$	$L1_0$	$B2$
Crystal structure ● Pd ○ Cu ◐ Cu or Pd	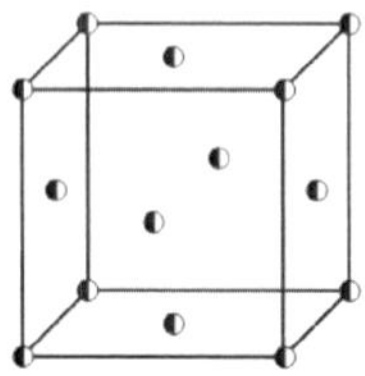	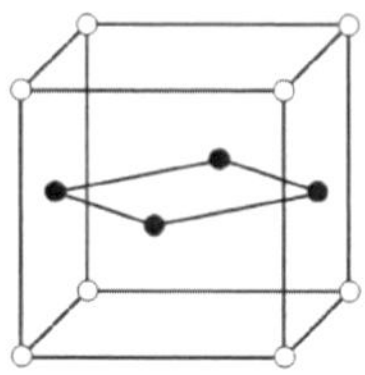	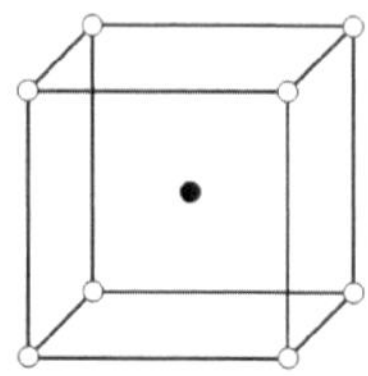

Note that the $L1_0$ sur-structure is given only for comparison since "CuPd(α)" phase does not exist for energetic reasons [17].

sheet, Cu electro-deposit) into a *fresh acidic* solution of a palladium salt. Other competitive methods are more complicated and usually involve the polyvinylpyrrolidone (PVP) as a stabilizer to ensure the formation of nanoparticles. Basically, the alloy at nanometric size is obtained from Esumi's method (or adaptations) by thermal decomposition of mixtures of copper and palladium precursors in high boiling organic solvents [18, 20–22] or by condensing Pd and Cu atoms at 350°C under UHV [23, 24]. Interestingly, Cu–Pd alloys are also known to be attractive as catalysts in gas phase to enhance the selectivity in the hydrogenation of dienes [24] and the reduction of NO by CO [23, 25].

Thus, this chapter concludes that a simple redox displacement reaction between Cu^0 and Pd^{2+}, carefully managed in acidic solution at room temperature, offers the opportunity to produce a thin and very stable layer characterized by a well-crystallized nanometric CuPd alloy. Such a displacement can be simply achieved in the presence of various Pd(II)-based salts like sulfate, nitrate, and chloride. However, formation of a pure CuPd alloy could be observed only with palladium sulfate and nitrate. Besides, electro-chemical data obtained from the reduction of a large series of organic halides (mainly iodides and bromides) showed that cathode materials very strongly activate the scission of the carbon–halides bond, occurring sometimes at –1V, to more positive potentials. Additionally, there was no strong passivating phenomena during electrolyses even if a moderate decay of the cathodic current could be observed in the course of electrolyses. Deposits were shown to act as porous materials and their structures may change with time; this could contest the assumption that palladium reacts (as catalyst) with alkyl halides [16]. It is worth stressing about the use of palladium deposits as modifier of the cathode surface (e.g., deposits onto platinum or glassy carbon). The catalytic action of palladium probably lies in the use of finely divided deposits of nanosized particles. The mode of catalysis is not yet fully determined, but it is rather conceivable that the insertion of palladium inside the carbon–halide bond gives rise to a chain such that a strongly adsorbed C–Pd–X linkage could occur. The insertion may justify the catalysis, especially owing to the constant regeneration of the copper–palladium alloy (see Scheme 3.1) supported by the strong electronic interaction between Pd and Cu upon alloying with a specific feature [25, 26]. In the proposed process, the adsorption rate and the insertion of palladium inside the C–halogen bond would be fast toward diffusion of the electroactive species. However, as stressed within the described work, catalysis by Cu–Pd surface is of huge interest for C–Br bond cleavage: the C–iodide cleavage reaction is also made very easy, but data are quite similar to those already observed with all kinds of palladized surfaces. Very often (but not systematically), the potential shift is so large that the cathodic reaction is fundamentally changed, and turns out to be mono-electronic: the method therefore can be seen as an efficient source of free radicals (with a possible coupling reaction outside the cathodic layer) more or less strongly adsorbed at the interface. These results agree with estimations given by Lund [27–29] about standard potentials that correspond to the reduction of free alkyl radicals in DMF between –1.39 and –1.72 V vs. SCE under quite similar experimental conditions.

References

[1] D.G. Peters, in *Organic Electrochemistry*, eds. H. Lund and O. Hammerich, M.M. Dekker, New York, Basel, p. 341 (2001).

[2] O.R. Brown, in *Physical Chemistry of Organic Solvent Systems*, ed. A.K. Covington, Plenum Press, New York, p. 747 (1973).

[3] E. Coulon, J. Pinson, J.-D. Bourzat, A. Commerçon, J.P. Pulicani, *Langmuir*, **17** (2001) 7102.

[4] E. Steckhan, in *Organic Electrochemistry*, eds. H. Lund and O. Hammerich, M.M. Dekker, New York, Basel, p. 1103 (2001).

[5] S.B. Rondinini, P.R. Mussini, F. Crippa, G. Sello, *Electrochem. Commun.* **2**(7) (2000) 491.

[6] G. Kokkinidis, *J. Electrochem. Chem.* **201**(2) (1986) 217.

[7] R. Kossai, J. Simonet, G. Jeminet, *Tetrahedron Lett.* **20** (1979) 1059.

[8] P. Hapiot, J. Simonet, *Electroanal. Chem.* **23** (2010) 105.

[9] C. Cougnon, J. Simonet, *Platin. Met. Rev.* **6** (2002) 9.

[10] J. Simonet, P. Poizot, L. Laffont, *J. Electroanal. Chem.* **591** (2006) 19.

[11] J.M. Savéant, *J. Am. Chem. Soc.* **114** (1992) 10595.

[12] C. Andrieux, I. Gallardo, J.M. Savéant, *J. Am. Chem. Soc.* **111** (1989) 1620.

[13] J.M. Savéant, in *Advances in Physical Organic*, ed. T.T. Tidwell, Academic Press, New York, p. 177 (2000).

[14] J. Grimshaw, J.R. Langan, G.A. Salmon, *J. Chem. Soc. Faraday Trans.* **90** (1994) 75.

[15] J. Simonet, *Electrochem. Commun.* **7** (2005) 619.

[16] J. Simonet, *J. Electroanal. Chem.* **7** (2005) 34.

[17] P. Poizot, L. Laffont-Dandras, J. Simonet, *Platin. Met. Rev.* **52** (2008) 84.

[18] I. Nekrasov, V. Ustinov, *Dokl. Acad. Sci. USSR, Earth Sci. Sec. (Engl. Transl.)* **328** (1993) 128.

[19] L. Zhu, K.S. Liang, B. Zhang, J.S. Bradley, A.E. DePristo, *J. Catal.* **167** (1997) 412.

[20] K. Esumi, T. Tano, K. Torigoe, K. Meguro, *Chem. Mater.* **2** (1990) 564.

[21] J.S. Bradley, E.W. Hill, C. Klein, B. Chaudret, A. Duteil, *Chem. Mater.* **5** (1993) 254.

[22] N. Toshima, Y. Wang, *Langmuir* **10** (1994) 4574.

[23] S. Giorgio, C. Henry, *Microsc. Microanal. Microstruct.* **8** (1997) 379.

[24] S. Giorgio, H. Graoui, C. Chapon, C. Henry, in *Metal Clusters in Chemistry*, eds. P. Braunstein, L.A. Oro and P.R. Raithby, Wiley-VCH, New York, Chapter 2, p. 1194 (1999).

[25] Y. Debauge, M. Abon, J.C. Bertolini, J. Massardier, A. Rochefort, *Appl. Surf. Sci.* **90** (1995) 15.

[26] A. Rochefort, M. Abon, P. Delichère, J. C. Bertolini, *Surf. Sci.* **294** (1993) 43.

[27] D. Occhialini, S.U. Pedersen, H. Lund, *Acta Chem. Scand.* **44** (1990) 715.

[28] D. Occhialini, J.S. Kristensen, K. Daasberg, H. Lund, *Acta Chem. Scand.* **46** (1992) 474.

[29] D. Occhialini, K. Daasberg, H. Lund, *Acta Chem. Scand.* **47** (1993) 1100.

[30] J. Simonet, Unpublished results.

electrode surfaces, onium salts were considered [6–9]. More specifically, persubstituted ammonium, iodonium or sulfonium ions were used as convenient sources of free radicals under one-electron transfer. Until now, the most suitable reaction for generating aryl radicals is quite certainly the reduction (chemical or electro-chemical) of aryldiazonium salts. Thus, aryl radicals can be formed through a one-electron process at very moderate reducing potentials and this concomitantly permits the addition of those radicals onto solid cathode surfaces. This process has been frequently used for grafting these radicals onto carbons [10, 11]. Due to the fact that transient radicals are readily reduced under conventional conditions [12–14], a need for specific electrodes with electro-catalytic properties could redirect electro-chemical processes in a different way. Thus, chemically modified electrode materials could change the nature of the reduction process favoring the formation of free alkyl or aryl radicals within potential ranges where these species may show some stability (toward the electron, the solution bulk, and the solid material at which they are formed). Recent discoveries on electrodes giving priority to transition metals (such as Ag, Pd and Cu) used as the cathode material have enabled the development of new techniques and reactions that were unknown so far [15–17]. Therefore, the growing interest in copper, palladium, silver, gold, iron, nickel, and tin, within the cathodic domain, certainly offers new perspectives. In particular, new methods with regard to transition metals as cathode materials for reducing non-activated organic halides such as alkyl-iodides (RIs) were successfully demonstrated at palladium and palladized interfaces [18]. Under such conditions, large energetic gains in these processes were already observed; in this context, the additive catalytic properties of some materials such as nickel–silver, nickel–palladium, or gold–silver appear to be of interest [19].

Given the great interest in simple and efficient methods of grafting various functional groups to graphitic materials and in light of the above discussion, the aim of this presentation is to focus on the facile one-electron cleavage of alkyl iodides (and further of ω-iodoalkylferrocenes $Fc–C_n–I$ with $n = 3, 4, 6$ and 12) at

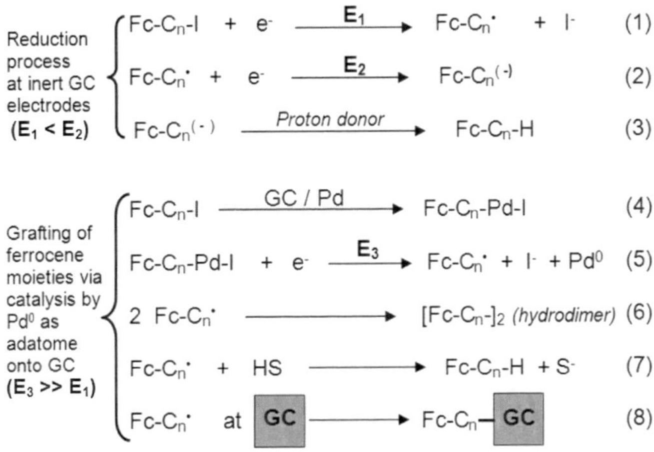

Scheme 4.1. Reduction of marked alkyl-iodides RIs at inert and metal-doped glassy carbon electrode: an overview in the case of palladium Ref. [32].

carbon electrodes doped by very thin electro-catalytic layers of transition metals, allowing one to modify these surfaces through covalent grafting of ω-ferrocenylalkanes specifically used as marked reagents (Scheme 4.1) in order to quantify *in situ* the reactivity and/or the grafting activity of possible transients. This will enrich the pool of precursors for radical grafting processes with commercially available or easy-to-prepare haloalkyl derivatives with redox active or other groups with desired physico-chemical response(s); in addition, very moderate potentials required for such grafting allow high functional group tolerance. Galvanostatic deposits easily prepared permit the preparation of metal-doped electrodes with extremely small amounts of electricity ($\approx 10^{-6}$ coulombs per mm^2 for deposition from bivalent cations). It was deliberately chosen to preferentially run electrolyses in strongly polar solvents such as propylene carbonate (PC) possibly mixed with ethylene carbonate (dielectric constants of 65 and 90, respectively) for providing strong C–I bond polarization and then still improving their cathodic reactivity. Electro-catalytic reactions are easily performed with deposits of palladium, gold, platinum, rhodium, and also with copper, nickel, and cobalt. Potential ranges were carefully limited and cathodic boundaries definitively less negative than −1.7 V vs. Ag/AgCl electrode. Not exceeding this potential was considered as crucial in order to avoid carbon charge when this material is employed as a substrate for achieving metal deposits. Problems bound to the nucleophilicity of the electrode material would induce different reactivity, especially, the alkyl grafting via S_N2-like reactions on alkyl halides [20, 21].

4.2. Primary Alkyl Iodides Electro-Chemically Reduced at GC Doped with Palladium

The reduction of 1-iodoalkanes (RIs) at GC in nonaqueous polar solvents has been discussed previously [20]. In particular, in DMF containing tetraalkylammonium salts, it was established that recurrent scans at the level of the main reduction step (Figure 4.1, peak I, curve A) leads to a progressive cathodic shift of the two-electron process resulting in immobilization of alkyl groups at the carbon surface. The key step of this immobilization was assigned [21] to the cathodic charge of GC. Consequently, nucleophilic carbon sp3 centers capable of reacting with RI were formed at the GC surface through carbon sp^2 sites via the reduction of polyaromatic or graphitized inclusions in the material, especially when rather negative potentials ($E \ll -1.7$ V) are achieved (see Chapter 12). On the contrary, doping of GC with very small amounts of palladium at freshly polished carbon (average deposit of 0.4×10^{-9} atom cm^{-2}) permits one to split the two-electron peak into two reduction steps. The first step, now mono-electronic (peak II, curve B, $E_{p/2} = -0.98$ V), is assigned to the formation of the alkyl radical $R^{\bullet}$. The reduction of alkyl iodides at GC covered with Pd nanoparticles was also achieved [22]. These free alkyl radicals were characterized by ESR spectroscopy as their spin-adducts with α-phenyl-*N*-*t*-butylnitrone used as *in situ* radical trap (for $C_{10}H_{23}^{\bullet}$, $g = 2.0065$ with *hfc* constants $a_N = 14.54$ G, $a_H = 2.39$ G). The test of the oxidation of potassium ferrocyanide performed after sonication of this electrode confirms

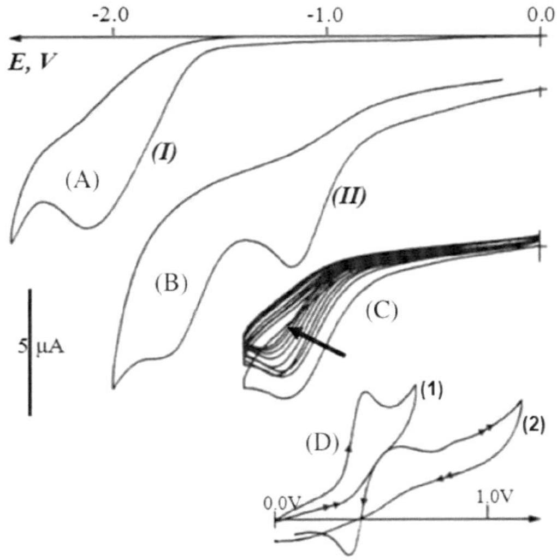

Figure 4.1. 1-iododecane (11 mmol L^{-1}): voltammetries in PC + $TBABF_4$. Working electrode: GC, area: 0.8 mm^2. Scan rate: 50 mV s^{-1}. (A) Simple reductive cyclic voltammogram at bare GC (two-electron step). (B) Use of the same cathode after a galvanostatic deposit of Pd (total charge: 10^{-6} C mm^{-2}). (C) Same experiment with nine successive scans. (D) Response at a bare GC/Pd electrode (first curve) compared to that of the electrode modified according to (C) in a solution of potassium ferrocyanide in a KCl saturated aqueous solution (second scan) [32].

the formation of a superficial layer strongly inhibiting electron transfer as shown in Figure 4.1, curve D.

4.2.1. *One-electron deposition from ω-iodoalkyferrocenes by means of palladium catalysis*

The method described above with primary alkyl iodides RIs was applied to ω-iodoalkyferrocenes when GC surface was partially covered with palladium particles. The efficiency of this process (still carried out in PC containing 0.1M $TBABF_4$) is well attested by Figure 4.2. While the reduction of ω-iodohexylferrocene needs the potentials $\ll -1.8$ V (Figure 4.2, curves A), the Pd-catalyzed

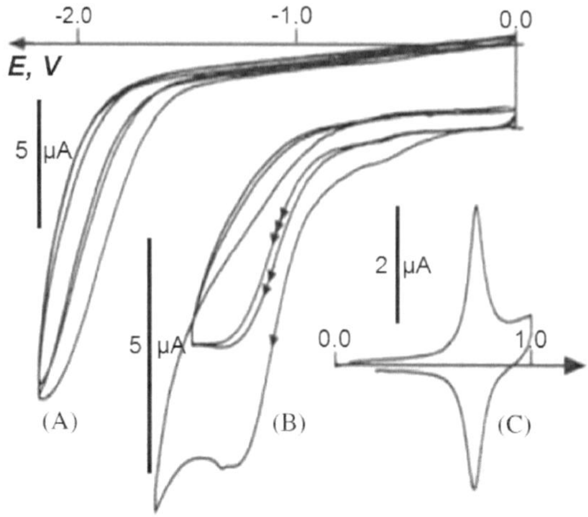

Figure 4.2. Voltammetries of Fc–C_6–I (12 mmol L¯1) in PC + 0.1 M $TBABF_4$ at a smooth GC electrode. Scan rate: 50 mV s^{-1}. Area of GC electrodes: 0.8 mm^2. (A) At a bare GC electrode: first three scans. (B) After coverage with a thin palladium deposit (1.4×10^{-6} C cm^{-2}): first three scans. (C) Response of the electrode obtained in (B) after a 30 s hold at −1.2 V. Cycling in DMF + $TBABF_4$ [32].

process is displayed in B with quite a large anodic shift of the potential by almost 1 V. Three successive scans between 0 and −1.5 V provoke a dramatic decay of the catalyzed step. The results of this grafting, assigned to the radical immobilization, are shown in Figure 4.2(C). A clear response of immobilized ferrocene at GC is seen through an almost perfectly reversible adsorption-like peak (curve C, anodic and cathodic currents are equal, E_p = +0.64 and +0.66 V, respectively) corresponding to the superficial concentration $\Gamma_{Fc} = 4.2\times10^{-9}$ mol cm^{-2}. Variation of the scan rate (v) does not visibly change the shape of this peak when v values remain moderate (curves D). The half-peak width at moderate scan rates is very close to 90 mV, which corresponds to an one-electron process for the cathodic immobilization process. Additionally, it can be seen that the variation of peak currents with v is very linear (Figure 4.3, inset) than the attesting

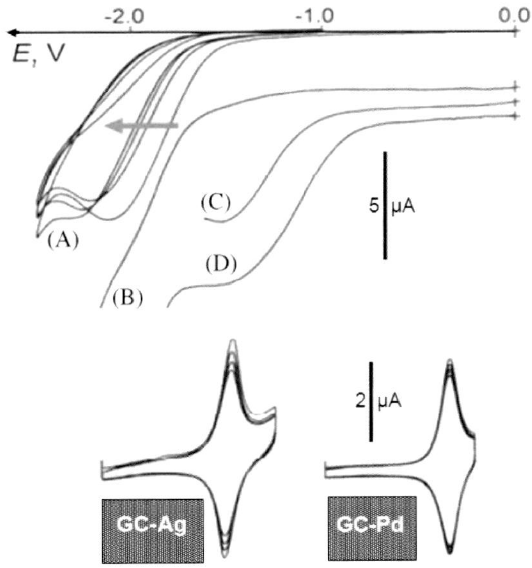

Figure 4.3. Voltammetries of Fc–C_3–I (15 mmol L^{-1}) in PC/ethylene carbonate (50:50 v/v) + $TBABF_4$. Use of smooth GC electrode. Scan rate: 50 mV s^{-1}. (A) Four first scans at a GC electrode. (B) Coverage of GC by gold (deposit: 1.25×10^{-6} C cm^{-2}). (C) Coverage of GC by silver as a catalyst (deposit: 1.25 C cm^{-2}). (D) After a galvanostatic deposition of palladium (charge: 1.25 C cm^{-2}). Below: cycling (in DMF + $TBABF_4$ solutions) of two GC/metal electrodes after holds of 30 s at −1.5 V attesting a dense grafting of ferrocene [32].

immobilized redox systems. Moreover, superficial concentration of ferrocene appears to be quite large, but such values of Γ_{Fc} values could indicate a kind of radical ablation of carbon, which was already similarly observed in processes involving silver nanoparticles. Also, the roughness factor of the carbon surface (obtained by the polishing procedure) is quite certainly higher than 5. The formed alkylferrocene layer is expected to be almost compact if one refers to the literature data [23–25] for gold substrates (for example, ferrocene moieties branched onto long alkyl chains ending with a mercaptan function to ensure the gold–thiol attachment).

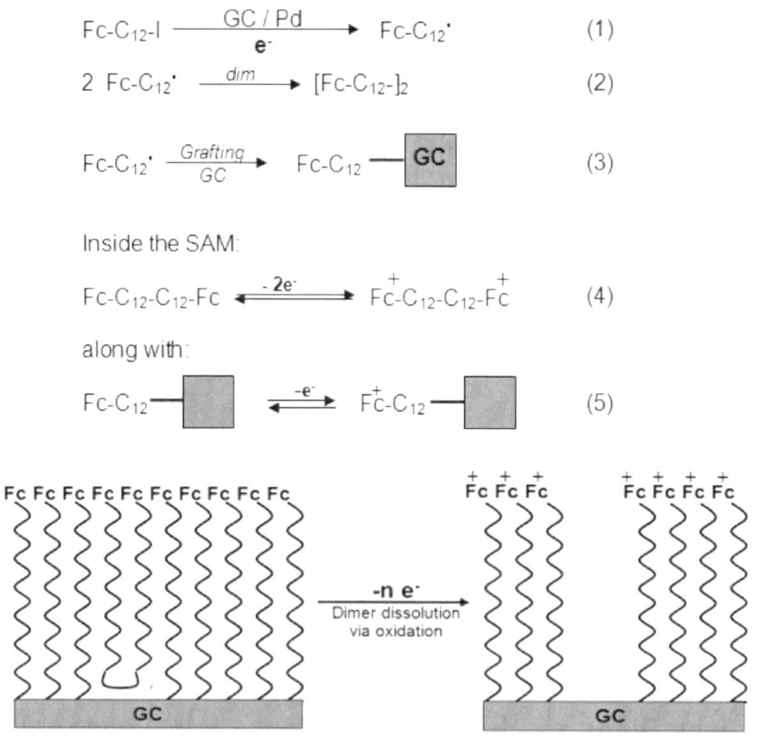

Scheme 4.3. Grafting of dodecylferrocene to GC with parallel coupling of $Fc–C_{12}^{\bullet}$ radicals and evolution (rate of the homo-coupling reaction of transient radical) of primarily formed monolayer during anodic cycling [32].

4.2.4. *Generation of three-dimensional (3D) electrodes made of GC, graphene, and transition metal microparticles*

While still employing RI substrates is a convenient model in indirect scissions of C–X bonds (here, those of carbon–iodide) at carbons, we further imagined GC electrodes modified by layer of natural graphite (NG) or/and graphene, those deposits being mixed to fine particles of transition metal (Pd, Ni, Au, Ag, Rh, etc.) either by insertion of nanoparticles or extremely thin potentiostatic deposits from acidic solution of salts (Scheme 4.4). In the present chapter, we focus only on the catalytic effect of deposited palladium, nickel, and gold particle on and into thin layers of graphene.

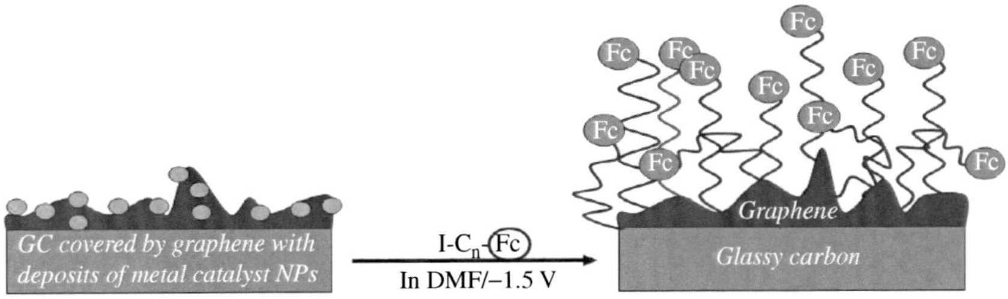

Scheme 4.4. Example of composite cathode materials used for deposit of ferrocene moieties present on alkyl chains immobilized at graphene [19].

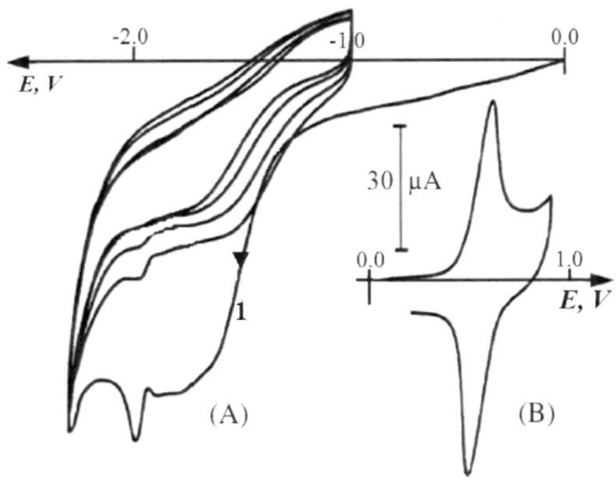

Figure 4.6. Reduction of Fc–C_{12}–I (5.3 mmol L^{-1}) at smooth GC electrode covered by a thin deposit of natural graphite and galvanostatically doped by gold (from $AuCl_3$). Doping coverage: $\Gamma_{Au} = 2\times10^{-10}$ mol cm^{-2}. Apparent surface area of the electrode: 7 mm^2. Scan rate: 50 mV s^{-1} [30].

It appears that the concept is quite valuable since it allies the catalytic capacity of graphene and its ability to act as a radicalophilic material toward free radicals generated by cathodic scission of the RI at about −1.2 V (case of Pd^0) under general conditions already described. Thus, as stressed in Figure 4.6, GC–graphite electrodes modified by very thin coverages of transition metals may induce profound discrepancies: as a matter of fact, Ni^0 deposits exhibit a rather poor ability in achieving a catalytic R–I bond scission, while beyond −1.7 V, palladium is extremely efficient (scission of the R–I bonds from −1 V). With a same electrode surface area, it is,

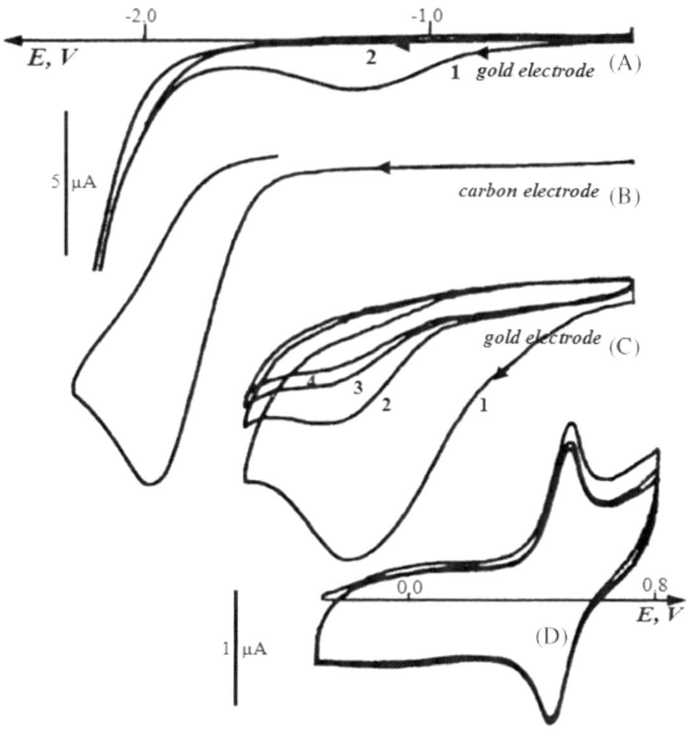

Figure 4.7. For comparison, the reduction of marked alkyl iodides was studied at smooth gold electrode as exhibited above (curve A). A reduction peak is observed at about −1.2 V assigned to the reduction of the adsorbed RI onto Au. This is totally different at GC electrodes (curve B). Voltammetries depend on the standby length at 0 V or the scan rate at gold to fill the surface by alkyl radicals (curve C). Results obtained via (A) or (D) after standby are coherent and correspond (via the grafting of $-C_{12}$–Fc and oxidation step of Fc) to a coverage onto gold of $\Gamma_R \sim 1.0 \times 10^{-9}$ mol cm^{-2}.

however, worth noting that the ferrocene surfacial concentration is roughly the same, and catalyst coverage is close to a monomolecular layer.

Micro-depositions of gold onto GC–graphene electrodes allowed one to get catalytic steps at about −1.5 V (Figure 4.7). Very unexpectedly, the specific use of smooth Au to catalytically cleave C–I bonds was recently found [31] possibly via the pre-adsorption (in organic solvent) of alkyl iodides followed by a fast reduction that leads to the robust immobilization of alkyl chains, leading then to a dense coverage at around −1.2 V as shown in Figure 4.7 curve A, first scan. A large difference with GC (shown in curve B) should be emphasized. For the moment, the scope of this highly unexpected catalytic phenomenon remains limited to primary alkyl-iodides (RIs). Coverage levels were found to be dense and corresponded (rough gold surfaces) to Γ_R or $\Gamma_{Fc} = 1.0\times10^{-9}$ mol cm^{-2} (as pointed out in Figures 4.7(C) and 4.7(D)).

4.3. Merging and Growing of SAMs from GC: Concept of Self-Elaborated 3D Interfaces

Lastly, there is an attempt to develop in this chapter the unexpected swelling of GC through graphite particles contained in this material often in a large amount, which depends on the pyrolysis conditions used for its synthesis. Also, as already stressed before, the role of the GC electro-chemical charge at < −1.7 V and its effects (charge of residual graphite followed by a dense grafting of long chains onto graphene network) may produce tremendous modifications of the external zones of the material. Such effects associated with the swelling of the surface via exfoliation are discussed elsewhere in the book.

Swelling procedures require the insertion of tetraalkylammonium cations (TAA^+). The use of the electro-chemical technique and the subsequent charge process of the material strongly influence the amplitude of the phenomenon. Thus, graphite nanoflakes present at the GC surface (or imbedded in this material) can be exfoliated through swelling and react via an SN process with long chain electrophiles such as RIs. If the GC surface is covered by Pd^0 particles as a catalyst, the process is quite similar, but it occurs from −1.5 V via the addition of alkyl radicals onto the graphene (or graphite) surface. Such an additional event (often ignored) corresponds to the dissociation of graphite plans during the charge until more or less an almost total material degradation via its exfoliation. In the present situation, the effect is dramatically amplified owing to the roughness

Chapter 5

Electro-Chemical Generation of Silver Nanoparticles: Catalytic Reactions at Solid Conductor Interfaces

5.1. Introduction

Easy and stable chemical modifications applicable to solid electrode surfaces still remain a challenging domain in the field of interfaces and electro-chemistry [1–3]. Then, studies devoted to solid surfaces derived from carbon have been intensively achieved by means of several modes of grafting: use of existing carboxylic groups [4] or created by chemical or electro-chemical oxidation of carbons, additions of free aryl radicals [5, 6].

Another way of producing modified electrodes that lies in the localized generation of very small-sized precious metal particles possibly used as potential catalysts remains a highly challenging field and electro-chemistry may appear to play a foremost role. Then, solid surfaces such as graphite and glassy carbon (GC) have been intensively modified by means of several modes of grafting [5–7]. Electro-deposition of functionalized polymers [7] appear to be promising to modify interfaces, especially when those polymers can themselves be decorated by redox groups or organometallic centers, strategically important in the field of electro-catalysis [8]. Mostly, for more than 10 years, silver metal was found to attract major attention

(A) (B)

Figure 5.1. Morphology of modified silver grains (size: 250 μm) *after* reaction with $PdSO_4/H_2SO_4$ leading to coverage or fragmentation of micro-particles superficially tailored from Ag–Pd [30].

Ag–Pd powder considered now as a new substrate was rinsed with water, then alcohol, and finally acetone. In the end, the powder was dried in an oven at 60°C for 1 h. Under such conditions, the Ag–Pd powder was easily stocked for months, reused, and easily reactivated into with a Pd–Ag alloy simply by repeating the steps described above. The thickness of the Ag–Pd capping at silver grains was not precisely determined, but an assessment of a few micrometers is reasonable. Obviously, the thickness of the additional layer depends on the contact time with the Pd^{++} solution. Figure 5.1 displays the morphology of modified silver grains (extremely small, mainly <0.5 μm and associated with amorphous grains of large active area) that appears like a concretion of small size elements apparently amorphous. These particles, having an average size of the order of 1 μm, are assigned to Ag–Pd microstructures highly reactive toward alkyl iodides.

Alternatively, it appeared to be of interest to produce solid electrodes modified by Ag–Pd particles. These electrodes were obtained at different metallic substrates by galvanostatic primary deposits of silver from a solution of $AgNO_3$ (10 g L^{-1} in a 0.1 N HNO_3 solution).

The imposed current was of the order of 0.5 mA mm^{-2}, and the average amount of electricity was in all cases 1.2 millicoulombs mm^{-2}. The silvered deposit was transformed into an Ag–Pd layer by contact with a palladium salt solution (see above).

5.2.2. *Electro-chemical procedure: Solvents, electrolytes, and electrodes*

In this study, the results mainly concern 0.1 M solutions of tetra-*n*-butylammonium tetrafluoroborate ($TBABF_4$) and tetra-*n*-butylammonium hexafluorophosphate ($TBAPF_6$) into different aprotic dipolar solvents. The solvents used were dimethylformamide (DMF) and acetonitrile (quality for analyses) purchased from SDS and propylene carbonate (quality for syntheses) purchased from Merck. It is worth mentioning that procedures given hereafter do not require especially dry solutions.

In this chapter, all potentials are given versus aqueous Ag/AgCl. The electro-chemical instrumentation is described throughout this book.

Electrodes used in voltammetry had an apparent surface area of 0.8 mm^2. GC, palladium, nickel, copper, iron, and platinum disks when used as smooth substrates were always carefully polished with silicon carbide paper (Struer) or with Norton polishing paper (types 02 and 03) before a silver deposit was made. Before use, the electrodes were twice rinsed with water, then alcohol, and finally acetone. Lastly, they were dried with a hot air stream.

5.2.3. *Coulometries and electrolyses*

Electrolyses and coulometric measurements relative to the cathodic activity of alkyl iodides in the presence of the Ag–Pd powder were carried out in three-electrode system cells with a catholyte volume of about 5–10 mL. The two compartments are separated by a glass frit (weak porosity). Experiments here described were completed on quite small substrate amounts of RI (typically, a concentration in the cathodic bulk <1 mmol L^{-1}). The building of silver electrode was achieved under argon atmospheres.

5.3. Methodology

Thin silver layer deposition method onto solid conductors (carbon, graphite, platinum, gold, and many other metals) is derived from a previous report devoted to the cathodic reduction of primary alkyl iodides (RI) at a massive silver electrode covered with a fine layer of Ag–Pd [28]. In the present study, a fine Ag–Pd powder is efficiently dispersed (homogenous suspension) in a liquid electrolyte. The first evidence can be simply visualized by the experiment displayed in Figure 5.2 that demonstrates the "migration" of silver from an Ag–Pd sample disposed at the bottom of the cell to a polarized solid electrode placed in the center of the cell by means of RI considered as a reactive vector toward silver.

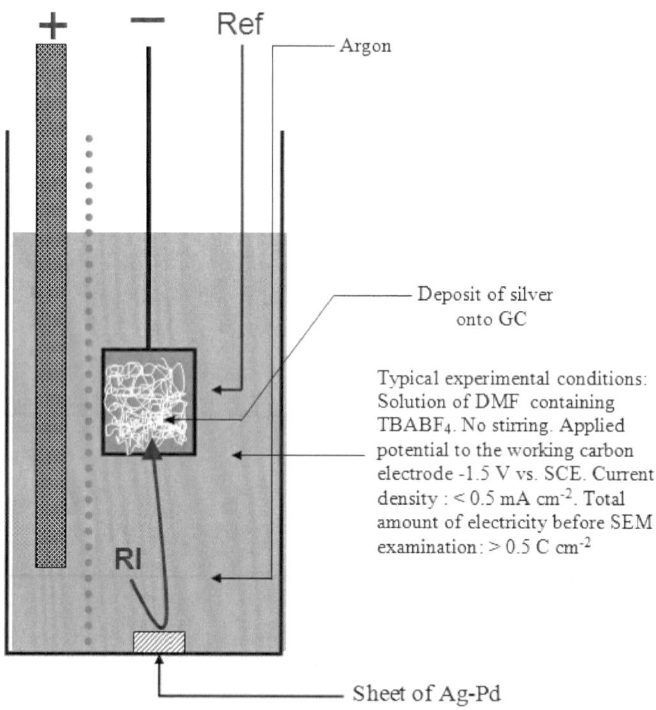

Figure 5.2. Experimental device establishing the migration of Ag from an Ag–Pd plate (bottom of the cell) to the polarized cathode whatever be the nature of the conductor (from an organic conducting polymer to iron or tin plates) (from Ref. [28]).

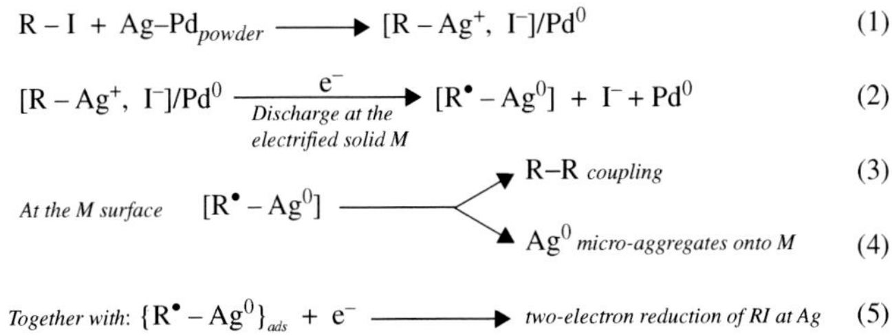

Scheme 5.2. Proposed mechanism for the deposition of Ag^0 NPs onto an electrified metal **M** through the activation of RI at Ag–Pd (from Ref. [27]).

Therefore, a solid conductor (carbon or metal) covered by a layer of Ag–Pd contributes to the activation of primary RI compounds (owing to a high activation energy) and a catalytic process takes place (Scheme 5.2, reaction (3)), allowing both the homo-coupling of R, and the deposition of Ag^0 at any electrified conductor.

The apparent role of Pd only permits one to strongly activate the reaction of silver toward the C–I bond. The contact with the solid electrode permits the discharge of the superficial reactive layer. At this stage, the free alkyl radical $R^{\bullet}$ is formed and the potential applied to the solid fixed electrode is not negative enough to allow its fast reduction. Under these conditions, the homo-coupling was pointed out (Eq. (3)). In the meanwhile, a chemisorption of the radical is also believed to happen. The radical formation appears to be the key point of the process: it permits one (Eq. (4)) to liberate an atom of silver at (or in the immediate vicinity of) the electrode. If Scheme 5.2 is valid, one may expect that a silvered electrode (in principle *only* covered by very small silver aggregates) may be produced. As a matter of fact, after a certain amount of electricity has passed, the regular reduction of RI at a silvered cathode (Eq. (5)) turns to be the major process. Experiments described hereafter allow confirming these reactions and also show important properties relative to the nature of the silver layers produced as well as the capability to show unforeseen electro-catalytic properties (when compared to those of smooth silver).

5.4. Silvering Processes

5.4.1. *Voltammetries*

The process described below in Scheme 5.3 depicts the silvering of most of the solid conductors. Thus, 1-iodooctane first reduced (Figure 5.3) at the GC micro-electrode exhibits (curve I) a two-electron step as expected [21] at a potential close to −2 V. Let us point out that the half-peak potential $E_{p/2}$ is not constant upon repetitive scans. If some Ag–Pd powder is added to the solution and efficiently stirred before scans, the voltammetric peak at GC progressively moves toward less negative potentials (curves II). *In situ* "metalized" electrodes (like iron and gold, curves III and IV, respectively) display a behavior similar to that of GC. A shift of about 1 V could be observed after several scans and this enables a well-defined and totally reproducible step: it is the case of iron, presumably because the discharge of Ag–Pd grains (Eq. (2), Scheme 5.2) is made difficult owing to the presence of superficial oxides and requires quite negative potentials to trigger silver deposition. Additionally, iron cannot be used for galvanostatic Ag^0 depositions from acidic solution of $AgNO_3$. On the contrary, smooth gold obviously generates a silver deposit from the first voltammetric sweep.

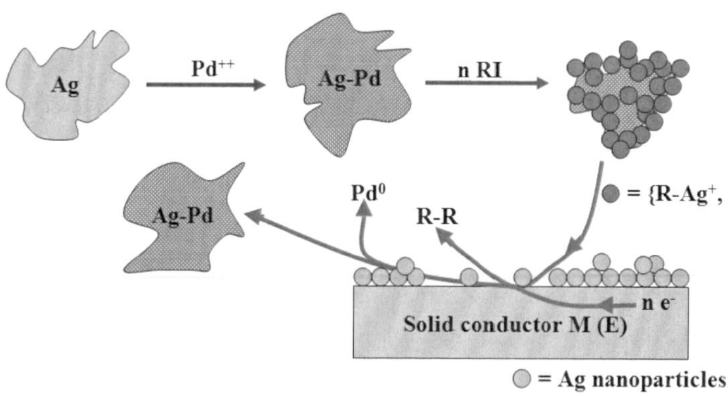

Scheme 5.3. The coverage of Ag–Pd micro-particles and its reaction with alkyl-iodide RI permits an Ag^0 deposit at polarized conductors (applied potential = E) (From Ref. [27]).

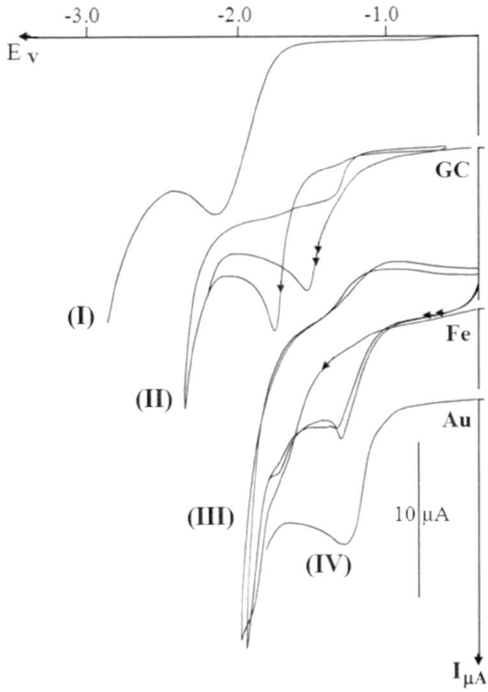

Figure 5.3. Voltammetric responses of 1-iodooctane (concentration: 6 mmol L^{-1}) in 0.1 M $TBABF_4$/DMF. Scan rate: 50 mV s^{-1}. Reference electrode: aqueous SCE. All cathode surface areas: 0.8 mm^2. (I) Response at a GC electrode (first scan). (II) Response at the same electrode after dispersion of an Ag–Pd powder (20 mg) in the electrolytic solution. Scan performed *after* stirring. (III) and (IV) responses at iron (three scans) and gold (one scan) substrates, respectively, after a mechanical dispersion of the Ag–Pd powder in the solution [27].

It has been established (Chapter 2) that a deposit of Ag–Pd at a silver surface displays the best activation for C–I bond cleavage at a potential close to −1.1 V. If now potentiostatic reductions of RIs are achieved at GC electrode (under conditions of Figure 5.3 that imply a remnant suspension of Ag–Pd in the electrolyte) and achieved at −1.3 V, any primary RI compound appears totally reduced after 1.2 F mol^{-1} (while no cathodic response for the RI compound at a not-modified carbon electrode is obtained at $E > -1.8$ V).

The analysis by means of GC/MS of the resulting solution shows the formation of the homo-dimer in a quasi-quantitative way. Thus, the indirect reduction of RI by means of Ag–Pd used as a mediator (powder in suspension) does not show a quantitative electrical yield while the Ag–Pd alloy, if directly deposited onto the GC interface, allows the process to be totally completed after an electricity consumption of sharp 1 F mol^{-1} (simultaneous occurrence of reactions (3) and (4) are given in Scheme 5.2).

5.4.2. *Metalizing procedures*

The procedure for silvering different solid conductors is easily applied to depositions of silver onto iron, copper, and gold chosen as typical substrates. The procedure is very simple: a solution of a tetraalkylammonium salt in DMF, a common 1-iodoalkane like 1-iodobutane (concentration < 10 mmol L^{-1}), a small amount of Ag–Pd powder (about 20 mg dispersed in 6 mL) and an applied potential of −1.0 V. Under efficient stirring, the current at the microelectrode (area: 0.8 mm^2) easily reaches 50 μA. In principle, the rise in current depends on the heterogeneous deposit rate of silver that makes the electrode more and more reactive toward the alkyl iodide. Silvering procedure (Scheme 5.3) was stopped after total electricity amount of 2×10^{-2} C. Electrodes modified in this way were sonicated in water for 10 min, and then rinsed with alcohol, acetone, and dried under hot air stream. It is remarkable that upon time, current limits are equal, but peak potentials appear slightly different. In the given examples, the "catalytic" effect decreases according to the following order: Au/Ag > Cu/Ag > Fe/Ag [29, 30]. However, the implication of the solid substrate in the electrocatalytic activity has not been demonstrated so far.

5.5. Structural Evidence of the Deposit: A Layer made of Pure Silver Nanoparticles

Until now, only silver deposits onto large areas (> 5 cm^2) of GC and copper were achieved. With small amounts of electricity, shiny white

(A) (B)

Figure 5.4. (A) and (B): Silvering procedure of solid conductors by reduction of 1-iodobutane in the presence of Ag–Pd powder. See text. Applied potential: −1.3 V. (A) Deposit of Ag nanoparticles onto copper (silver indirect deposit of the left part of the sample). Amount of electricity: 2 C cm^{-2}. (B) Extremely thin deposits of silver onto pieces of GC. The shiny aspect of the produced silvering obviously depends on the amount of electricity. The right sample corresponds to an amount of electricity of 1 C cm^{-2} while the left one used 2.5 C cm^{-2} (from Ref. [27]).

deposits were obtained (Figure 5.4). The mass increase is, in all cases, extremely small and suggests that the thickness of the layer is, at the maximum, of a few tens of nanometers. EDS analyses of the external layers revealed the exclusive presence of silver: no trace [29] of palladium being detected.

5.6. Building of New Electro-Catalytic Surfaces

Lastly, these electrodes Metal-Ag NP were preliminary tested from comparative studies relative to alkyl-bromides (RBr), known to be moderately active [25] at most of solid electrodes. The example is displayed in Figure 5.5 with 1-bromobutane when reduced at several solid electrodes. Potentials and peak currents obtained at a GC electrode ($E_{p/2}$ = −2.23 V, curve I) were compared to different kinds of silvered surfaces. Under the conditions of Figure 5.5, smooth silver (curve II) does not exhibit a large catalytic effect ($E_{p/2}$ = −1.99 V) while the silvered gold electrode covered with Ag nanoparticles (Au–Ag NP) appears to be much more efficient ($E_{p/2}$ = −1.53 V). If the latter surface shows a nice efficiency to reduce RBr compounds, the

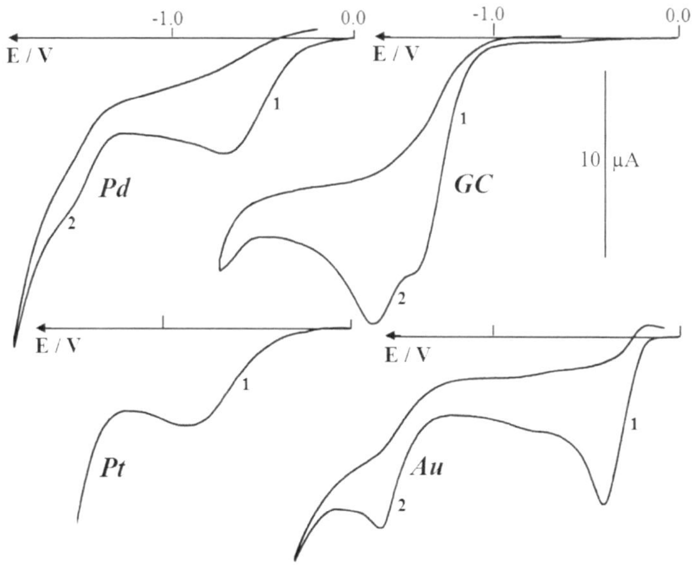

Figure 6.1. Voltammetries of AllBr (concentration: 11 mmol L^{-1}) in PC–EC (50:50 v:v)/0.1M TBAI at several stationary solid electrodes (Pd, GC, Pt and Au). Scan rate: 50 mV s^{-1}. Apparent surface areas of all electrodes: 0.8 mm^2 [10].

and certainly underlines the catalytic influence of different cathode materials. In any case, current densities progressively decrease and electrolysis turns to being quite slow after 0.5 F mol^{-1} have passed. A continuous metal surface allylation is expected to progressively prevent the catalytic reaction. This proposal is supported by the study of a rapid system (oxidation of potassium ferrocyanide) in testing Au and Pt electrodes after reduction of AllBr at 1 min holds at peak potentials. Also, a free allyl radical generation within 0 V and –0.5 V has been checked by a trapping procedure using nitrosodurene: Electron Spin Resonance (ESR) coupling constants fit well with those already reported in the case of cathodic degradation at GC of allyldimethylsulfonium [11] under quite close experimental conditions.

Moreover, similar results could be noted by using GC, graphite, and graphene deposited onto several conducting support systems. These different substrates were beforehand efficiently covered with

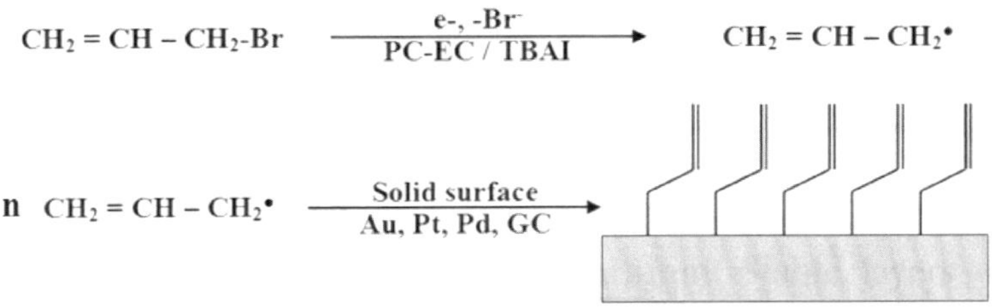

Scheme 6.1. A general mode of grafting allyl radicals onto solid surfaces [10].

very thin deposits (less than 0.1 μm, average layer thickness) of Au, Pd, Pt, and Rh. With as-metalized carbon layers, an electro-catalytic step assigned to such allyl radical was confirmed. However, potentials found for the first step and the associated currents could differ depending on the nature of the deposited metal and the thickness of the deposit.

Evidences given above were applied to the cathodic immobilization of allyl radicals at carbon surfaces (Scheme 6.1). For this, fixed potential electrolyses were achieved at GC (apparent surface area of electrodes: 7 mm^2) at a potential of > -1.2 V (level of the first step) by using PC–EC/0.1 M TBAI solutions containing small amounts of AllBr. After an amount of $\leq 4\times10^{-3}$ C of electricity was passed, electrodes were rinsed and sonicated in water for 10 min. In order to quantify the number of unsaturated bonds present at the solid surface, a simple halogenation procedure of the double bond of fixed allylic layer was employed. This process implies saturating the double bond with I_2 or Br_2 (diluted in alcohol (5%) or in cyclohexane (3%), respectively). This reaction well known in organic chemistry [14] is applied here for readily transforming the allylic deposit layer into either α,β-di-iodo or α,β-di-bromopropane layers. The EDX analysis after iodination process confirms the presence of iodine. A basic electro-chemical two-electron elimination (well known for C–C bonds substituted by two leaving groups in vicinal position [1]) permits one to completely regenerate the allyl group and concomitantly quantify the global amount of allyl groups immobilized at the carbon surface by coulometry. For this, the

Furthermore, electro-catalysis and solid grafting at gold surfaces was also achieved [16]. Therefore, this model for preparing new surfaces by a ***simple*** cathodic alkynylation obviously expands possibilities and the method can also be used for substituted compounds (let us recall that propargylic derivatives are commercially available or easy to prepare).

As a matter of fact, the cathodic reduction of propargyl bromide has been carried out in DMF containing $TBABF_4$ supporting electrolyte. At a stationary GC electrode (Figure 6.3(A)), a diffusion-controlled two-electron process ($E_{p/2} = -1.52$ V) was obtained. Alternatively, using a gold electrode, a quite similar two-electron step ($E_{p/2} = -1.83$ V) was obtained, which does not exhibit any

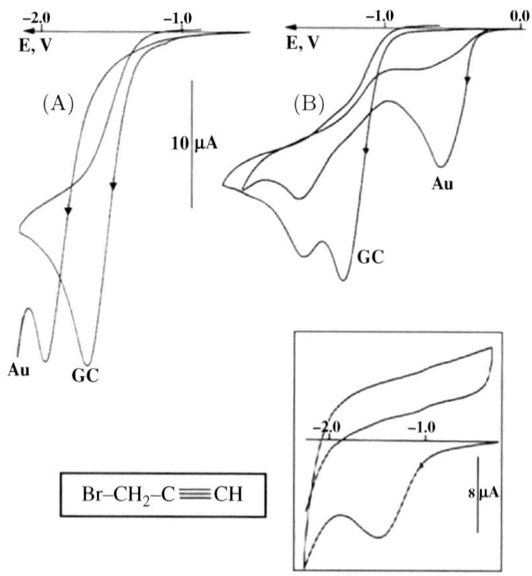

Figure 6.3. Propargyl bromide [15]. Voltammetries at different solid electrodes. Surface area: 0.8 mm^2. $v = 50$ mV s^{-1}. (A) Solvent: DMF. Response at GC and smooth gold electrodes. Concentration 20 mmol L^{-1}. Electrolyte: 0.1 M TBAI. (B) Voltammetric responses in PC containing 0.1M TBAI. Concentration: 17 mmol L^{-1}. *Inset*: Deposition of Pg$^•$ radical (from (B) solution) at a GC electrode (7 mm^2) (charge: 2.5×10^{-2} C at −1.2 V) sonicated and rinsed with EtOH and acetone and then reacted with Br_2 (3% in cyclohexane). Cycling in DMF + 0.1M $TBABF_4$ (first and second scans).

catalytic acceleration of the process. The reason would be that at very negative potentials, an anionic polymerization, via the electro-generated propargyl carbanion, may lead to the electrode coverage.

On the contrary, the use of TBAI as electrolyte (in PC or PC/EC) allows a rather fast nucleophilic substitution of PgBr by I^- (method already described above with AllBr in the same chapter). Now, voltammetries at different solid electrodes are quite different (Figure 6. 3(B)) due to two well-separated steps, particularly clear at the gold surface. Let us add that the large viscosity increase of PC solutions provokes a neat decay of peak currents.

At gold, the first one-electron step is clearly assigned to the formation of a free propargylic radical ($E_{p/2}$ = −0.41V). The second step, at quite negative potentials, corresponds to the reduction of the propargyl radical into propargylic anion ($E_{p/2}$ = −1.44 V). Also worth noting is the ***huge*** potential difference (ΔE > 1.0 V) between these two steps, which stresses the strong catalytic effect brought about by gold. Other metals tested as solid substrates led also to appreciable electro-catalytic reactions although with smaller efficiencies. Thus, both copper and platinum exhibit similar surface reactions. With all these substrates, the second step is located within a narrow potential range (peak potentials between −1.5 and −1.65 V). This observation would support that the second step corresponds to the reduction of $Pg^•$ radical whatever be the nature of the cathode material. The reductive integration of the steps obtained at GC surfaces revealed the electrode coverage as corresponding to $\Gamma = 5.1 \pm 0.8\times10^{-9}$ mol cm^{-2}, which sounds to be adequate with a dense radical grafting onto a carbon of high roughness (Scheme 6.2). The addition of $Pg^•$ onto the primary alkynyl layer (to form a kind of "multi-layer") and the anionic polymerization of PgBr should also be certainly mentioned.

Macro-electrolyses of propargyl bromide at several electrodes were carried out in the presence of spin traps such as *N-t*-Bu-α-phenylnitrone. Thus, the radical addition onto nitrone in solution has been verified, and this resulted in a six-line spectrum (g = 2.0067, a_N = 14.11 G and a_H = 3.13 G at 22°C, EC/PC mixture).

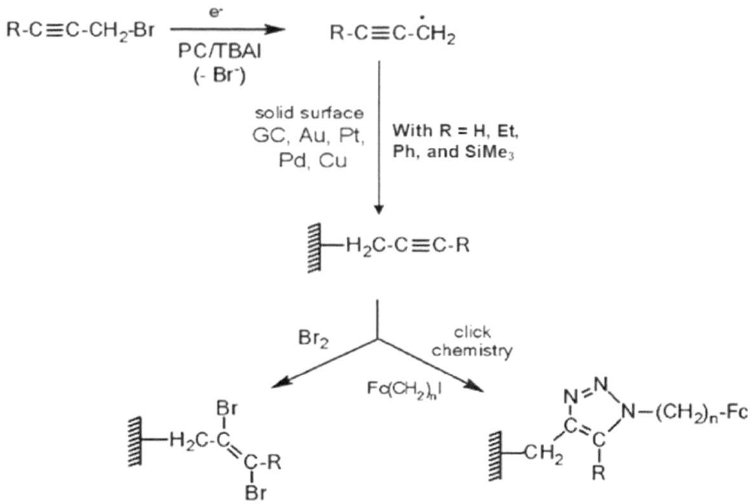

Scheme 6.2. Formation and deposition of propargyl radical at solid electrodes (click chemistry in the presence of NaN_3/Cu(I)) [15].

Additionally, fixed potential coulometries achieved at the first step ($E > -1.0$ V) confirmed the one-electron process. Substitutions of the acetylenic proton by Et, Ph, and $SiMe_3$ groups confirm that the intermediate radicals lead to surface additions [16].

6.4. One-Electron Cleavage of Benzylic Bromides at Palladized Substrates

A large series of substituted benzyl halides Ar–CH_2–X were reduced at solid electrodes in PC and dimethylformamide (DMF) containing especially tetraalkylammonium iodides (TAAI) for permitting the conversion into benzylic iodides. Alternatively, this approach privileged the use of palladium electrodes and especially those covered with layers of Cu–Pd, Ag–Pd, Au–Pd, and Ni–Pd were found to favor the one-electron scission of the C–Br bond. Under these conditions, benzyl bromides are reduced in two separated steps assigned to the transient formation of a free radical capable of coupling or adding onto the cathode material. The electrolytic generation of varied benzyl radicals allows the addition *in situ* onto unsaturated organic systems (mono and di-benzylations that may take place in solution or at the interface toward immobilized unsaturated reagents). Experiments performed for testing grafting of benzyl radicals onto palladized interfaces were successful and underlined the interest of such interfaces in the future.

This contribution specifically shows the interest in palladized electrodes via the transient generation of benzyl radicals. The formation of paramagnetic species at the catalytic interface could be confirmed by means of ESR studies performed at much less negative potentials. Thus, a bibenzyl generation occurs according to a homo-coupling and not by means of an S_N2 displacement from the benzylate (Wurth reaction). Moreover, our interest in producing benzyl radicals directly at the polarized metal surface prompted us to propose a new mode for immobilizing such transients at electrode surfaces (let us recall the grafting of benzyl radical has already been developed through corresponding diazonium ion reduction [9]).

Voltammetries of benzyl bromide classically achieved in PC and DMF in different salts are strongly dependent on the cathode material. As shown in Figure 6.4, the two-electron reduction step (E_p = −1.60 V, formation of the benzyl anion) observed at GC is generally shifted toward less negative potentials when using palladium and palladized surfaces [16, 17].

Thus, employing palladium alloys (such as Cu–Pd, Ag–Pd and palladized palladium) efficiently splits the global reduction of the C–Br bond for efficiently separating the two-electron transfers. More precisely, using Ag–Pd as the cathode material permits obtaining two separated steps (peak potentials E_{p1} = −0.97 V and E_{p2} = −1.35 V assigned to the formation of the free benzyl radical (step (I)) and the second one (step (II)) to its reduction). The sum of currents ($i_I + i_{II}$) is more or less equivalent to the total current obtained at GC. The global current is proportional to benzyl bromide concentration. The current inequality ($i_{II} < i_I$) is presumably due to the fast coupling of the benzyl radical at the interface. Lastly, the overall processes with all tested palladized electrodes were found to be diffusion controlled.

The potential of step (II) (assigned to the radical reduction) fits well with values previously published [18] for the standard reduction potential of benzyl radical (−1.215 V). Therefore,

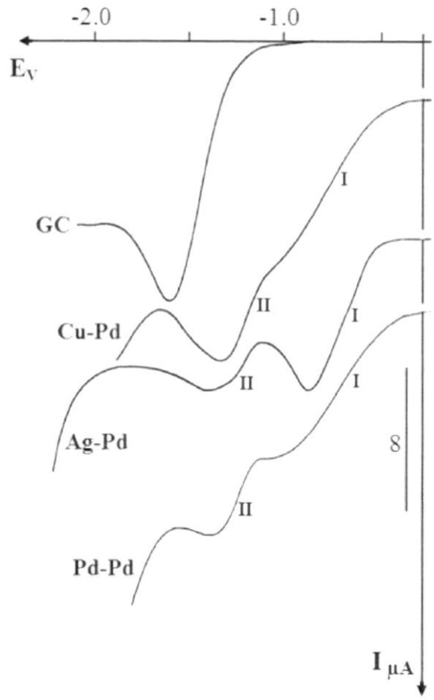

Figure 6.4. Voltammetries of benzyl bromide (concentration: 8.6 mmol L^{-1}) at several solid electrodes. Reference electrode: ECS. Solvent: PC + 0.1 M TBABr. Scan rate: 50 mV s^{-1}. Used cathodes materials: GC. Copper–Palladium, Silver–Palladium and Palladized palladium (average thickness of Pd^0: 5 10^{-2} μm) [16].

reductions of benzyl bromide by macro-electrolyses achieved at potentials > −1.0 V may be suitable for specifically producing this radical. The intermediate formation of benzyl radical was obtained via the ESR technique by using *N*-*tert*-butyl-α-phenylnitrone (TBPN) as trap in excess, allowing one to obtain a paramagnetic adduct (Scheme 6.3). The six-ray ESR response obtained at Ag–Pd and Ag surfaces, respectively) matches with a nitroxide already reported in the literature implying the benzyl radical [18]. Additionally, *in situ* electro-generated benzyl radicals can be added to unsaturated organic radicalophilic systems. More concretely, generation of benzyl radical at Ag–Pd (fixed potential: −0.8 V) in the presence of 2,5-dimethyloctadiene yields exclusively bibenzyl adduct (28%) together with the expected bibenzyl. Its formation from the benzyl radical was reported [19] to be fast (10^9 M^{-1} s^{-1}). Lastly, the addition of benzyl radical to acenaphthylene reaches

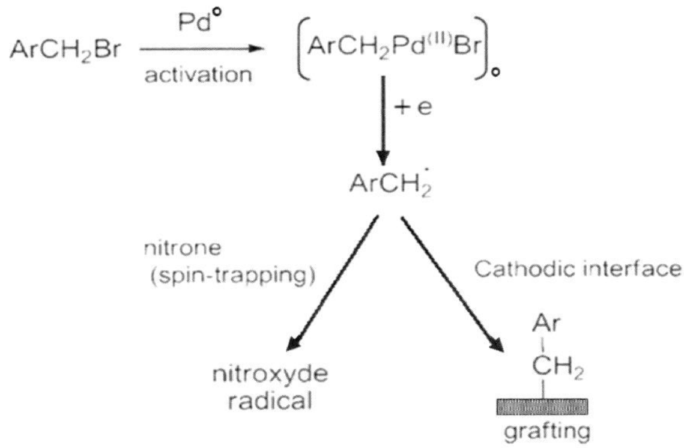

Scheme 6.3. Benzyl bromide: the one-electron catalytic scission of the C–Br bond through Pd^0 activation [16].

10% of mono-addition and 7% of di-addition. Lastly, radical addition to phenyl acetylene affords monobenzyl-styrene (3%) and dibenzyl-styrene (7%).

Additionally, several substituted benzyl bromides could be efficiently grafted at different electrode materials (Scheme 6.3) and 4-nitro, 4-iodo, 4-bromo, 4-methyl, and 4-fluorobenzyl bromides were successfully produced [16].

6.5. 9-Fluorenyl Radical

6.5.1. *Generation*

Very similar to other activated bromides of which the scission process was found to be strongly activated, 9-bromofluorene (FluHBr) was in the same way electro-chemically reduced at solid electrodes such as GC and smooth transition metal surfaces (gold, palladium, platinum) [24]. Actually, at smooth palladium and gold, two well-separated steps are obtained and are assigned (i) to the electro-catalytic formation of the radical and then (ii) to its reduction. Additionally, galvanostatic depositions of different transition metals (Pd, Au, Ni, Co, Rh) onto solid substrates such as GC, platinum, and gold lead to comparable results and allow the grafting of $FluH^{\bullet}$ onto those heterogeneous surfaces.

It is worth noting that the potential of the second reduction step remains the same irrespective of the metal. Fixed potential electrolyses at the level of the first step permitted one to obtain (especially at Au and GC surfaces) cathodic signals assigned to the grafting of 9-fluorenyl radical FluH•. The transient formation of this radical has been established by means of ESR (electrolysis at low temperature or trapping by a nitrone).

Thus, among organic halides, quite a reactive carbon–bromine bond is present in 9-bromofluorene. Quite astonishingly, a sole study [19] in organic solvents led to indicating its two-electron electro-chemical cleavage. A convenient way for producing free 9-fluorenyl radicals is of huge interest (e.g., grafting of a conducting polymer like polyfluorene [20] because of the presence of anchoring sites at a platinum surface). Processes that permit the facile and effective immobilization of 9-fluorenyl radical onto solid different surfaces are briefly presented below:

EPR spectro-electro-chemistry (at a GC/Pt electrode in a three-electrode cell) similarly was carried out in PC/THF (1:2 v/v) using, as supporting electrolyte, a 1:3 molar mixture of TBAI with $TBABF_4$ or $Bu_4N[F^-@Si_8C_{48}H_{40}]$ which was previously described [21–23].

6.5.2. *Fluorenation of solid surfaces*

The reduction of FluHBr in an aprotic solvent appears strongly dependent on the electrode material [24]. Thus, at GC cathode, a sole irreversible two-electron step (first scan, Figure 6.5(A)) is observed ($E_{p/2}$ = –0.83 V) in a PC-AN mixture in the presence of ammonium salts, and it appears to be of limited interest in the present goal. On the contrary, at metallic electrodes (palladium, gold, platinum, rhodium, nickel, or cobalt), the reduction of FluHBr is strongly facilitated and occurs in two well-separated steps. For example, the use of GC activated with a very thin layer of Pd gives two peaks I and II (Figure 6.5(B)). It was found that half-peak potentials are equal to –0.14 V and –0.62 V respectively. The step (I) is — quite obviously — assigned to the one-electron cleavage of the C–Br bond

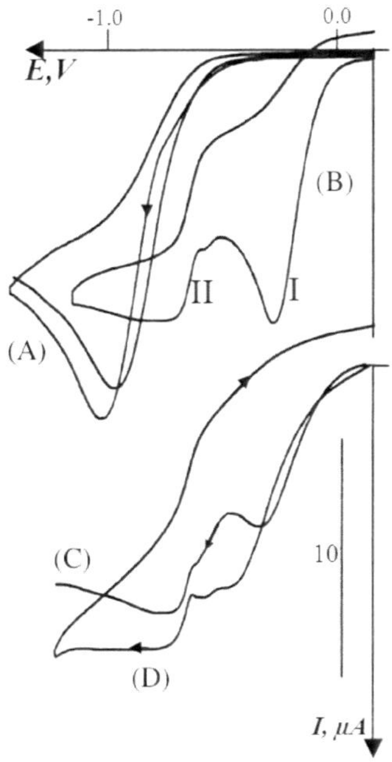

Figure 6.5. Voltammetries of FluHBr at different solid electrodes [23]. Scan rate: 50 mV s^{-1}. Apparent surface area of all used electrodes: 0.8 mm^2. (A) Electrolyte: PC–AN (80:20 v/v) containing $TBABF_4$. Concentration: 6.1 mmol L^{-1} at GC (first two scans). (B) The same GC electrode now covered with a layer of Pd (average thickness: 0.1 μm). (C) Smooth gold. (D) Smooth palladium.

of FluHBr with the formation of 9-fluorenyl free radical while the second peak II corresponds to the reduction of the radical (Scheme 6.4: reactions (1) and (2)). It is noticeable that first step strongly depends on the cathodic material whereas step (II) is practically invariant to the nature of the electrode. Curves C and D, drawn at smooth platinum and smooth gold, do confirm this point. Step (II) always exhibits a kind of pre-wave that corresponds to the reduction of the radical in the presence of a proton donor (acidic impurities present in the solvent). This step totally disappears here possibly in the course of the second scan, or by using a super-dry solution (e.g., by adding activated neutral alumina to the solution).

was confirmed by the spin-trapping technique at GC and gold surfaces.

The reduction of FluHBr, cathodically achieved at metallic electrodes in polar organic solvents, appears particularly powerful to split the carbon–bromine bond in a one-electron way permitting the easy formation of the transient 9-fluorenyl radical (Scheme 6.4). The use of several electrode materials, such as palladium, gold, platinum, nickel, cobalt, or rhodium, led to comparable results [24]. Quite specifically, the use of tetraalkylammonium iodides as the electrolyte led to the convenient *in situ* conversion of FluHBr into FluHI. This method permits (without catalysis at adequate metals) generalizing the one-electron scission at many solid surfaces. The transient formation of FluH$^\bullet$ was confirmed by EPR technique and attempts to immobilize this radical onto solid surfaces were reported. This was successfully done onto gold and carbon surfaces.

The additional interest of the fluorene grafting onto metals concerns the immobilization of electro-generated poly-fluorene known as an organic conducting polymer [25]. After achieving the grafting of the fluorenyl moiety at GC covered by gold or platinum, the polymerization of fluorene in solution may be obtained [26] for fixing the entire polymeric matrix at the electrode (see Scheme 6.5).

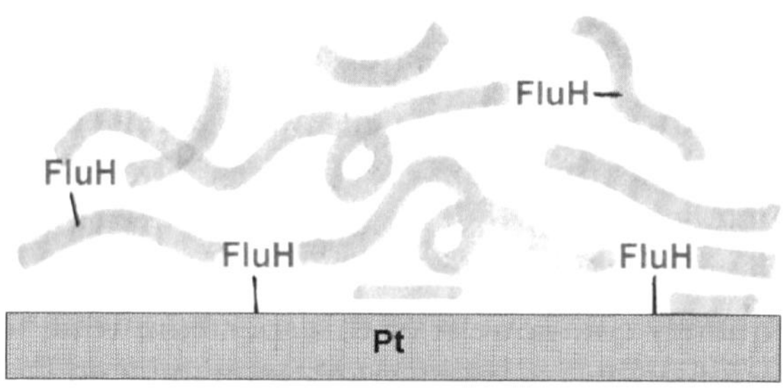

Scheme 6.5. After immobilization of FluH$^\bullet$ radicals at a solid electrode, the polymerization of fluorene (oxidation in acetonitrile or methylene chloride) may be achieved allowing the formation of $-(FluH_2)_n-$chains (gray ribbons = electro-generated polyfluorene) quite possibly reticulated by attached fluorenyl moieties. An additional treatment (reduction of the polymeric layer) may produce supplementary reticulations between chains [26].

References

[1] D.G. Peters, in *Organic Electrochemistry*, eds. H. Lund and O. Hammerich, M.M. Dekker, New York, Basel, p. 341 (2001).

[2] P. Poizot, O. Durand-Drouhin, M. Lejeune, J. Simonet, *Carbon* **50** (2012) 73.

[3] P. Poizot, L. Laffont-Dantras, J. Simonet, *J. Electroanal. Chem.* **624** (2008) 52.

[4] A.A. Isse, A. De Guisty, A. Gennaro, *Tetrahedron lett.* **47** (2006) 7735.

[5] V. Jouikov, J. Simonet, *Electrochemistry Commun.* **12** (2010) 331.

[6] F. Hui, J.-M. Noel, P. Poizot, P. Hapiot, J. Simonet, *Langmuir* **27** (2011) 5119.

[7] C. Lagrost, M. Guilloux-Vitry, C. Mangeney, P. Hapiot, J. Guilane, *J. Phys. Chem. C.* **111** (2007) 5701.

[8] J. Pinson, F. Podvorica, *Chem. Soc. Rev.* **34** (2005) 429.

[9] D. Belanger, J. Pinson, *Chem. Soc. Rev.* **40** (2011)3995.

[10] J. Simonet, *Electrochem. Commun.* **13** (2011) 1417.

[11] P. Martigny, J. Simonet, G. Mousset, J. Vigneron, *Nouv. J. Chim.* **7** (1983) 299.

[12] The grafting is then achieved via a preliminary treatment at GC beyond –1.7 V and substitution with 4-bromo-1-butene according to a process analogous to a nucleophilic substitution may occur. Note that a similar mode of deposition cannot be expected with AllBr ($E_2 > -1.7$ V).

[13] T. Brinck, A.G. Larsen, M.K. Kenneth, K. Daasbjerg, *J. Phys. Chem. B.* **104** (2000) 9887.

[14] J. March, *Advanced Organic Chemistry, Reactions, Mechanisms and Structures,* John Wiley, Chichester, Third Edition, p. 724 (1985).

[15] V. Jouikov, J. Simonet, *Electrochem. Commun.* **15** (2012) 93.

[16] V. Jouikov, J. Simonet, unpublished observations.

[17] F. Hui, J.M. Noel, P. Hapiot, J. Simonet, *Langmuir* **27** (2011) 5119.

[18] F.J. Gonzalez, P. Hapiot, V. Konovalov, J.M. Saveant, *J. Am. Chem. Soc.* **120** (1998) 10171.

[19] F.M. Triebe, K.J. Bohrani, M.D. Hawley, *J. Am. Chem. Soc.* **101** (1979) 4637.

[20] J. Rault-Berthelot, J. Simonet, *J. Electroanal. Chem.* **182** (1985) 187.

[21] J. Zeitouny, V. Jouikov, *Phys. Chem. Chem. Phys.* **11/33** (2009) 7161.
[22] A.R. Bassindale, M. Pourny, P. G. Taylor, M.B. Hursthouse, M.E. Light, *Angew. Chem. Int. Ed.* **42** (2003) 3488.
[23] A. Atto, A. Hudson, R.A. Jackson, N.P.C. Simmons, *Chem. Phys. Lett.* **33** (1975) 477.
[24] V. Jouikov, J. Simonet, *Electrochem. Commun.* **13** (2011) 254, and unpublished works.
[25] J. Rault-Berthelot, J. Simonet, *Synthetic Metals* **33** (1989) 329.
[26] J. Simonet, Unpublished observations.

Chapter 7

Cathodic Reduction of Aryl Halides in Organic Carbonates for Achieving Aryl–Aryl Couplings

7.1. Introduction

The cathodic reduction of some aryl-halides (ArX) is achieved essentially in propylene carbonate (PC) considered for its high dielectric permittivity. Different electrode materials such as copper, silver, palladium, silver–palladium alloy, and nickel are concomitantly employed. Such conditions could appear quite unusual, but they permit an unforeseen activation of the C–X bond by the metal. The metal's behavior suggests an obvious similarity with Ullmann's reaction and conditions under certain aspects of an aryl–aryl bond generation. The use of organo-metallic intermediates deposited *in situ* at the metal interface demonstrates the existence of an activation process often enabling high yield Ar–Ar linkages, especially in the case of aryl iodides. Additionally, very novel and unexpected reactions achieved in the presence of PC/palladium catalysts are presented.

7.2. Ar–Ar Linkage Generated by Electro-Chemistry

Achieving aryl–aryl linkages is an important challenge in chemistry, particularly in supramolecular chemistry as well as in the building of redox oligomers. Efficient methods already exist for building aryl–aryl bonds. Metals catalysts (Mg^0, Pd^0 and Ni^0) may react with C–X bonds of aromatic halides for activating oxidative insertion (in the presence of appropriate ligands), and organo-metallic linkages (such as $C–Pd^{II}–X$) then provide wide possibilities for generating C–C couplings by the reaction with electrophiles like RX or Ar′X [1]. Additionally, the Gatterman's method of de-diazotation (actually, the reduction of diazonium salts by Cu in acidic media) remains a grand method for creating aryl radicals. Beyond the building of Ar–Ar′ bonds, the process is additionally employed for the functionalization of radicalophilic surfaces [2]. Similarly, ArX compounds (especially when X = I) can be converted into bi-aryls by using copper catalysts (Ullmann's reaction [3] essentially performed with activated ArI (Scheme 7.1, Eq. (2)) at high temperatures). The key intermediate would be an organometallic cuprous complex $ArCu^I,X^-$ being written as ArM^+,X^-. Analogous species were shown being formed [4] by contact of alkyl bromides (RBr) with highly activated copper (transient formation of RCu^I,Br^-). These intermediates are

Organic chemistry

$$ArX + M \longrightarrow ArM^+, X^- \quad (1)$$
$$ArM^+,X^- + Ar'X' \longrightarrow Ar\text{-}Ar' \quad (2)$$

Electrochemical activation

$$ArM^+,X^- + 2e \longrightarrow [ArM^o]^- + X^- \quad (3)$$
$$[ArM^o]^- + ArX \longrightarrow Ar\text{-}Ar + M^0 + X^- \quad (4)$$

together with:

$$[ArM^o]^- \longrightarrow Ar^- + M^0 \quad (5)$$
$$Ar^- + \text{proton donor} \longrightarrow ArH \quad (6)$$
$$Ar^- + PC \xrightarrow{H^+} Ar\text{-}C(O)\text{-}O \quad OH \quad (7)$$

Scheme 7.1. Formation of Ar–Ar linkages via chemistry (Ullmann's reaction using copper, M = Cu) and electrochemistry (possibly via some metallic cathodes made of metal M as a catalyst together with an organic carbonate such as PC as solvent). One may expect that native metal M° could also activate the overall reaction toward coupling achievement.

often considered as nucleophilic species for generating R–R bonds with organic halides. In the field of electro-chemistry, they should also be advantageously considered as possible intermediates in the cathodic reduction of alkyl bromides within the growing world of smooth copper electrodes.

Quite generally, a direct electro-chemical reduction of ArX does not afford Ar–Ar linkages [5]. Data concerning ArXs deal mostly with a two-electron scission when using carbon materials. Studies on dissociative electron transfer carried out in polar organic solvents [6] have found that the transient aryl radical $Ar^{\cdot}$ is very easy to reduce. In concrete terms, it was established that $E^0_{Ar/Ar-}$ in the simple example of phenyl radical is 0.05 V vs. SCE [7], which means the overall process should definitively yield the aryl anion (Ar^-) in most of the cases, whatever be the Ar substituent. Quite recently [8], voltammetric reduction of a large palette of ArBr compounds was achieved at smooth silver in non-aqueous polar solvents, and the interest in using transition metal as the electrode material was also discussed. The spectacular catalytic effect of silver on reduction potentials was attributed to the selective affinity of Br^- (generated by the cathodic scission) toward silver.

Additionally, studies using metals [9, 10] as cathode materials (copper, nickel, silver, palladium, either pure or included into different alloys) may boost reviews devoted to new methods on aryl-aryl linkage formations. This new approach fundamentally based on the possible oxidative insertion of metals used as a cathodic material is proposed: more precisely, electroactive species chemically formed according to Eq. (1) (Scheme 7.1) are chemically activated for creating efficient transient nucleophiles. Additionally, in order to boost the reactivity of ArX toward these nucleophiles, the use of solvents of quite high polarity is put forward for boosting the polarization of the C–X bond and then increasing its electrophilicity. For this, PC possibly mixed with ethylene carbonate (EC) leading to mixtures that possess dielectric constants $\varepsilon \geq 66$, much higher than that of DMF ($\varepsilon = 37$), was successfully used. This work is then based on the capacities of ArXs to form dimers under very unusual experimental conditions as solvents and electrode materials (Scheme 7.1).

7.2.1. *Experimental*

Electro-chemical studies were conducted with freshly prepared 0.1 M solutions of tetra *n*-butylammonium tetrafluoroborate ($TBABF_4$) principally in dimethylformamide (DMF), acetonitrile (AN), PC, and EC. All potentials are referred to the aqueous saturated calomel electrode (SCE).

Glassy carbon (GC), nickel, gold, copper, and silver disks used as solid substrates had a surface area of 0.8 mm^2. Prior to use, all electrodes were carefully polished with Norton polishing paper (types 02 and 03). Metals could also be deposited galvanostatically onto GC from acidic solutions of their salts. By this way, electrodes denoted as GC–M were easily obtained. Such depositions were carried out with an amount of 2×10^{-3} C mm^{-2} of electricity at a current density of 500 μA mm^{-2} (average thickness ≪ 0.1 μm). Before use, the electrodes were thoroughly rinsed with alcohol and acetone.

Ag–Pd alloy electrodes were prepared according to a specific procedure [10] fully described in Chapter 2.

7.2.2. *Cathodic generation of Ar–Ar bonds: Conditions and limitations*

The reduction of some aryl iodides in PC is presented by data in Figure 7.1. At a GC electrode (Curve A), 1-iodonaphthalene (NpI) undergoes two main reduction steps (I and II). At GC, they correspond to the two-electron cleavage of the C–I bond (step (I), $E_{p/2}$ = –1.78 V) and the reversible reduction of naphthalene (step (II), –2.48 V). With palladium, copper, and nickel electrodes, a catalytic step (IIIc) is then observed: all these materials display very large positive potential shifts relative to the reduction potentials at carbon. Thus, at Ag–Pd stationary electrode (GC surfaces modified by a very thin ($d < 0.05$ μm) deposit of this catalyst), half-peak potentials of –1.24 V is observed.

Similarly, the cathodic voltammetry of PhI confirms that the reduction process at a GC electrode ($E_{p/2}$ = –1.93 V) corresponds to the cleavage of the C–I bond. On the contrary, Ag–Pd (ultrathin

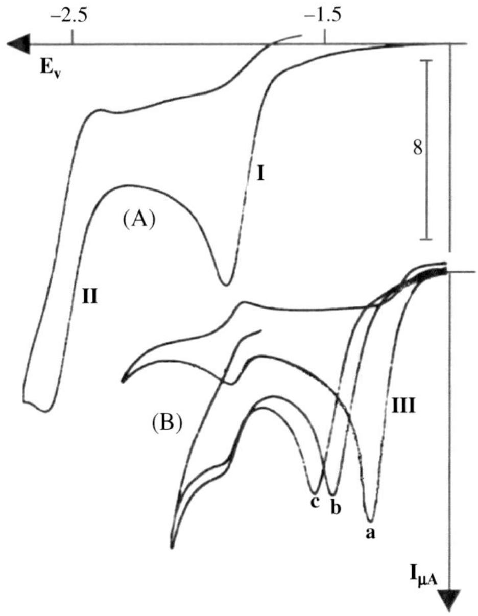

Figure 7.1. Voltammetries of 1-iodonaphthalene (concentration: 5,5 mmol L^{-1}) at solid electrodes in PC + $TBABF_4$. Scan rate: 50 mV s^{-1}. (A) 1-iodonaphthalene (NpI) at a GC electrode (surface area: 0.8 mm^2). (B) NpI reduced at (a) GC modified by Ag–Pd, (b) by Ni, and (c) modified by Cu [12].

deposits of Ag–Pd) triggers a catalytic system that was checked being diffusion controlled. Similarly, strong electrode catalytic effects were obtained at Cu, Ni, and Pd either as metal plates or galvanostatic metals deposited at GC permitting reductions of most ArIs at potentials > −1.5 V.

Aromatic bromides (ArBr) display similar results at transition metal surfaces. So at Ag–Pd electrode, NpBr exhibits the step (IIIc) at −1.38 V while PhBr shows this step at −1.40 V. Voltammetry data obtained at solid surfaces modified by the deposits of Cu, Ag, and Pd are rather close to those obtained with parent ArIs at GC surfaces.

Some cathode materials, especially in the case of aryl bromides, ArBrs, clearly lead to reduction potential shifts in voltammetry, in particular Ag, Ag–Pd, Cu, or Pd. Similar to ArIs, it is expected that aryl bromides also preliminarily react with the electrode material

Chapter 8

Large Immobilization of Organic Bielectrophilic Species at Carbon Surfaces

8.1. Introduction

Glassy carbon (GC) is a convenient but rather complex electrode material to carry out electrolyses within the cathodic range. Comments and discussions concerning the structure of GC [1–5] and its capacity to be electro-chemically charged in aprotic polar organic solvents [6–8] at potentials < −1.7 V vs. Ag/AgCl are stressed in Section 12.2. Thus, strongly polarized carbons may act as nucleophilic reagents and allow many surface reactions with appropriate organic electrophiles, and therefore lead to surface functionalizations [9].

Precisely, a part of this book is amply devoted to the reactivity of cathodically charged carbons (GC, graphite, graphene as well as most of carbonaceous conducting materials) toward mono-electrophilic molecules such as alkyl halides. Additionally, this chapter aims to discuss the cathodic reactivity of bielectrophilic species such as α,ω-dihaloalkanes (I–R–I) possibly modified through the alkyl chain by redox groups for playing the role of a catalyst or markers. This strategy may present a double interest: (i) one is producing polymeric materials through electron transfer by means of scission of carbon–halogen bond linkages (radical cleavage or S_N2

type reactions depending on the applied potential and the possible presence of adequate catalyst) and (ii) second concerns the ability of electrophilic or radical fragments to react with carbon interfaces. The proposed strategy should then appear as an original approach for grafting a conducting (or redox) polymer generated *in situ* at the carbon surface. By this way, ferrocene units are then used as efficient redox probes. This preliminary attempt should point out the formation of a chemically grafted redox polymer at carbon electrodes for quantifying the amount of immobilized alkyl chains as a function of their size. Coverage amounts are expected as high and apparent ferrocene surfacial concentrations (readily appreciated via a simple voltammetry of ferrocene) are currently superior to 10^{-8} mol cm^{-2}. Lastly, ferrocene layers deposited accordingly were found to be chemically and electro-chemically stable.

This contribution is first focused on the "reductive attachment" of α,ω-diiodoalkanes possibly containing ferrocene moieties and their ability to generate branched redox polymers at interfaces and bring some further information to preceding reports [10–19]. Also, in the hypothesis, charged fine carbon particles are employed as polynucleophilic entities for generating highly versatile coverages for electrodes and materials devoted to specific use (for example, large modified carbon plates). Briefly, the cathodic reactivity of I–R–Is was only considered in polar organic aprotic solvents containing tetraalkylammonium salts as the electrolyte. It is well known that these compounds in contact with GC, polarized at potentials more negative than −1.7 V vs. Ag/AgCl, may, at least partly, lead to mono-iodoalkyl chain immobilization onto carbon. Such process should correspond to the cathodic charge of graphitized and fully-realized zones present in carbon followed by a displacement reaction (Chapter 12). This mode of grafting is here applied to the reaction of charged carbons toward bis-(ω-iodoalkyl)ferrocenes. Therefore, ferrocene units are then used as efficient redox probes. This preliminary approach should point to quantifying the amount of immobilized alkyl chains as a function of their length. Coverage amounts were found as high and apparent ferrocene surfacial concentrations are superior to 10^{-8} mol cm^{-2}. Lastly, ferrocene layers

$$\mathrm{RI} + e^- \xrightarrow{e^-} \mathrm{R}^\bullet + \mathrm{I}^- \quad (1)$$

$$\mathrm{R}^\bullet \xrightarrow{e^-} \mathrm{R}^- \xrightarrow{\text{Proton source}} \mathrm{RH} \quad (2)$$

$$2\ \mathrm{R}^\bullet \xrightarrow{\text{Dim}} \mathrm{R-R} \quad (3)$$

$$\mathrm{R}^\bullet + \text{Surface material} \longrightarrow \text{Surface grafting} \quad (4)$$

$$\text{If material = carbon} \xrightarrow{ne^-} \text{Charged carbon} \quad (5)$$

$$\text{Charged carbon} \xrightarrow{\mathrm{RI}} \text{Surface grafting via SN} \quad (6)$$

Scheme 8.1. Electro-chemical reduction of 1-iodoalkanes: some different possible pathways occurring at polarized carbon electrodes.

deposited accordingly were found to be chemically and electro-chemically stable.

Specifically, alkyl iodide RIs and diiodides (I–R–Is) belong to peculiar series of electrophilic species and many electro-chemical surveys concern these species [20, 21]. As shown in Scheme 8.1, 1-iodoalkanes RIs are well known to be electro-chemically reduced at GC according to a two-electron bond cleavage process with the major formation of the corresponding alkane (RH). Using this cathode material, the two-electron process occurs due to a large cathodic instability of the free radical $R^\bullet$ (issued from a dissociative electron transfer according to reaction (1) and reduced at a potential less negative than that required for the two-electron bond scission, (Eq. (2)) [22]. In aprotic organic solvents, some amounts of homo-dimers could be formed either by homo-coupling (reaction (3)), but more likely obtained through a Wurtz process that implies organic carbanions, which may act as nucleophilic transients. Practically, reactions (1)–(4) occur at an ideal electrode. Reactions (5)–(6) should be considered as side reactions that depend partly on (i) the capability of the free radical $R^\bullet$ to get some affinity with the solid interface (reaction (5)) or/and on (ii) the electro-chemical charge (reaction (6)) of the cathode material that leads to a classical SN displacement.

Until now, it was expected that radical additions and SN displacement reactions may take place and depend on the nature of the concerned primary alkyl halide and the potential applied to the

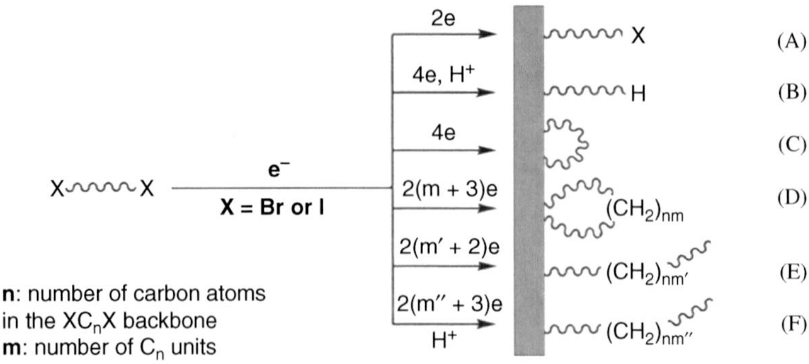

Scheme 8.2. Different kinds of attachments expected as possible with α,ω-di-electrophilic species at carbon polarized surfaces.

carbon cathode. Interestingly, cathodic voltammetric experiments achieved with 1-iodohexane showed a progressive inhibition of the electrode, but definite evidence concerning the exact nature of the organic deposit onto the GC surface [23] was brought about by using ω-iodoalkylferrocenes (II) and the efficiency of charged carbon achieved *independently* of its reaction toward alkyl iodides was proved by using a glove box. Concerning α,ω-diiodoalkanes, many possible reactions may occur at the carbon interface and are summarized in Scheme 8.2. Focusing on immobilization processes, the modification of carbon surface is certainly complex and difficult to foresee and analyze.

8.2. Indirect Grafting of Ferrocene by Means of Diiodoalkanes at Charged Carbons

The reduction of α,ω-diiodoalkanes (model substrate: α,ω-diiodooctane (I)) is now considered for comparing electrode capacities of electro-chemically modified by structures like (III) bearing a ferrocene moiety branched on their aliphatic backbone as displayed in Scheme 8.3. For that, studies achieved in aprotic polar solvents were undertaken with a special focus on researching modifications of GC and graphite surfaces. The building of compact films from primary and secondary RXs has already been described as a function of

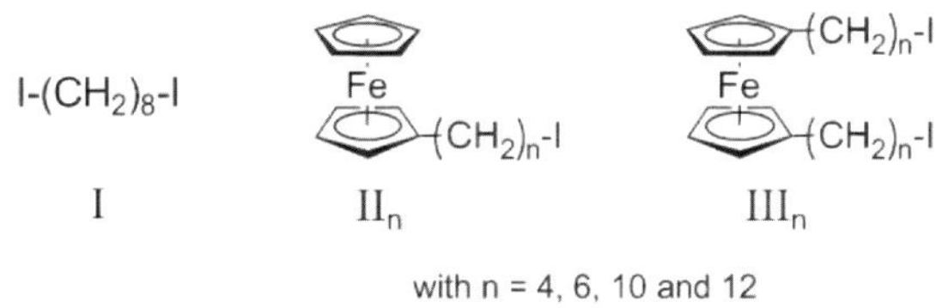

Scheme 8.3. Different electrophilic species studied in this chapter.

the chain length [24–26]. Quantification of surface deposits is worth being considered using ferrocene moieties. On the other hand, ω-iodoalkylferrocenes used as a probe showed that ferrocene was directly immobilized in large amounts onto carbon [23]. In the redox polymers marked with ferrocene generated by Wurtz coupling in basic media, the influence of linker length and the anodic response of immobilized ferrocene (either fixed onto carbon matrix or incorporated in the polymeric bulk) are evidenced. Therefore, a finer structural knowledge of redox layers immobilized at GC surface should be gained and high-density coverage then obtained allows envisaging the process as particularly promising for specific functionalizations in an aim to build new types of electrodes.

8.2.1. *Experimental section*

In the present study, voltammetric results were obtained using electrolytic solutions of tetra-*n*-butylammonium tetrafluoroborate ($TBABF_4$) in acetonitrile (ACN) and *N,N*-dimethylformamide (DMF). All these chemicals were obtained from Aldrich and used as such. Supporting salt concentration was in all cases fixed at 0.1 M. Experiments described in this work did not need special treatment of the electrolytic solutions. In particular, there was no special care for getting extremely dry solutions. The average amount of water found in solutions was checked to be about 1000 ppm.

All potentials are referred to the aqueous $Ag/AgCl/KCl_{(sat)}$ placed in a bridge filled with the corresponding solvent/supporting salt. Electrodes used in voltammetry had an apparent surface area

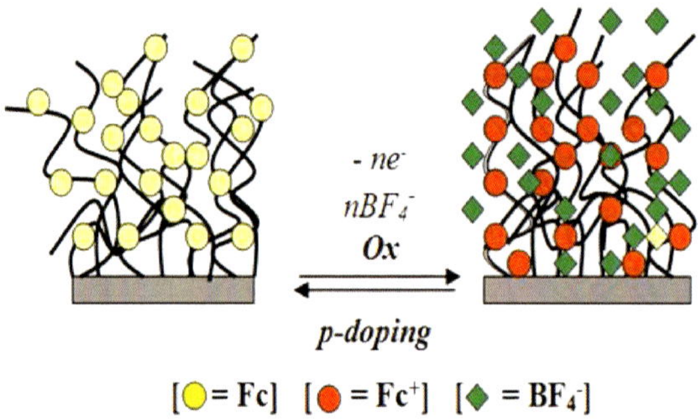

Scheme 8.5. Doping processes of the attached ferrocene layers from long chain diiodoalkylferrocenes that may induce irreversible ion motions [33].

A theoretical approach on the coverage of ferrocene stems onto gold surfaces (compact layers of ferrocene-terminated self-assembled SAM mono-layers immobilized by mean of thiolate functions) was verified by experiments [29–32]: $\Gamma_{Fc} = 1.2\times10^{-10}$ mol cm^{-2}.

8.3. Remarks about the Incorporation of Redox Moieties inside Hydrophobic Polymeric Layers at Solid Surfaces

Then, the cathodic activation of bis(ω-iodoalkyl)ferrocenes may lead to an interesting source of new organometallic redox polymer possibly possessing a random distribution of ferrocene moieties in matrices. Similarly, comparable immobilization of tetrathiafulvalenes (TTF) using this procedure can be imagined as well. The building of the interfacial material in the case of Fc occurs apparently in two stages: (i) the grafting of a first Fc through an initial electro-chemical activation of the carbon surface for creating nucleophilic sites capable of displacing iodine at least from one of the terminal chains of (III_n) species as already recently described with (II_n) series [23] and (ii) the subsequent reactions of the second C–I bond possibly oriented according to at least three modes of reactivity as sketched in Scheme 8.6.

The first pathway **(a)** should correspond to the simple hydrogenolysis of the C–I bond through an initial electron transfer. The

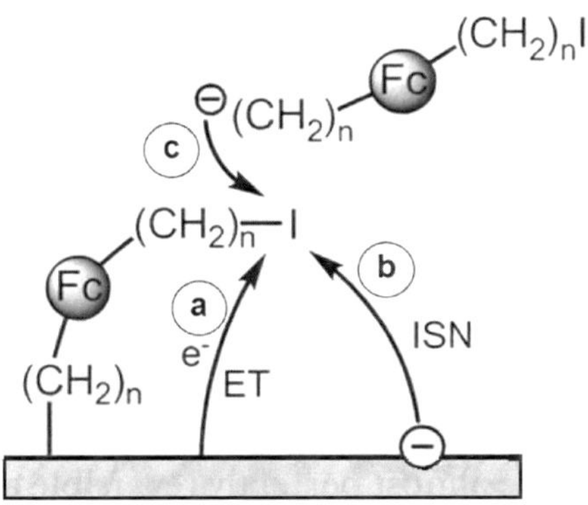

Scheme 8.6. Visualization of three possible pathways for a redox polymer formation at a GC cathode. (a) Heterogeneous electron transfer to the remnant C–I bond. (b) Nucleophilic attack (ISN, interfacial nucleophilic substitution) owing to a specific charge of carbon that depends on the applied potential. (c) Homogeneous S_N2 obeying a Wurtz reaction process. The chain length may condition the rate of the pathway (for example, if $n = 10$, process c > a or b).

alternative pathway **(b)** may occur from the solid surface when the layer is not too thick and the carbon surface is sufficiently charged (a process analogous to a SN substitution). Quite probably, the latter situation is not likely with starting monomers III_n (short alkyl chains) leading to double grafting, apparently a consequence of their chain flexibility. The third pathway **(c)** could fit with the partial heterogeneous two-electron reduction of III_n at the covered interface, provided that alkyl chains bound to Fc moieties are not too short. That would imply that carbon–carbon bond formation through a Wurtz-type reaction is most probably the key of the polymerization and accumulation of immobilized Fc moieties at the polymer–electrolyte interface. Consequently, it is not likely that the number of Fc units per grafted chain could be quite large, especially with bi-electrophilic species dotted with long alkyl chains ($n \geq 10$). The produced films would turn being too thick to allow a fast electron transfer to III_n. On the contrary, with short chains (small n values), the direct reduction of the second C–I looks easy and this possibility does not favor such anionic polymerization. In summary, it could be expected that the capping process is controlled by the length of alkyl chains (path **(b)** vs. path **(c)**), meaning that the lengthening of shift is strongly conditioned by a standby time for at least 1 min before starting the catalytic reduction in the presence of Au, Pt, and Rh. This point supports the existence of an antecedent electrode reaction between **DB3** and the metal. This important point had been already mentioned with alkyl halides [11] reduced at Pd and Cu and interpreted by a chemical reaction (in fact, rather similar to an oxidative insertion within the C–halide bond) being at the basis of catalysis. Reactions of primary alkyl iodides with Cu, Pd, and Ag were already described [14] at GC cathodes for allowing organometallic transients to be generated at disconnected metal interfaces.

At the scale of the electro-chemical cell, the reaction of **DB3** with metals can be well followed using the process schematized in Scheme 9.2. Thus, the voltammetric response of **DB3** upon time shows the progressive merging of a pre-peak growing fast in intensity and constantly shifting toward less negative potentials (Figure 9.1 (B))

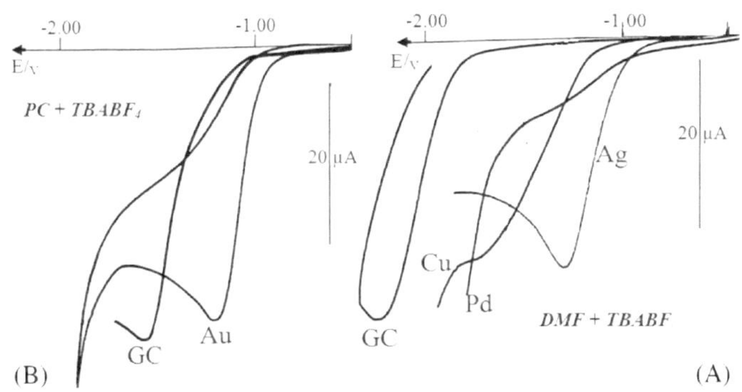

Figure 9.1. Voltammetries of **DB3**. Concentration: 12 mmol L^{-1}. Voltammetries displayed (DMF + $TBABF_4$) on the right part of (A) exhibit a direct catalytic effect brought by different electrode materials in the order Ag>Pd≫Cu>Pt and compared to smooth GC supposed to not possess any catalytic activity. Alternatively, in PC + $TBABF_4$, as displayed in the left part (B) when placing a piece of gold (area ≈ 2 cm^2) in the bottom in the cell for a few minutes, GC is progressively covered with Au NPs, which then produces an unexpected catalytic effect detectable by a progressive shift of the reduction peak strongly depending on the contact time with growing NP surface density.

for the whole duration of the experiment. Within about 30 min, the overall voltammetry of **DB3**, first drawn at the GC electrode, was found to have deeply changed (if obeying the experimental conditions given in the caption of Figure 9.1). Additionally, it could be observed that the modified carbon electrode gains, by this way, strong catalytic abilities.

The status of composite material by deposition of NPs was verified by SEM and EDX experiments. It could be clearly seen that the contact duration of **DB3** solution in the catalytic source (metal piece, for example Pt deposited at the bottom of the cell) is a key factor for the deposition of NPs onto electrodes. The other important point is the reactivity of the metallic NPs toward chemically dissolved **DB3** (actually the reverse reaction of their formation). Thus, the contact of a polarized GC plate with **DB3** rapidly leads to the brightness of carbon assigned to the deposit of platinum NPs.

The uniform coverage of carbon surface (GC and graphite) by precious metal NPs, is illustrated by Figures 9.2 and 9.3 and the occurrence of metal deposition is fully supported.

Coverages of metal NPs onto carbon are based on the occurrence of an oxidative addition of the metal within the C–Br bond. The rate of this reaction is boosted both by the solvent polarity and the structure of α,ω-dibromoalkanes [18]. Thus, at a GC cathode, 1,2-dibromoethane would lead to the formation of ethylene while 1,4-dibromobutane does not exhibit any fast reactivity toward metals. Additionally, **DB3** certainly favors a mutual activation of the two C–Br bonds by electron-withdrawing effect. Palladium is deposited this way onto Au and Pt by short (< 30 s) fixed potential electrolysis at micro-electrodes. Moreover, the coverage of palladium onto platinum could be achieved at −1.2 V from a "mature" solution of palladium after 2 h of immersion of a Pd solid plate. The current was low (1.5 $\mu A\ mm^{-2}$) and the amount of generated electricity was 0.2×10^{-3} C mm^{-2}. The resulting modified electrode could be tested by means of the classical electro-catalytic reduction of 1-iodobutane in DMF containing $TBABF_4$. The one-electron diffusional process was fully observed at −1.24 V (obtaining of the butyl free radical)

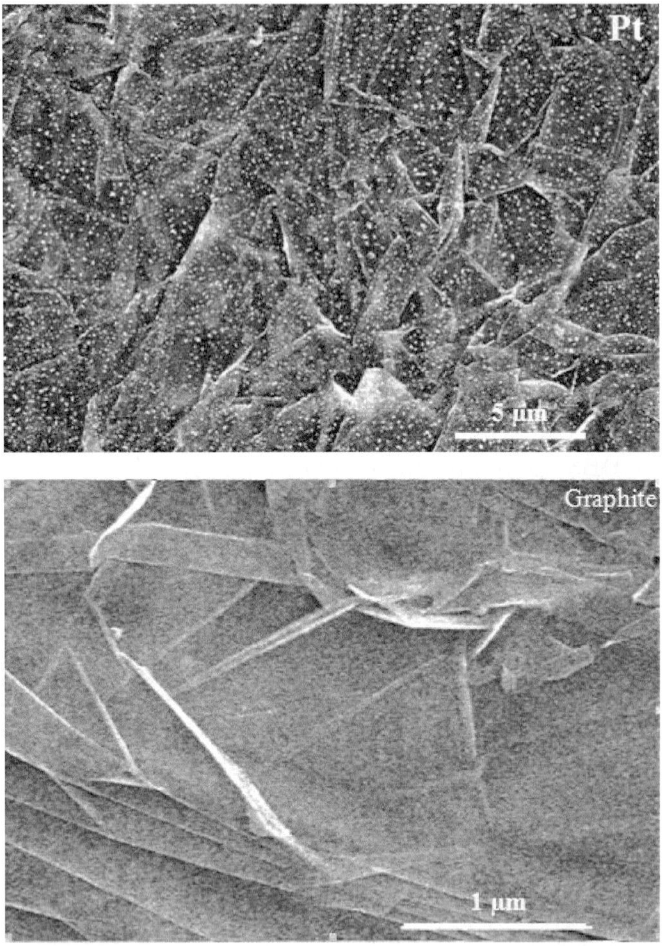

Figure 9.2. An example of platinum NP deposition onto graphite foils used as cathodic materials (apparent surface area: 1.6 cm^2). Concentration (top by comparison of the support shown at the bottom) of **DB3**: 16 mmol·L^{-1} in PC–EC (50–50) containing $TBABF_4$ 0.1 M. Applied potential: −1.4 V vs. Ag/AgCl. Source of platinum; a thin foil 0.8 × 0.6 cm (10 mg) placed at the bottom of the cell. Total amount of electricity: 1 C cm^{-2}. Time of contact: 30 min [19].

and then confirms the occurrence of an efficient platinum palladization, which can be also built by means of galvanostatic deposit. Similar palladizations were easily obtained at many conducting solid interfaces.

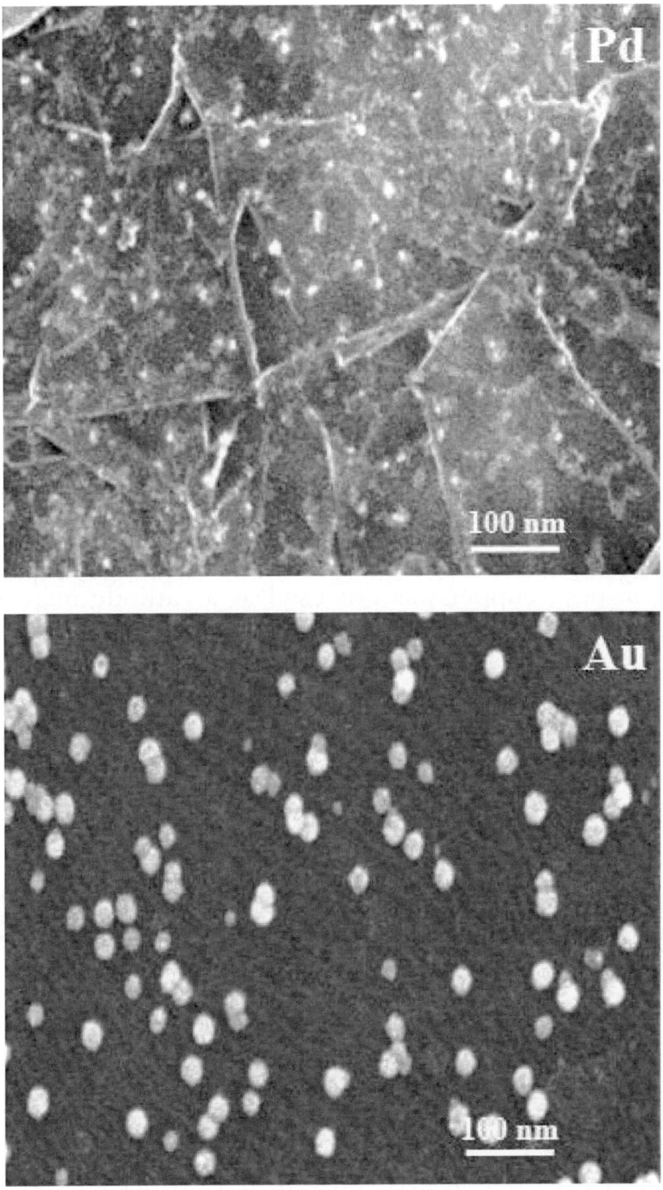

Figure 9.3. Deposition of palladium and gold NPs onto graphite and GC, respectively. In both cases, concentration of **DB3**: 18 mmol L^{-1}. Liquid electrolyte: PC/0.1 M $TBABF_4$. Applied potential: −1.5 V. Amount of electricity: 1.2 C cm^{-2} (graphite/Pd) and 1.8 C cm^{-2} (GC/Au) [19]. See conditions in text.

9.5. Copper: A Remarkable Cathode Material for Reduction of α,ω-dibromoalkanes

Copper electrodes (made of smooth metal plates, copper powder suspensions, or galvanostatic deposits from acidic solutions of Cu^{2+} onto several conducting substrates like GC, platinum, and nickel) may exhibit exceptional catalytic properties, in particular with a series of α,ω-dibromoalkanes $Br–(CH_2)_n–Br$ [19]. Thus, the use of all coppered surfaces leads to spectacular reduction potential shifts ($\Delta E \approx 1$ V) compared to the use of GC taken as a standard interface. Copper was also found to chemically react with dibromoalkanes especially when $n = 3$ (the case of **DB3** described in this chapter). Thus, the preliminary complexes of organo-copper formed are destroyed by a two-electron cathodic discharge. It is stressed here that the participation of copper in the cathodic process gives rise to tremendous surface structural changes.

In the past, copper was not used as a cathode material. This apparent lack of interest could be justified by its ability to spontaneously produce a layer of copper oxide (CuO). Quite similarly, the attention of organic chemists (formation of organo-coppers) appears mostly troubled by the presence of the oxide: for this, sophisticated methods were developed to create very reactive Cu^0 species [20–21], especially by reduction of Cu(I) phosphine complexes. Subsequently, the interest in Cu^0 should be more or less regarded as a reducing reagent. Specifically, aryl iodides may produce in the presence of Cu^0, aryl–aryl bonds, in what is called the Ullmann's reaction [22–24]. Zero copper may also react with alkyl halides [25] through oxidative insertions (formation of RCu(I) intermediates). These latter reactions are certainly of great interest for a specific use in electro-chemistry, especially when carbon–halogen bonds are involved and aprotic solvents of high polarity are employed.

Most of the catalysts currently effective in electro-chemistry do not strongly activate alkyl bromide scission. Even silver remains of limited interest. More complex palladium surfaces (in particular utilized as alloys with copper or silver [26]) could be more efficiently achieved by the reductive cleavage of C–Br bonds (see Chapters 2 and 3). Organic dibromides such as α,ω-dibromoalkanes Br–$(CH)_n$–Br (DB*n*) react easily at copper electrodes in aprotic solvents (in DMF) in the presence of TAAX salts, from **DB3** (1,3-dibromopropane) to **DB12** (1,12-dibromododecane).

9.6. Reactivity of DB3 at Copper Surfaces

Let us recall that **DB3** is reduced at a GC electrode according to an irreversible single two-electron step (electrolyte used: TAAX salts in DMF) within the potential range from −2.3 to −2.5 V. Thus, the two C–Br linkages are expected to behave identically under those conditions, leading to a two-electron process [13]. The main product obtained by potentiostatic electrolysis was cyclopropane (Scheme 9.3). One notices the progressive shift of the reduction step to more negative potentials under recurrent scans. This effect is presumably provoked by the *in situ* generation of polymers such as polypropane.

Most of electrode materials like smooth palladium and platinum do not bring much improvement in the ease of reduction of **DB*n***. On the contrary, a great change was observed by introducing copper metal onto the electrode interface. Thus, catalytic interfaces made of copper deposited onto platinum and copper–palladium alloy exhibited huge discrepancies toward GC. As a matter of fact, a very large potential shift could be observed between carbon (E_p = 2.31 V) and galvanostatic thin deposit copper coverage onto platinum that positively shifts the peak potential to E_p = −1.39 V. This strongly catalytic

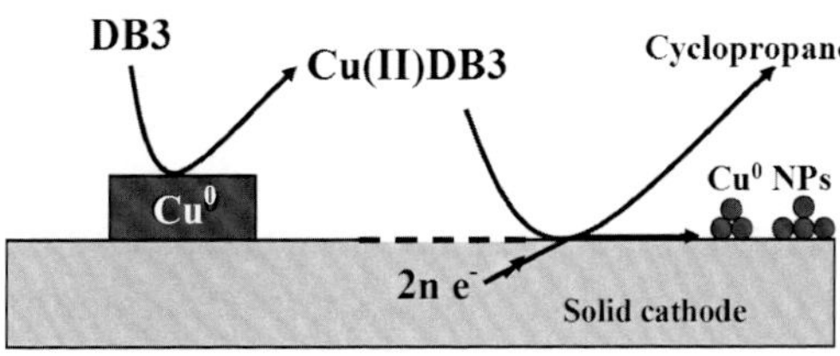

Scheme 9.3. The electro-chemical reactivity of 1,3-dibromopropane at copper electrodes [19].

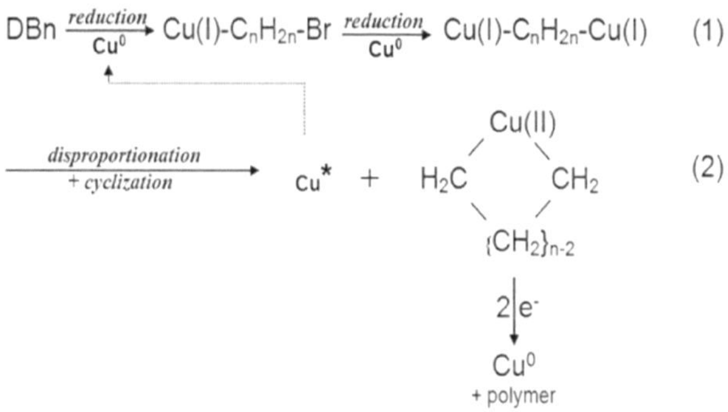

Scheme 9.4. The reactivity of α,ω-dibromoalkanes at polarized copper [10].

reduction of **DB3** is exclusively obtained at a smooth copper electrode in the absence of copper oxide. This main cathodic peak was always found to be diffusion controlled and peak currents were satisfactorily proportional to **DB3** concentration. Coulometric measurements confirmed the two-electron reduction process and robustly suggested the major formation of cyclopropane like previously reported at GC [18] as well at other electrode materials.

Why does a large electro-catalysis specifically obtained by copper with **DB3** for primary alkyl bromides (RBr) not exhibit any reduction process at a potential close to −1.5 V under similar experimental conditions? It is believed that **DB3** reacts with copper according to Scheme 9.4. Each C–Br bond would (separately?) react with copper (reaction (1)) and lead, through disproportionation, a cyclic transient (reaction (2)) more strongly electroactive, especially if the ring is tense ($n = 3$).

In order to clarify **DB3**'s reactivity toward copper, electrolyses were performed in the presence of strongly stirred copper suspension (size of Cu particles: 1000 mesh in fluidized bed system). The main electrode (used here for the electrical contact) was either a GC plate (no reduction could occur at this material at the applied potential of −1.6 V vs. SCE) or more simply a copper grid for activating the surface by Cu NPs deposit). The interest of this process lies in nature of the electrified plate (GC or platinum), which gets progressively covered by

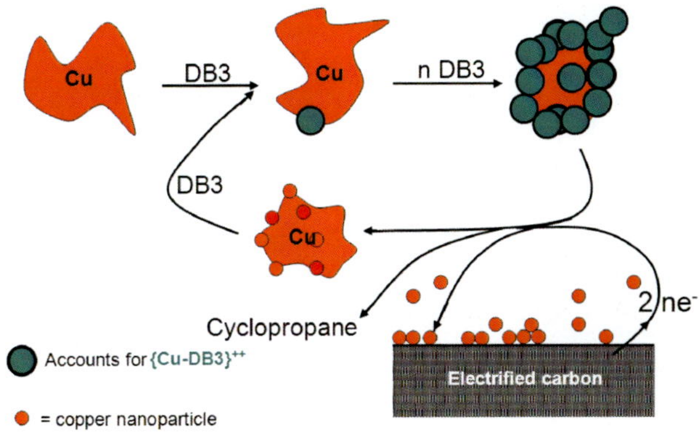

Scheme 9.5. Cathodic surfacial reactivity of **DB3** in contact with a suspension of Cu micro-particles for leading to Cu^0 NPs deposits.

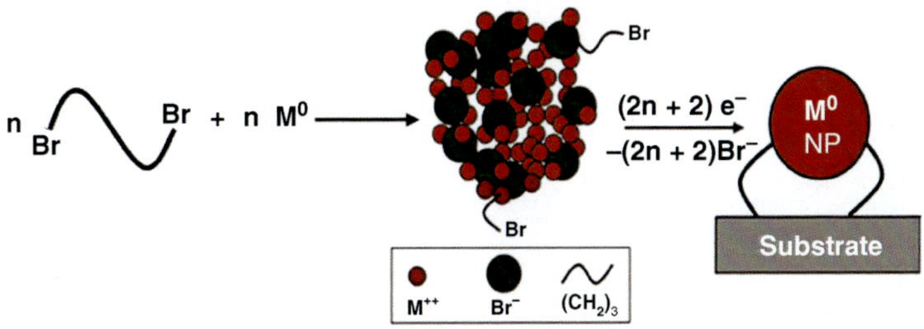

Scheme 9.6. The generation of metal M NPs (possibly attached to the solid substrate via cathodic one-electron cleavage of bromo-alkyl terminal links) via organometallic polymerization followed by an electro-chemical reduction by scission of the two terminal C–Br links (see Scheme 9.2) [17].

copper while electrolysis progresses (Scheme 9.5). The grafting in some cases can be expected with α,ω-dibromoalkanes (as depicted in Scheme 9.6). A strong stirring permitted a continuous tight contact between copper particles and the electrified plate. Moreover, such experiments permit a quantitative confirmation (amount of deposited copper upon time) according to reaction 2 (Scheme 9.4). Practically, it was established that 60–75% of the theoretical quantity of electricity (close to one Cu^0 per **DB3**) was enough to obtain an almost compact layer.

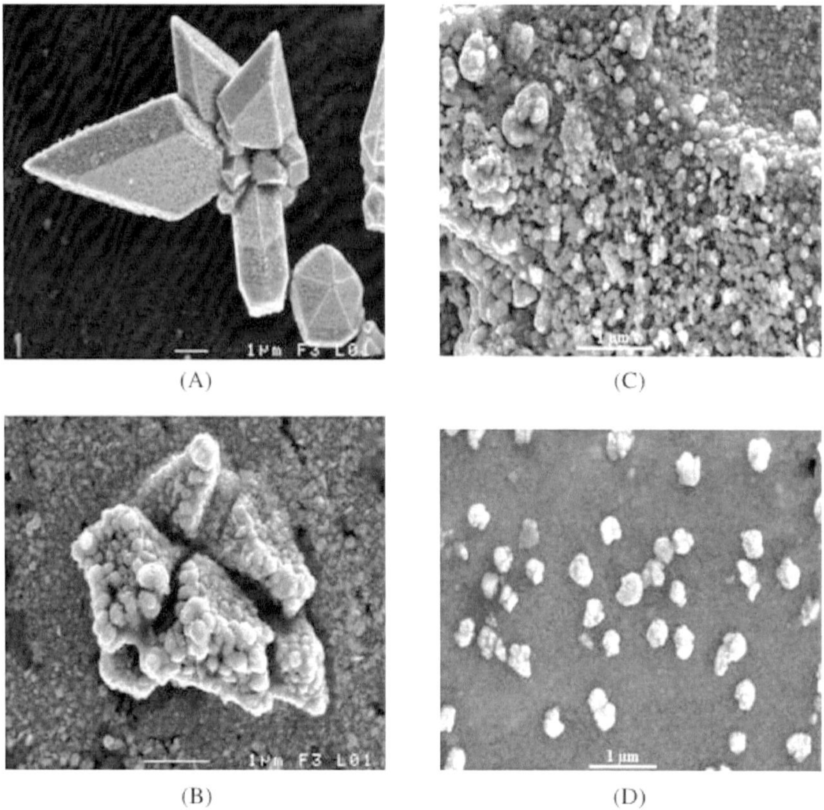

Figure 9.4. SEM images showing the reactivity of copper toward **DB3**. (A) Cu crystal onto a GC surface from a galvanostatic deposit from Cu(II) acetate that corresponds to an electricity of 0.7 C cm^{-2}. (B) SEM image of a copper crystal sample after reaction (5 min) with pure **DB3** and then rinsing with alcohol and acetone. (C) The surface according to (A) after contact and then reduction of **DB3** (concentration: 10 mM). Electrolysis at −1.5 V. Electrolyte $TBAPF_6$ in DMF. Amount of electricity: 2 C cm^{-2} (D) Nanostructural morphology of a copper deposit at a GC surface, using the method **DB3**/copper suspension. Applied potential: −1.8 V. Total electricity amount: 0.1 C cm^{-2}.

A large and fast reactivity of **DB3** toward copper is easily demonstrated by visualizing the corrosion of Cu micro-crystallites in real time. Accordingly, one notices the concomitant deposition of Cu^0 NPs on the conductive support (Pt or carbon) after a contact of 10 min (Figure 9.4).

GC electrodes covered by copper deposits, using the procedure presented above, were employed for generating catalytic materials. It then appeared that modified GC surfaces by activated Cu^* deposit (Scheme 9.4) became extremely efficient afterward. Also the progressive influence of deposits similarly obtained at bare GC from (copper–**DB3**) complexes (easily achieved by stirring a copper powder in the presence of any α,ω-dibromoalkane) could be obtained as well. Additionally, such transfer of copper in the presence of **DB3** is well evidenced by the progressive corrosion of copper crystals (e.g., deposited onto a GC surface) depicted in Figure 9.4. SEM images show that copper metal is progressively displaced from crystals (chemical reaction) to the carbon surface (electro-deposition according to the reaction (2), Scheme 9.4). An almost total lack of prominent copper crystals could be then obtained with large amounts of electricity. Lastly, the direct contact of smooth copper with pure **DB3** may confirm the spectacular corrosion of the metal. This fast reaction was totally unexpected and particularly efficient with **DB3**. A slow evaporation of the liquid phase allows observing the formation of copper microcrystals at the surface. Those copper nanostructures were found to present a large proportion of bromine atoms at their surface (possibly via attached structures) depicted in Scheme 9.6.

The definitive structure of a complex between copper and **DB3** has not been determined so far. Definitively, there are no Cu^{2+} ions present in solution as shown in UV–Vis spectroscopy (no absorption between 300 and 700 nm) and in voltammetry (absence of redox systems specific of copper cations). It is presently expected that a (copper–**DB3**) complex is formed and mainly stays as an adsorbed species at the interface of copper NPs at the interface and in suspension in the liquid bulk. Fine filtrations of solutions of **DB3** in the contact of copper permitted the total elimination of any copper deposit and therefore excluded the existence of a soluble copper complex. These NPs covered by organic functions mimic cationic micelles and therefore could be considered as electro-active organometallic species. Their electro-chemical discharge could then concomitantly contribute to copper deposits onto electrified surfaces like GC, platinum, gold, and nickel.

polar aprotic organic solvents such as DMF containing tetraalkylammonium or alkali metals salts) eventually deposited onto solid conductors (glassy carbon (GC), silver, gold) lead to intensive graphite exfoliation, which can be easily monitored through gentle voltammetric sweeps or by means of fixed-potential electrolyses at moderate potentials (at the threshold of the cathodic charging, usually at $E < -1.8$ V). Such simple and spectacularly efficient processes may produce a quite stable modified composite solid; its surface may exhibit astonishingly reproducible electro-chemical responses (both in nature and intensity). Of course, those responses strongly depend on the nature of the ions (principally cations) that are implied in charge–discharge processes of graphite and of the resulting graphene. Voltammetric reduction peaks assigned principally to the progressive insertion of TAA^+ cations into the material (arising from the reduction of graphene planes) may depend on the ion diffusion rate through the carbon sheets or the pores as well as through graphene planes; the whole process can be perceived as a complex phenomenon, especially with natural graphite given the heterogeneity of its structure. Lastly, it is clear that a few carbon layers remain at the solid surface at the state of cathodically doped graphene planes or flakes being tightly held at the conductor surface by π,π-interactions. Almost always, the cathodic charge is associated with a symmetric process (that of an anodic discharge). Alternatively, the mixed injection of electrons and cations in the materials is highly remarkable in many aspects: in particular, for generating such charged materials as reducing species and using these species to modify graphite (or graphene) planes using adequate organic electrophilic reagents for building new modified interfaces; and carrying out further investigations aiming for the creation of new composite and reactive materials. Stable attachment processes of graphene layers are mostly described at electrode materials such as carbon, gold, and silver. Especially, profound modifications of these graphene-like surface structures were deposited not only at carbon, but also at transition metals possessing catalytic capacities could, in principle, lead to a large palette of conducting composite materials. Different reactions are expected to occur: with a series of organic electrophiles (in the reaction with doped graphene acting then as a reducing species or as a polynucleophile) or *in situ* produced free radicals successfully generated with the aim of building a new layer in close contact with the substrate.

More classically, glassy carbon (GC) is widely used as a convenient electrode material [9, 10] in the presence of polar aprotic organic solvents at the potentials more reducing than −1.8 V. The interest in such materials in various forms (rods, plates, microelectrodes) allowed the prospects for a large variety of electro-organic reactions [10]. However, recent studies revealed a high complexity of GC related to the existence of inclusions of graphite-like structures in this material, which is no longer electro-chemically inert at potentials $E < -1.7$ V. These carbon {sp2}-inclusions [11–13] are also candidates to be charged and therefore they may confer GC with particular properties. The charge occurs via the insertion of organic or inorganic cations and concomitantly generates a number of polynucleophilic zones, which are reactive by a simple contact with a large series of electrophilic reagents [14–16]. Additionally, these unsaturated inclusions certainly account for high reactivity of such carbons, in particular, toward radicals generated at higher potentials (when GC is not electro-chemically charged) for decorating unsaturated solid materials.

Thus, in most of the polar organic solvents, the cathodic charge of varieties of graphite [17, 18], such as highly oriented pyrolytic graphite (HOPG), can afford rather well-organized {carbon–cations} materials strongly depending on the applied potential and resulting from the progressive transient of the form $[C_n^{-}TAA^{+}]$ through the concomitant addition of electrons and tetraalkylammonium (TAAX) salts. The addition of inorganic salts into graphite has also been explored [13]. At any time, the stoichiometry of these added salts depends on the applied potential and the amount of electricity consumed [20]. Constant-potential reduction providing a first evidence of exfoliation of natural graphite by ammonium salts have been reported [8]. With HOPG, less sensitive to exfoliation, such electro-chemically prepared "graphite salts", in fact quite

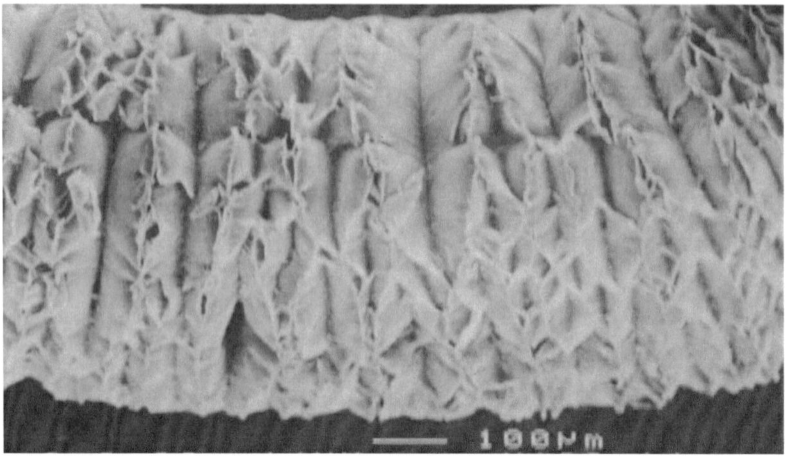

Figure 10.1. Electro-chemical pre-exfoliation [8] of a natural (Ceylon) graphite flake through fixed potential electrolysis (contact with a stirred mercury pool, dimethyl formamide(DMF)/tetramethylammonium perchlorate, $E = -2.1$ V).

akin to {TAA–amalgams}, have to be considered both as a novel class of reducing agents and as nucleophiles reactive toward electrophilic species like some π-acceptors or organic derivatives possessing a strongly polarized carbon–heteroatom bond. Therefore, charged graphites are of great interest at least for two reasons: (i) their capacity to stock electricity (conversion of the material surface to a polynucleophilic reagent) and (ii) with larger amounts of electricity, to split under the pressure created by the insertion of bulky TAA^+ cations. The bursting of this material was seen for quite a long time (Figure 10.1). In this chapter, the important point is the new process permitting to cap GC with very thin deposits of graphene via exfoliation, which allows to boost its nucleophilic properties at the potentials < -1.8 V. Such modification of GC implies considering this material as containing large amounts of graphite nanoparticles that allow, under cathodic charge, creation of polynucleophilic centers [18].

Graphene witnessed a huge development in chemistry and physics [19]. In particular, many ingenious approaches were reported showing that the recent research is highly focused on organic methods (total syntheses) [20,21], while other methods rely on processes of solution-based, chemical [6], and electro-chemical exfoliation [2,3]. Methods based on graphite exfoliation in ionic liquids deserve mention [22]. Thus, among the other processes, the electro-chemical method based on an artificial disaggregating via graphite exfoliation may appear quite helpful. Presently, the main electro-chemical approach for obtaining graphene principally remains the reduction of graphene oxide [23–27]. Additionally, graphene is commercially available and attempts to modify it chemically and electro-chemically were successfully achieved. Some tri-dimensional {graphene-GC} electrodes modified with redox units (such as ferrocenyl, anthraquinoyl, or aromatic units) were tested in catalysis. When using fluidized bed processes, graphene suspensions could permit huge electro-chemical functionalization. In particular (but not limited to), carboxylation of graphene was achieved at a high level (reaching one carboxylate function per 100 carbons) [28].

Coming back to exfoliation, it was mentioned that the first evidence of electro-chemical splitting of graphitic materials by the insertion of TAA^+ was reported four decades ago [8]. Based on this finding, the present chapter develops an original method allowing a quasi-total disaggregation of graphene layers from graphite mechanically deposited onto a solid conducting support. Such graphite–carbon interfaces appear rather stable and the method has been reported in detail. As explained hereafter, it allows a facile and reliable method for obtaining modified solid interfaces covered with very thin layers of graphene. Different TAAX salts were tested and compared in this process. Such recipe is facile and efficient in non-aqueous solvents at about −2.0 V vs. Ag/AgCl and it certainly looks as a fascinating method for *in situ* production of thin graphene layers at electrified solid substrates. Moreover, it can be easily applied to a large panel of solid conductors, conducting polymers, and metals (gold, silver, rhodium, copper, platinum, and palladium). This chapter is limited to the composite GC–graphene material.

Furthermore, some examples of grafting of chemical functions to these graphene overlays were realized using *in situ* electro-generated radicals. Under precise conditions, the exfoliation and the addition of these radicals can be concomitantly achieved. Novel processes exploiting the capability of the deposited graphene layers to be charged (at $E < -2$ V) were developed: as-doped materials transformed into polynucleophilic species provide avenues for new organic reactions for many kinds of carbons (see Chapter 11). Examples of graphene modification with redox groups and aromatics are provided in this contribution. Globally, this general method allows an electro-chemical tailoring of many solid conductors primarily overlaid with graphene.

10.2. Exfoliation Procedure Applied to GC–Graphite Surfaces

10.2.1. *An analytical evidence for the presence of graphite nanoinclusions in GC*

Before speaking about GC modification, it is important to draw attention to electro-chemical activation, or pre-exfoliation, of natural graphite. SEM images of thus activated graphite (Figure 10.1) compared to non-treated graphite samples make clear the interest in this procedure for a deep exfoliation of graphite (as depicted in Scheme 10.1). Fine pieces of this preliminarily "expanded" material (preferably obtained using the procedure already described in [8]) were deposited under gentle pressure onto a GC disk and this procedure guarantees an efficient and quite well-reproducible capping of the solid surface. In most cases, an additional final exfoliation (realized during the first scan in voltammetry) is necessary in order to obtain a stable (and definitive) coverage of the solid conductor.

However, under specific conditions (in the presence of certain electrolytes, in particular tetramethylammonium salts TMA^+X^-), it is necessary to remember that depending on its mode of preparation, GC is not always inert toward TMA^+. In fact, carbon samples furnished by Tokai carbon C° (GC rod) gave totally flat responses from 0 to −2.1 V to be qualified as very weakly reactive toward free

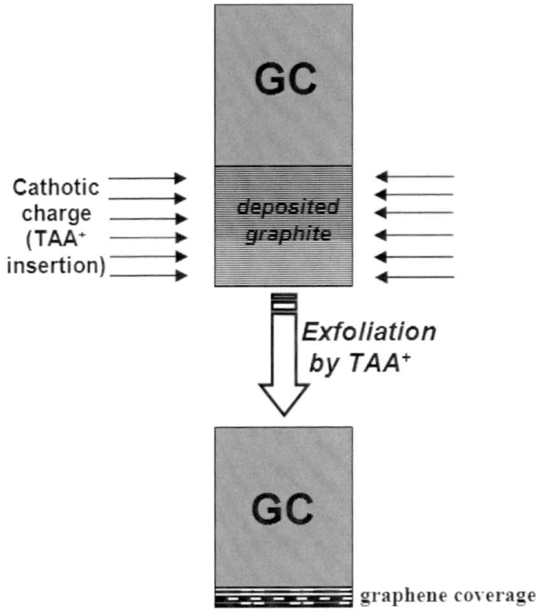

Scheme 10.1. Schematic representation of the exfoliation process based on the electrolyte insertion (TAA^+) into mechanically deposited natural graphite. The example concerns GC, but all conducting substrates are, in principle, suitable for such deposition.

radicals. On the contrary, the capping with pre-deposition of pre-exfoliated graphite, as shown in Figure 10.2(A), provoked a massive exfoliation of graphite after an additional sharp peak (−1.7 V). Afterward, the carbon surface appears to be definitively modified by a thin graphene deposit detected in the course of a first scan by the specific charge of graphene owing to the reaction (insertion?) of TMA^+ ions. This specific voltammetric pattern appears to be totally specific for each TAA^+ as inferred from the voltammetric responses (reduction + oxidation) of the graphene layer. Thus, the signals obtained with TEA^+ (tetraethylammonium cation) as displayed in (Figure 10.2(B)) are very reproducible both in potentials and in peak currents. One may notice that the discharge occurs in several steps. It is worth noting that the shape of the overall electro-chemical response is not at all altered during the multiscan voltammetry.

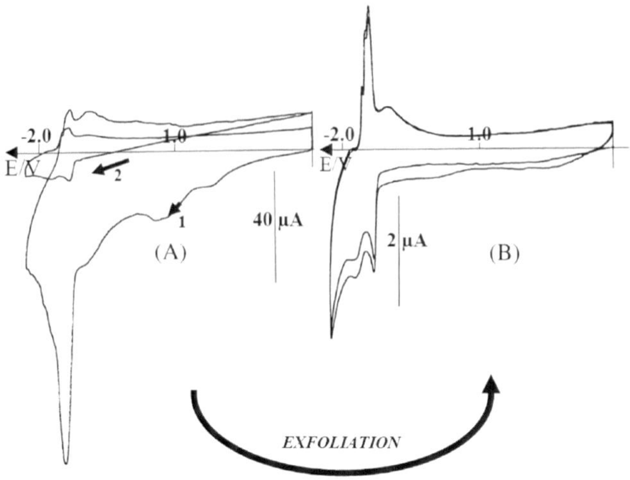

Figure 10.2. Voltammetric responses of a carbon disk (0.8 mm^2, Tokai Carbon C°) covered with natural graphite (in A). DMF + 0.1 M $TMAClO_4$. Scan rate 50 mV s^{-1}. Two first scans that display (from the second scan) the exfoliation of the graphite layer by insertion of TMA^+ cations. (B) Cycling of the produced electrode in DMF + $TEABF_4$ under the same conditions. Notice the difference of sensitivity between the two voltammetric responses [19].

On the contrary, the testing of other GC produced under different conditions may present diverse media with a much higher level of fullerenes and nanoinclusions of graphite (e.g., [11–13]). That could be due to different vitrification temperatures of phenolic resins during the preparation of GC. Actually, the use of TMA salts could be considered to be the most reactive toward charged graphite compared to bulkier electrolytes (such as tetrabutylammonium iodide or tetrafluoroborate). Afterward, these GC–graphite electrodes were electrified within a quite large range of potentials. Typically, voltammetry with Carbone Lorraine carbon (pyrolysis of phenolic resins at 1500°C) using several TMA^+ ions revealed that the electro-chemical responses are rather close, but not identical to those obtained after the graphite deposit. The example displayed in Figure 10.3 is strongly indicative of a quasi-similarity of the responses obtained without and with graphene. The constant anodic "triplet"

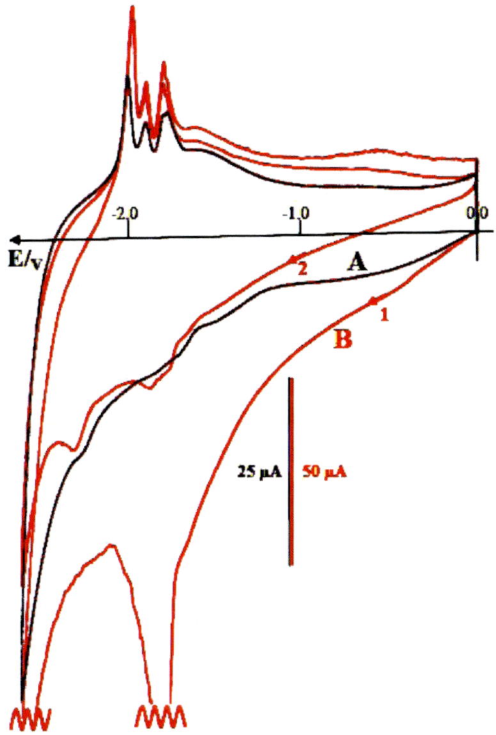

Figure 10.3. Comparative voltammetric responses of a GC electrode (Carbone Lorraine GC). Surface area: 7 mm^2, $TMABF_4$. (A: black) Smooth and (red, two scans) covered with graphite. Scan rate 50 mV s^{-1}. Note the scale difference and the rough similarity of charge–discharge curves after exfoliation [19].

(between −1.65 and −2.06 V) is specifically assigned to the discharge of the oxidation of the material and the reversible insertion of TAA^+. As emphasized later, tetraethyl, tetrabutyl, and tetrahexylammonium cations behave differently and those peaks should be associated with specific level of discharge.

After graphite deposition (curves in red), a huge exfoliation is observed while the second scan is quite similar to that already observed with GC (black curve) substrate. The main difference is the intensity of the process since the currents are at least two times those obtained with uncapped GC. We notice several reduction and oxidation signals assigned to charge and discharge steps. In the present case of TMA^+, an anodic large "triplet" should be considered as a kind of signature both of graphene and of the associated cation. Additionally, the amount of electricity obviously depends on the radius of the cation residing at the surface or inserted in the material in the case of multi-graphene overlays.

10.2.2. *Exfoliation processes and cathodic charge by insertion of several TAAX at GC–graphene materials*

Several salts were tested for identifying their specificity to influence charge and discharge processes, principally identified by means of voltammetric experiments in DMF/TAAX electrolytic solutions, always at room temperature and at 0.1 M concentration of the salts. With different salts, it is worth mentioning that many repetitive scans performed between 0 and –2.4 V do not globally affect the electro-chemical stability of the graphene interface despite those recurrent charge–discharge processes. Possibilities could lie on stabilization due to a kind of π,π-interaction, especially with already filled GC of many graphite nanoparticles due to the conditions of its preparation. As mentioned above, graphene deposits display stable and reproducible electro-chemical responses without any noticeable current decay during repetitive cycles. Furthermore, any deposit first exfoliated with TAA^+ can then be cycled in another salt at once, leading to the specific response proper to the nature of a cation without any apparent pollution brought from the preceding experiment. This could also signify that any charge–discharge is relatively clean, possibly because of the graphene layer thinness. For example, the deposit of TMA^+ can be cycled in tetraethylammonium tetrafluoroborate (displayed in Figure 10.4). With a cathodic scan limited to –2.2 V, three steps are obtained and they correspond to the reversible charge of an ultrathin layer of graphene. Sequenced scans permit to demonstrate that each cathodic step is associated with a corresponding anodic discharge step. The peak currents are linearly proportional to scan rates that arise from the absence of solution-related diffusion of cations in the charge process (quite possibly

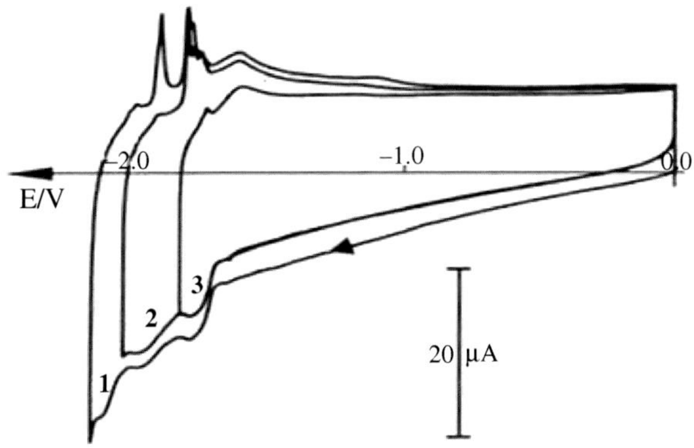

Figure 10.4. Building of GC electrodes first modified by means of graphite exfoliation using voltammetry in DMF/TMABF$_4$. Scan rate v = 50 mV s^{-1}. Results with a GC electrode (7 mm^2) covered with graphite via a procedure mentioned in the text. The resulting voltammograms in DMF/TEABF$_4$ successively drawn with cathodic boundaries: (1) −2.1 V, (2) −2.0 V, and (3) −1.8 V [19].

excluding a significantly fast diffusion of TEA$^+$ inside of remnant pieces of natural graphite). Lastly, it is mentioned that graphene deposits obtained this way are definitively stable toward air and moisture and can be kept in the laboratory for several days without any alteration in their state.

The procedure of exfoliation is easily applicable to all ammonium salts, in particular with bulkiest cations like tetrahexylammonium (THA$^+$) and tetraoctylammonium (TOA$^+$) salts. However, while TEA$^+$ enables an efficient exfoliation quite similarly as TMA$^+$, the possibility of alternatively employing other cations of larger radii is dramatically reduced. Under such conditions, the insertion is too slow to provoke any exfoliation. On the contrary, as mentioned above, the exfoliation completed with TMA$^+$ may induce the formation of graphene layers able to react, under charge, with TAA$^+$ cations (as featured in Figure 10.5). This figure shows the ease of charging graphene with different cations TAA$^+$ in the order: TMA$^+$ > TBA$^+$ > TOA$^+$. An inverse order is noticed for the discharge. Also, with a given scan rate, the average currents used within the charge–discharge processes are much higher with small size cations, which

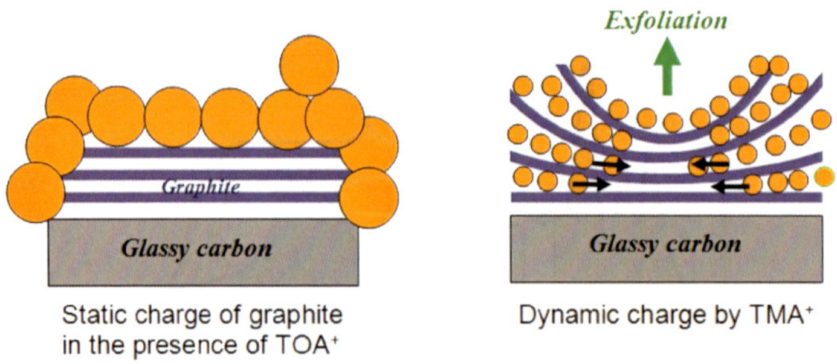

Scheme 10.2. Stability of the ionic deposition during the charge depending on the cation (tetraoctylamamonium vs. tetramethylammonium) involved into the cathodic charge process (from Ref. [19]).

chemical and electro-chemical stability of charged graphene layers has to be taken into account, especially in instances when subsequent chemical reactions may occur (e.g., addition of free radicals to graphene, or specific reactions due to the negative charge of graphene acting as a polynucleophilic material toward adequate reagents).

When the cathodic potential reaches −2.3 V, several charge peaks (their number is ≤4 depending on the TAA^+ electrolyte) are observed. When efficient exfoliation processes of graphite are achieved, the electro-reduction (or charging) concerns the few remaining layers or graphite microparticles that are immobilized at the GC interface presumably by π,π-interactions.

The reduction of graphite deposits onto GC (without exfoliation: first scan in the presence of TBA^+) may reveal different features. Sometimes, in the course of the first scan, some small reduction peaks were observed around −1 V, which could be assigned to electroactive edge functions appearing upon aging of graphite in the contact with air (e.g., ketone and quinone carbonyl groups, aromatic carboxylic acids). After exfoliation, reversible electro-chemical steps were observed at $E < -1.8$ V, assigned to the reduction of ultra-thin graphite (or graphene) layer(s). For example, in the case of TEA salts (triflate, tetrafluoroborate, or perchlorate) in DMF, several specific well-defined reversible steps were obtained (Figure 10.6). The discrepancies mentioned above appear quite similar to those already reported for HOPG under very comparable conditions (chronopotentiometry showing progressive intercalation of the graphite layers with TMA^+ [8, 19]). GC electrodes modified using commercial graphenes have been prepared for comparison, but their voltammetry unfortunately did not show reliable results in terms of reproducibility, stability, and intensity. The analytical characterization of synthesized macro polyaromatic species, expected to be close to graphene structure, gave comparable successive reversible cathodic peaks at carbon electrodes.

After a long series of experiments on the cathodic charge of natural graphite, one can rationalize assigning the observed cathodic peaks to the progressive charge of graphene, leading to polyanionic forms depending upon a number of factors such as the applied potential, the counter-cation present in the vicinity of the charging interface, the nature of solvent as well as the procedure of substrate preparation. Electronic charges are delocalized in mono-layers, but rearrangements of those layers into a "graphite-like" system by π,π-interactions may undoubtedly occur. Moreover, such interactions might stabilize graphene layers at GC interfaces. The described exfoliation seems to be of high interest since it could be considered as superficial doping processes for GC that may boost its reactivity.

10.3. Reactivity and Modification of Carbon–Graphene Interfaces

10.3.1. *Supported charged graphene layer as an efficient polynucleophile*

As pointed out above, overlaid with thin graphene layers, GC surfaces are good candidates for original procedures permitting elaboration of new interfaces based on simple deposition of graphite followed by an easy cathodic polarization procedure. Thus, building and reactivity of such interfaces are visualized by highly reproducible electro-chemical responses scanning the potential beyond −1.7 V, which is considered to be the charge threshold potential of the whole GC composite. For instance, with a given electrode size, astonishingly reproducible amounts of electricity were observed with TMA^+ and TEA^+ (total cathodic charge within a defined potential range remains almost constant with an average error much less than 10%). Additionally, recurrent scans did not bring any noticeable discrepancies, which imply that charge–discharge processes do not affect the stability of graphene layers deposited onto GC. At the scan rates below 0.1 V s^{-1}, the currents do not change along the repetitive scans. On the contrary, this observation is no more valid with bulky TAA^+ cations and the attempts at exfoliation (under the conditions of graphite immobilized at a GC surface) were disappointing, ineffective, or poorly reproducible.

Although the exact mode of exfoliation as a function of the TAA^+ size is not yet totally clarified, it is highly expected that graphite exfoliation likely goes on until only very few graphene layers remain at the interface. The rather good reproducibility relative to the amounts of charge observed after exfoliation would suggest that a small number of graphene layers is concerned (one to three if one refers to the gold–graphene system under similar conditions) in the final coverage process as developed in Chapter 15. However, the outer layer (or possibly graphene stacks or fragments) is expected to be disposed at the interface and to be exposed to the solution. These interfacial species remain practically the only ones further reacting when short potential sweeps are imposed. This could imply that diffusion of TAA^+ cations through the graphene stacks is quasi-absent within the quite short time laps corresponding to the whole sequence of scans in voltammetry. Additionally, the role of the GC substrate itself would not be significant, which supports the idea that graphite inclusions in GC are small and are much less reactive than the deposited graphene itself. Furthermore, the negligible role of the substrate is observed when metallic substrates (silver or gold) are employed instead (see Chapter 15). Subsequently, the experiments on grafting the redox groups may exhibit significantly different reactivity when both GC and graphene are simultaneously subject to charging and to the resulting nucleophilic substitution toward electrophiles such as organic RX compounds. Preliminary experiments support this to be a plausible model. Therefore, the reduction peaks observed with a given TAA^+ salt ($E < -1.8$ V) could be regarded first as related to the superficial formation of a polyanionic graphene, which corresponds to surface chemical activation, i.e., its reduced form continuously evolving while potential decreases and diffusion of TAA^+ inside graphene planes progressively take place. Thus, each cathodic peak would correspond not only to a specific partial coverage but to thermodynamic insertion already studied in detail [13]. Localized repartition of TAA^+ cations related to the insertion could be seen as a kind of soft digging since, at a given potential, their surface concentration cannot be established without a fixed setting in the graphene network. The total cathodic charge could be essentially calculated through the average amount of the inserted TAA^+ cations (very close to that of that of the number of electrons globally involved). For example, the charge integration found in a given process (case of the three charge steps of $TMABF_4$, between −1.7 and −2.3 V) was found to be 4.8×10^{-4} C for a surface of 7 mm^2. The resulting value (2.2×10^{-8} ions cm^{-2}) seems high, but one should take into account the large roughness of the interface along with partial involvement of the ions toward the internal layers, which could induce graphene deformations and globally increase the active surface substantially affected by the cathodic charge. Generally, weak electrophilicity of bulky TAA^+ cations is expected to favor the formation of ion multi-layers. On the contrary, the values obtained with $TOABF_4$, based on the probable formation of ion mono-layers (2.5×10^{-9} ions per cm^{-2}), would fit better the above proposed model.

10.3.2. *Specific surface modifications induced by deposited charged graphene*

Still using the operating mode developed above, a large series of experiments dealing with the electro-chemical doping of graphene were carried out. Many chemical modifications and functionalizations of graphene layers could be undertaken using pre-doping

C_{sp3} within the graphene network. Different modifications of the external graphene layer (preliminarily formed in the presence of different TAA^+ cations by cathodic charging at $E < -1.8$ V) contrary to the follow-up processes using, in principle, free radicals in the absence of graphene reduction are discussed in detail below.

First of all, the electro-chemical generation of 2-methylanthraquinone radical (from 2-bromomethyl-anthraquinone, AQ–CH_2Br) was carried out by simple electron transfer at GC–graphene electrodes. After one scan between 0 and −1.6 V (in DMF/$TMABF_4$ at v = 50 mV s^{-1}), the voltammetric response of the modified GC–graphene electrode was as shown in Figure 10.6, stressing out two large but totally different zones (a yellow and a blue one) that clearly show diverse charge–discharge systems. More specifically, the example is of interest because it demonstrates that different models of charging may coexist at the electrode: the yellow zone displays the global charge of graphene (from about −1.7 V) while the blue one (that emphasizes, in aprotic conditions, the surface charge of immobilized anthraquinone via external radical addition) features two reversible one-electron steps between −0.7 and −1.5 V.

The amount of electricity consumed by these two different kinds of charge–discharge systems are roughly equivalent. Graphene charge corresponds to about 0.3×10^{-3} C. The two-electron reduction of grafted anthraquinone consumes about 0.4×10^{-3} C. It is interesting to imagine that one electron implied in the graphene charge could be (also) involved in the grafting of a single AQ group. A simple experiment based on a charge of 0.5 millicoulomb at −2 V applied to a GC–graphene electrode and immediately transferred into a solution of AQ–CH_2Br led to a quite similar result. Note that the amount of AQ immobilized this way is extremely large, which supports a reaction of free radicals with sp^2 carbons belonging to graphene planes. The conditions reported in Figure 10.7 with just one scan allow obtaining superficial concentrations of $\Gamma_{AQ} \approx 3 \times 10^{-8}$ mol cm^{-2}, which cannot be justified only by edge effects. It is expected that such a large level of immobilization is unlikely with a simple multi-layer effect: this multi-layer form is quite certainly provoked by the existence of several planes of graphene highly reactive

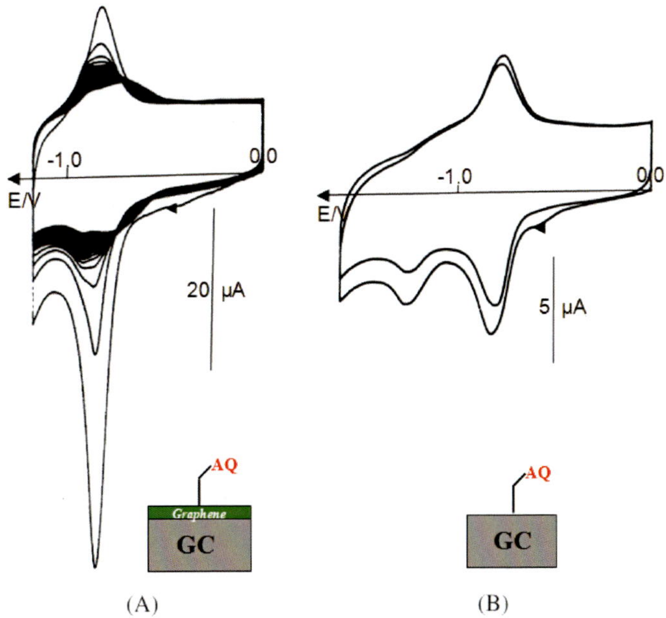

Figure 10.7. A fundamental difference between the radical addition ($AQ–CH_2^{\bullet}$), occurring at GC–graphene (A), vs. smooth GC (B). Deposits are achieved from a solution of $AQ–CH_2Br$ (2×10^{-2} M L^{-1} in aprotic DMF, reduction at −0.7 V). Cycling of the produced surfaces in DMF-$TEABF_4$ at scan rate 50 mV s^{-1}. It is certainly important to question about the instability of the grafting of {2-methylAQ} onto the graphene network. The low stability of the grafting would promote a tendency to regenerate the graphene network. Also, it is remarkable that the half-peak width (first scan with **A**, very sharp peak) is very close to the theoretical value of 90 mV from [19]).

with free radicals, the external layer being concerned first by the radical addition.

In order to bring a definitive support about the nature and the location of functionalization via *in situ* produced free radicals, intensities of the first step (in the case of the grafting of $AQCH_2^{\bullet}$ radicals) were compared: GC–graphene vs. smooth GC (Figure 10.7 curves A and B, respectively). The difference between the two responses (both relative to the first scan) are huge (the ratio of almost 10); accurate values with experimental superficial concentrations are $\Gamma_{AQ} = 3.0\times10^{-8}$ mol cm^{-2} (with A) vs. 3.3×10^{-9} mol cm^{-2} (case B). Average values with smooth GC should approximately agree with the roughness of a surface polished with a 800 μm silicon paper. In contrast, the coverage with graphene significantly increases, at least at the first sweep, the efficiency of the grafting level. The progressive decay of reduction current in the course of successive scans could be provoked by the bulkiness of species intercalated in the graphene matrix during the charge–discharge processes. Degradation would be therefore analogous to an ultimate exfoliation of the functionalized layers and lead to an almost total destruction of the coverage.

10.4. Conclusion

Direct exfoliation of natural graphite applicable to many carbons containing graphitic inclusions is described both using the voltammetric method (usable with microelectrodes) and specifically adapted fixed-potential macro-electrolysis at large GC plates. It offers a hard-to-underestimate possibility for easily creating graphene-overlaid GC interfaces and new materials based on them. These interfaces possess specific reactivity, at least within the cathodic range, toward cations and organic electrophilic species. Doping such interfaces at $E < -1.8$ V provides immense possibilities for realizing chemical reactions involving redox catalysis by the deposited graphene (easily cathodically doped within this range of potentials). Those produced materials are stable, have good electronic conductivity and are suitable for versatile surface modifications that quite certainly deserve further studies and development.

References

[1] K. Yan, H. Peng, Y. Zu, H. Li, Z. Liu, *Nano Lett.* **11** (2011) 1106.

[2] Y.L. Zhong, T.M. Swager, *J. Am. Chem. Soc.* **134** (2012) 17896.

[3] W. Sirisaksoontorn, A.A. Adenuga, V.T. Remcho, M.M. Lerner, *J. Am. Chem. Soc.* **133** (2011) 12436.

[4] H.-L. Guo, X.-F. Wang, G.-Y. Qian, F.-B. Wang, X.-H. Xia, *ACS Nano* **3**(9) (2009) 2653.

[5] A.G. Güell, A.S. Cuharuc, Y.-R. Kim, G. Zhang, S.-Y. Tan, N. Ebejer, P.R. Unwin, *ACS Nano* **9**(4) (2015) 3558.
[6] A. Ambrosi, C.C. Chua, A. Bonanini, M. Pumera, *Chem. Rev.* **114** (2014) 71.
[7] M. Pumera, *Electrochem. Commun.* **36** (2013) 14.
[8] J. Simonet, H. Lund, *J. Electroanal. Chem.* **75** (1977) 719.
[9] R. L. McCreery, *Chem. Rev.* **108** (2008) 2646.
[10] H.J. Schäfer, in *Comprehensive Organic Synthesis*, eds. B.M. Trost and I. Fleming, Vol. 3, Pergamon Press, New York p. 633, (1991).
[11] G.M. Jenkins, K. Kawamura, *Nature* **231** (1971) 175.
[12] P.J.F. Harris, *Philosophical. Mag.* **84** (2004) 3159.
[13] P. Bernier, S. Lefrant, *Le carbone dans tous ses états*, Gordon and Breach Science Publishers, Amsterdam (1997).
[14] V. Jouikov, J. Simonet, *Langmuir* **28** (2012) 931.
[15] V. Jouikov, J. Simonet, *Electrochem. Commun.* **42** (2014) 34.
[16] D. Lorcy, K.-S. Shin, M. Guerro, J. Simonet, *Electrochim. Acta* **89** (2013) 784.
[17] C. Dano, J. Simonet, *J. Electroanal. Chem.* **564** (2004) 115.
[18] V. Jouikov, J. Simonet, *Electrochem. Commun.* **45** (2014) 32.
[19] V. Jouikov, J. Simonet, *Electrochim. Acta* **172** (2015) 28.
[20] K.S. Novoselov, A.K. Geim, S.V. Morozov, D. Jiang, Y. Zhang, S.V. Dubonos, I.V. Grigorieva, A.A. Firsov, *Science* **306** (2004) 666.
[21] X. Y. Yang, X. Dou, A. Rouhanipour, L. J. Zhi, H. J. Rader, K. Mullen, *J. Amer. Chem. Soc.* **130** (2008) 4216
[22] Y . Yang, F. Lu, Z. Zhou, W. Song, Q. Chen, X. Ji, *Electrochim. Acta* **113** (2013) 9.
[23] A.Y.S. Eng, C.K. Chua, M. Pumera, *Electrochem. Commun.* **59** (2015) 86.
[24] W.Y.K. Khoo, M. Pumera, A. Bonanni, *Anal. Chim. Acta* **804** (2013) 92.
[25] A. Bonanni, A.H. Loo, M. Pumera, *Trends Anal.* **37** (2012) 12.
[26] C.T.J. Low, F.C. Walsh, M.H. Chakrabarti, M.A. Hashim, M.A. Hussain, *Carbon* **54** (2013) 1.
[27] A.Y.S. Eng, M. Pumera, *Electrochem. Commun.* **43** (2014) 87.
[28] V. Jouikov, J. Simonet, *Electrochem. Commun.* **43** (2014) 67.
[29] S. Sarkar, E. Bekyarova, R.C. Haddon, *Angew. Chem. Int.* **51** (2012) 4201.
[30] D. Bélanger, J. Pinson, *Chem. Soc. Rev.* **40** (2011) 3995.

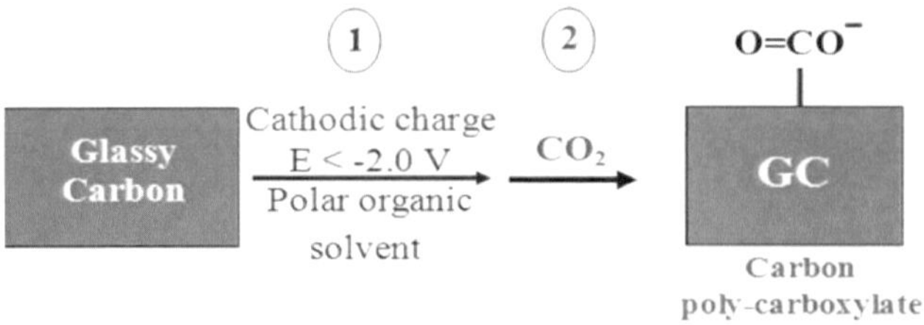

Scheme 11.1. Carboxylation of GC via electro-chemical doping of the material [44].

Scheme 11.1 via the electro-chemical generation of anionic sites at carbon surfaces. For this purpose, a typical activation of a GC cathode in contact with a saturated solution of CO_2 in *N,N*-dimethylformamide (DMF) saturated with a tetraalkylammonium (TAAX) salt is usually conducted. Only under these conditions, a large reduction step (threshold at about −1.7 V vs. Ag/AgCl) is observed. This step quite rapidly vanishes upon recurrent scans. First of all, it could be expected that the main reduction step is assigned to the specific reduction of CO_2. As a matter of fact, as discussed further, this reaction could correspond to the cathodic charge of GC and is followed by the reaction of this material with carbon dioxide acting as an electrophile. The reaction leads globally to the carboxylation of carbon. Moreover, the destruction of the carboxylate layer can be simply achieved by anodic oxidation of this material (analogous to a non-Kolbe reaction with evolution of CO_2). This supports a very large multi-layer character of carboxylation deeply affecting the carbon bulk.

This chapter is devoted to cathodic carboxylation of carbon and is especially focused on such materials as GC, graphite, and graphene.

11.2. Technical Methods Applied to Carbon Carboxylation

Voltammetry and electrolyses were principally performed in 0.1 M solutions of TAAX in *N,N′*-dimethylformamide DMF. Solvents such as acetonitrile (ACN) and propylene carbonate (PC) were also used in this study. All experiments are very simple to perform and needed no special treatment of electrolytic solutions. Carbon dioxide was used as a saturated solution in DMF (0.2 mol L^{-1}) at room temperature, by bubbling it directly in the electro-chemical cell. There is no need for a prior de-aeration of the electrolytic solution. Potentials are referenced to the aqueous Ag/AgCl/ $KCl_{(sat)}$.

Used carbon electrodes had apparent surface areas of 0.8 mm^2 (microelectrodes) and 7 mm^2 (large electrodes). Glassy carbon samples (rods) were purchased from Tokai Carbon C° (code: GC Rod) for building GC stationary electrodes and from Carbone Lorraine (plates of area > 0.5 cm^2) for making large samples used in macroanalysis. HOPG was purchased from Union Carbide and commercial graphene obtained from XG Sciences as samples named XGnPGraphene nanoplatelets — Grade C.

Prior to use, carbon electrodes were carefully polished first with silicon carbide paper (Struers 1200) and then with Norton polishing paper (type 02). Rinsing with water and acetone was followed for every experiment. After CO_2 immobilization, the surfaces were sonicated (5 min). Coulometric measurements and electrolytic grafting were performed using three-electrode cells separated with a fritted glass.

11.3. On the Cathodic Reactivity of Carbons

Basically, let us recall that GC remains nowadays a convenient electrode material (with convenient imperfections) to carry out analyses and electrolyses, especially within a large cathodic range [1, 2]. The present approach only concerns aprotic media (aprotic polar solvents eventually mixed with activated neutral alumina) that allows one to reach the potentials of −2.8 V. Under such conditions, many kinds of electro-organic reactions can be performed at negatively polarized carbon without competing with the reduction of organic solvents [3]. A large series of organic reactions can be run since the role of remnant water in many cases is considerably minimized. Within the cathodic range, a large number of organic reductions was successfully accomplished where the use of metallic reducing agents (e.g., Li or Na in THF as solvent) is the classical approach. In the field of grafting for modifying or extending the functional properties of carbon electrodes (e.g., for electro-catalysis), different techniques can be efficiently employed. Therefore, many organic layer depositions by means of bond anchoring to the surface were reported principally by means of free radical attachment to the solid surface. Mostly, the generation of radicals via electro-chemical reduction of onium salts such as diazonium [4], iodonium [5], and ammonium [6] has been successfully investigated. Furthermore, cathodic scissions of activated carbon–halogen bonds (such as benzyl, propargyl, and allyl halides already mentioned in Chapter 6) result in the grafting of unsaturated radicals of many kinds onto carbon [7, 8], and they represent the real interest under specific conditions for additional functionalization of carbon surfaces.

On the other hand, it is essential here to consider GC as a reactive material in the cathodic charge. As a matter of fact, GC cannot be regarded as an "inert" material. Rather recent investigations [9, 10] on its real structure revealed that it contains quite generally the ribbons of graphite-like structures in discontinuity with "crystallite" boundaries expected to correspond to all types of unsaturated and/or aromatic forms of carbons. Therefore, GC could be represented as a complex solid that exhibits conducting zones (e.g., reticulated structures) in mixture with micro-graphitized and fullerenized [9] parts where sp^2 carbons certainly predominate. Subsequently, the cathodic reactivity would depend on the level of graphitic zones in relation with pyrolysis conditions used during its preparation (temperatures up to 2500°C).

Experimentally, preliminary experiments dealing with the GC cathodic polarization at $E < -1.7$ V vs. Ag/AgCl, as described in Chapter 12, demonstrate the reversible cathodic charge of commercial GC, resulting in the formation of nucleophilic material capable of inducing both internal and interfacial reactions. Thus, this is in full agreement with the cathodic polarization of graphite: in polar organic solvents, highly oriented pyrolytic graphite (HOPG) under cathodic charge [11] specifically leads to well-defined insertion stages $[C_n^-, TAA^+]$ via concomitant uptake of electrons and TAAX salts. Therefore, such "graphite salts", in a certain extent equivalent to TAAX amalgams, could be considered both as reducing species and polynucleophilic materials.

Now, carbon dioxide (well recognized as an "electrophilic" entity in organic chemistry) was recently [12] established as reactive toward negatively charged GC both for creating a new kind of modified materials (and interfaces) and as a new mode of sequestration of this gas in carbons using a very simple electro-chemical reaction. Indeed, it is well known to easily act as an electrophilic agent toward electro-generated anions [13, 14] of π-acceptors. With metallic electrodes, such as lead cathode, carbon dioxide was reported to generate [15, 16] oxalate ion via its anion radical as a reaction transient. The proton availability and the use of specific catalysts have recently enriched the list of reduction products electro-chemically derived from CO_2; methane, formic acid, methanol, carbone monoxide, formate, glyoxylate, formaldehyde, and glycolate can thus be formed. For example, the use of a silver cathode [17] led to a complex mixture of reduced forms of CO_2. A recent review by Pletcher [18] aims to point out the real assessment of so many attempts to valorize CO_2 by means of electro-reductions at new electrodes and catalysts.

The capacity of GC to easily form a polynucleophile under cathodic charge (exclusively in the presence of non electrophilic TAA^+ salts) was explored for the first time, electro-chemically charging this material in polar aprotic organic solvents (essentially *N,N*-DMF). Therefore, reduced GC may act as a macro-nucleophilic material toward carbon dioxide for readily building [carbon carboxylated] materials. Owing to the large amounts of charge injected, the global reaction is expected to affect the material to a large extent, not limited to the surface. Further functionalization reactions of as-carboxylates decorating carbon are presented here. This process should be seen as an easy way of building a new multi-purpose carbon material both in surface and in volume. The elemental carboxylation reaction was found to be also applicable to industrial cokes and carbon fibbers.

11.4. Electro-Generation of GC Polycarboxylates

With fairly important amounts of electricity injected during the charging of GC to render it polynucleophilic, the reaction was found to affect this material to a large extent, not at all restricted to the surface. The method may dramatically affect the roughness of the surface, so terms of "surface concentrations" estimated after cathodic treatment, quite obviously do not have a strict sense. The elemental carboxylation reaction was found to be also applicable to industrial cokes and carbon fibers.

At a GC electrode (Figure 11.1(A)) in DMF (water content < 1000 ppm)/$TBABF_4$, it is found that a saturated solution of carbon dioxide exhibits a large step that progressively vanishes upon recurrent scans totally disappearing after about 10 cycles. The coverage of the surface could be established as shown in Figure 11.1(B). So, the use of bulkier TAAX cations leads fast to very small currents in voltammetry.

Therefore, chemical carboxylation of GC occurs (Scheme 11.2) producing a polycarboxylate, at least superficially depending on salt bulkiness. The apparent surface coverage of such interfaces was found to be quite large. When the modified surface is reacted for a few minutes with a bulky electrophile such as a long chain RI (e.g., 1-iodotetradecane) dissolved in DMF, the blocking of the surface may happen as shown in Figure 11.1(B). This does confirm the reactivity of CO_2 toward GC through a global material activation.

In order to quantify grafting levels (obtained after polycarboxylation, Scheme 11.2), several methods were used. First is the formation of an activated ester (e.g., reacting 3-nitrophenol with the carboxylated GC), which is expected to lead to a two-electron reduction (E_p = −0.98 V, ester cleavage with expulsion of nitrobenzene reduced afterward at E_p = −1.32 V). The current integration (ester reduction) corresponds to the global immobilization of 1.0×10^{-8} mol cm^{-2}. Second, the addition of a redox-marked electrophile such as ω-iodohexylferrocene (already used as a redox probe [18]) successfully permits to verify the immobilization of ferrocene (Fc)

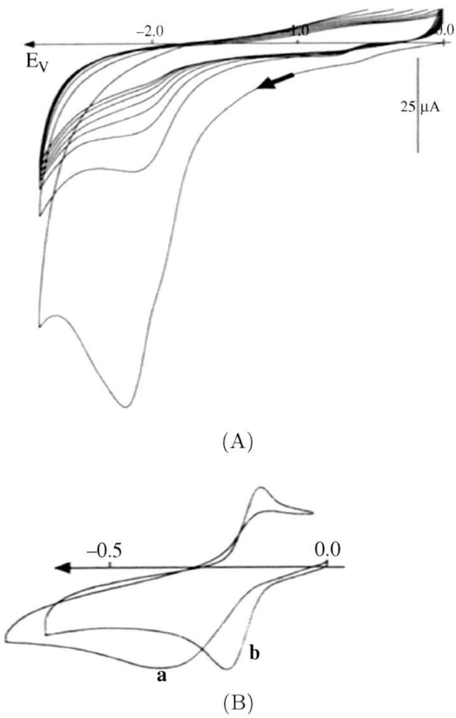

Figure 11.1. Recurrent voltammograms (A) at a GC electrode from Carbone Lorraine (7 mm^2) in contact with a DMF + $TBABF_4$ solution saturated by carbon dioxide at 25°C; Scan rate: 50 mVs^{-1}. (B) Voltammetry of chloranil (curve a) in DMF at a GC surface modified by a poly-carboxylate layer followed by an esterification with 1-iododecane. For comparison, curve b depicts the voltammetry at a smooth GC electrode [12].

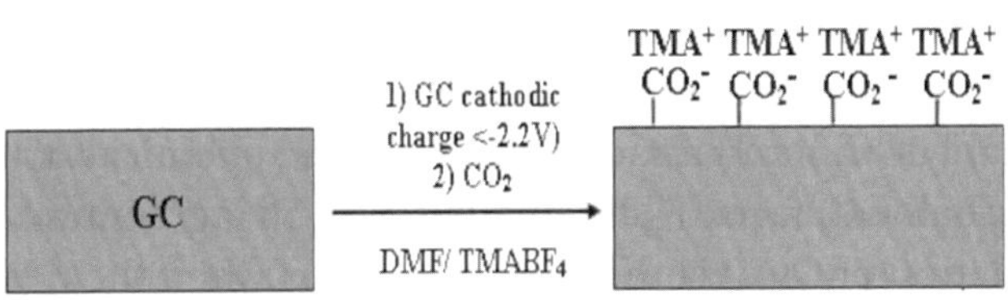

Scheme 11.2. Cathodic generation of a carbon–polycarboxylate [12].

From the electro-chemical point of view, it has been established that graphite could be anodically oxidized in aqueous acidic media. Thus, intercalation compounds were successfully produced in the presence of sulfuric acid, perchloric acid, and fluorosulphonic acid [24, 25].

Similarly, the use of non-aqueous solvents in the presence of anions stable to anodic oxidation such as ClO_4^- and BF_4^- led to ternary phases in which solvent was involved [26, 27]. Still in the anodic range, the intercalation kinetics was found to depend on the degree of crystallinity. For comparison, a theoretical model (graphite–H_2SO_4) was used [28] as a rapid characterization method for the host–lattice material. More specially and in relation to the present study within the cathodic range, the intercalation of metal cations (essentially alkali metal cations) [29, 30] and tetraalkylammoniumcations [31, 32] was successfully carried out in aprotic solvents like dimethylsulphoxide (DMSO), DMF, tetrahydrofuran,

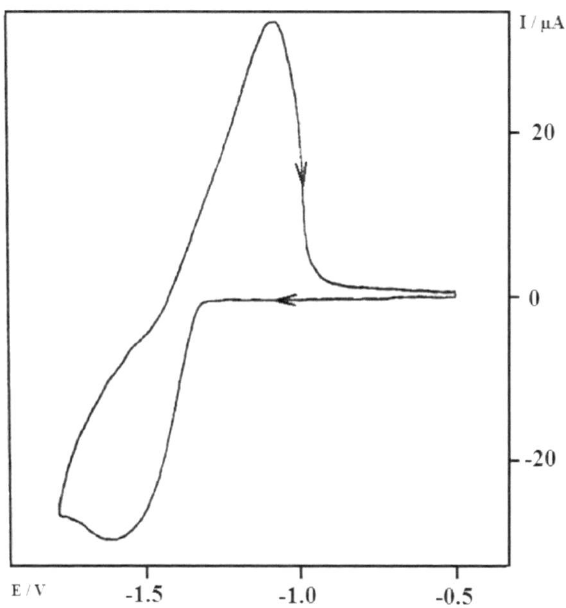

Figure 11.3. Voltammetric response of natural graphite (ca. 2 mg) fixed onto a GC electrode; the latter shows no activity within the explored potential range. Electrolyte: DMF + 0.1 M Bu_4NBF_4. Scan rate: 0.1 V s^{-1}. Reference electrode Ag/AgI/0.1 M I^- in DMF [45].

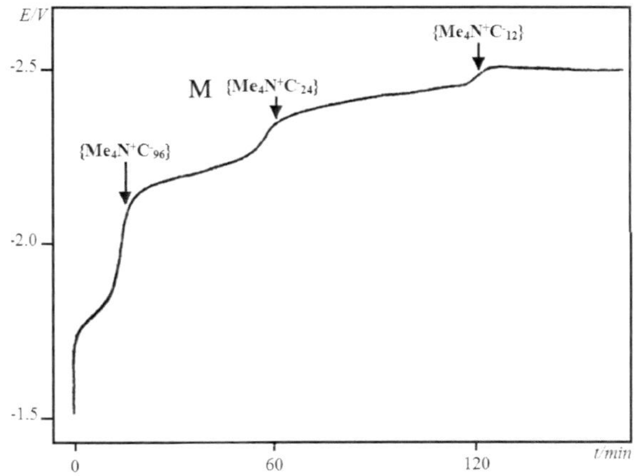

Figure 11.4. Chronopotentiometry of an HOPG sample (ca. 0.5 mg) fixed on a GC surface. DMF/0.1 M Me_4NBF_4. Current density: 375 μA cm^{-2} [41].

and 1,2-dimethoxyethane. The reversibility of the charge process was found to reach high levels (almost 100% in DMSO and DMF with TAAX cations). With metal salts MX, one has to bear in mind the relation between the redox potential of graphite E_G and $E_{M/M+}$ (E_G is the reduction potential of the graphite sample taken as its charging threshold potential). If $E_G < E_{M/M+}$, the deposit of metal M could be observed on the carbon surface. On the contrary, if $E_G > E_{M/M+}$, the intercalation of M^+ occurs, and the corresponding stoichiometry was determined by chronopotentiometric analysis. Therefore, reversible formation of reduced phases of graphite permitted them to be used as reducing reagents [33]. With bulky intercalated cations like TAAX ions, one may expect that ion pairing between these ions and the diffused negative charges distributed on reduced graphenes is not tidy. Under these conditions, the polyanionic carbon skeleton (Figures 11.3 and 11.4) could be considered as an efficient polynucleophilic material capable of reacting *in situ* with organic electrophilic species under the condition that these electrophiles are not electro-active at the reduction potential of graphite. Another possibility is to react *in situ* electro-chemically reduced graphite with electrophiles added afterward in the disconnected cell.

When modified electrodes were needed, graphite was often considered as an excellent and cheap material, especially due to its ability to form upon ageing oxides and carboxylic functions on its surface. Alternatively, a mild anodic carboxylation (via direct oxidation) of carbon felts in acid media has been described [34–39], which allowed chemical functions to be immobilized permitting a judicious and original way of preparing modified electrodes. However, primarily anodic carboxylation is a destructive method. Here, it is alternatively proposed to utilize reduced (via electron transfer i.e., without alteration of carbon skeleton) graphite as an electro-chemically generated polynucleophile and then to react it with carbon dioxide. This type of reaction is well known and fully described in organic electro-chemistry [40], especially for obtaining carboxylated π-acceptors. Thus, next task is to estimate the extent of carboxylation of graphite (especially HOPG) by CO_2 used under partial pressure. The stability of the graphite polycarboxylates was evaluated and the level of carboxylation under these experimental conditions was described.

The reactivity of CO_2 was confirmed by FTIR and functionalization of carboxylate groups principally with redox markers. The chemical versatility of this carboxylated material is obvious. Lastly, its anodic collapse was shown to be easily provoked by a non-Kolbe reaction followed both by the total loss of the chemically sequestrated CO_2 and possibly a "carbeniumization" of the GC that opens many possibilities of reactions with adequate nucleophiles.

Measurements described hereafter were performed exclusively on HOPG samples (from Union Carbide) of about 0.5 mg immobilized by means of a non-electroactive gel (epoxy glue) onto GC electrode considered as weakly reactive toward CO_2. Here, we focus on data related to the insertion of Me_4N^+. Cyclic voltammetry (Figure 11.3) showed a quasi-reversible step obtained with Bu_4NBF_4 in DMF with a threshold potential of −1.8 V vs. Ag/AgCl. Currents (cathodic = charge, and anodic = discharge) are diffusion controlled [41]. Current densities (the apparent area of the HOPG crystal was estimated to be about 0.12 cm^2) applied to the graphite within the working potential interval were between 80 and 250 μA cm^{-2} with Bu_4N^+ and between 150 and 500 μA cm^{-2} with Me_4N^+. Chronopotentiometry (Figure 11.4) in dry DMF (containing less than 250 ppm water and 0.1 mol L^{-1} of the supporting salt) showed several plateaus corresponding to the progressive filling of spans between graphene layers. Cations of radii larger than that of Bu_4N^+ were disregarded due to huge exfoliation processes already observed [11] with tetrahexylammonium and tetraoctylammonium salts. In the course of chronocoulometric experiments (with exclusive use of Me_4NBF_4), the HOPG sample is progressively swelling, but without compromising the overall electricity storage process that remains more or less reversible. The charge process exhibits several potential jumps that match the progressive filling of inter-layer by TAA^+ cations. From a large panel of data relative to different HOPG samples, stoichiometry of the most highly filled stages $C_n\{R_4N\}$ was determined:

with Me_4N^+: three potential jumps: n = 96 ± 20%, 24 ± 8% and 12 ± 8%.
with Bu_4N^+: two potential jumps: n=48±6% and 24±12%.

Then, when exposing reduced HOPG samples to a CO_2 atmosphere, a neat shift of the chronopotentiometric curves was noticed. These are ascribed to an effective reactivity of graphite with CO_2: indeed, the graphene plane carboxylation leads to the decay of the insertion efficiency. This point was checked by analysis (X-ray photoelectron spectroscopy) of the external faces of a sample (Figure 11.5). From a general point of view, it was found that the O/C ratio (due to the presence of C–O, C=O and O–C=O functions) increased significantly after the electro-chemical carboxylation of graphite. It was noticed, however, that the oxidation of the material, presumably at the edges and surface defects of the sample, was already significant without any treatment. However, the O/C ratio increased 2.5 times in the outer part of the HOPG crystal. After splitting the crystal, the material core (after reaction with CO_2) revealed a O/C ratio of 0.094, i.e., 3.5 times larger than that obtained by splitting the material without any treatment [41].

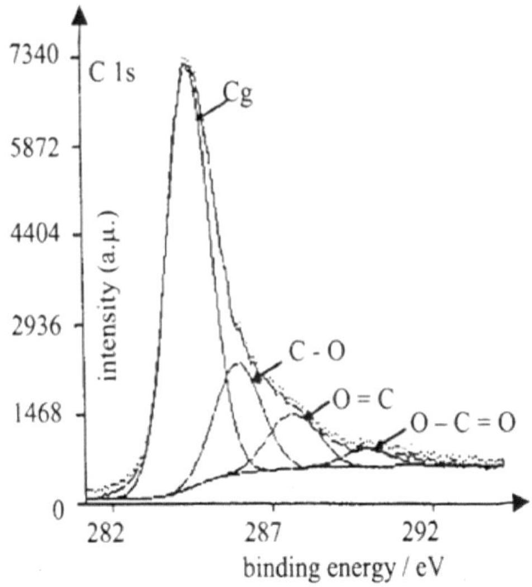

Figure 11.5. XPS analysis of HOPG samples modified by carboxylation. Response of the external surface showing the deconvolution of the C_{1s} signal into its four components [41].

11.5.2. *Procedure for a carboxylation of natural graphite*

A known amount of natural graphite was introduced into the cathodic compartment of an H-shaped cell equipped with a porous glass separator. The solvent was DMF containing either 0.1 M Bu_4NBF_4 or 0.1 M Me_4NBF_4. The cathodic compartment was maintained under an argon atmosphere. The graphite flakes were fixed and compressed using a platinum grid or let free in the solution over an efficiently stirred mercury pool. The potential applied to the Pt grid or to Hg pool was always set to −1.8 V. This method provided an electric contact between flakes and electrodes good enough to enable an efficient electron transfer. The apparent degree of reduction n as a function of the amount of electricity q injected during the process could be estimated from the expression

$$n = \frac{m \times 96{,}500}{12 \times q}$$

where *m* is the mass of graphite (in g) and *q* the charge (in C). It was assumed that the electric yield of charging was quantitative. Thus, with a tetramethylammonium salt (e.g., tetrafluoroborate or perchlorate), when *n* reaches a value of 12 (maximum of insertion), graphite flakes turned to a steel blue color, which instantaneously disappeared in contact with air or carbon dioxide.

When *n* reached the desired value, the reduction process was changed: the argon bubbling was replaced at once by saturation with CO_2 as fast as possible. Carbon dioxide was dried by passing through two successive traps containing concentrated H_2SO_4 and then through pure DMF. The contact of the reduced graphite with CO_2 was prolonged (at the applied potential of +0.5 V for up to 24 h for totally oxidizing still reduced zones remaining in the bulk of material). Since carboxylation is slow and therefore not quantitative, the residual charge remaining in the graphite sample was in all cases released by an additional electro-chemical discharge at 0 V in the same cell until the current dropped to zero. This treatment allows remaining TAAX cations to leave the material before the contact with air. This procedure was justified as it may prevent the oxidation of graphite via superoxide that can add to charged carbon yielding carbon peroxides [43].

Last, graphite flakes were filtered off and rinsed successively with DMF and acetone to eliminate the residual electrolyte. Since the present mode of carboxylation must produce immobilized TAAX carboxylate functions, acidification was performed with 0.1 M hydrochloric acid. Before drying at 60°C for 24 h, two rinsing cycles with distilled water were completed until neutral pH.

11.5.3. *The extent of carboxylation from analytical methods*

The extent of graphite carboxylation was determined by several methods. The first one deals with the acidification of carboxylate functions with diluted HCl and then titration of the remaining amount of the strong acid by diluted sodium hydroxide. The second efficient way is based on the evidence of the instability of the graphite–carboxylate linkage in the presence of NaOH (1 M).

Alternatively, carbon dioxide was allowed to react with barium hydroxide to precipitate barium carbonate. On the basis of a CO_2 evolution considered as quantitative, the gravimetric method was preferentially used owing to its high simplicity.

Last, adding organic electrophiles, symbolized by RX and containing specific elements different from carbon and oxygen (Si, N, Cl, Br and F) that are *a priori* supposed to be totally absent in the basic graphite used for analysis, esters and amides could be formed and the level of those elements can be determined (Scheme 11.3) by elemental analysis [41].

As a matter of fact, the average grafting of one carboxylic function per 200–250 carbon atoms was found to be rather general. Surprisingly, better carboxylation assays were found for larger amounts of graphite using the gravimetric method. The importance of a preliminary cathodic charge of the material (the higher the reduction stage, the larger is the rate of carboxylation) is also evident. It is amazing to note that the carboxylation rate also exhibits plateaus, suggesting that the different reduced stages could show a specific reactivity toward CO_2. Thus, a maximum carboxylation level of $C_{80}CO_2^-$ was reached (Figure 11.6) from over-reduced graphite ($C_{12}NMe_4$). The yield of carboxylation relative to the initially reduced form is 15% in contradiction with results obtained with much lower rates of reduction with potentials fixed at −1.8 V. Thus, using moderate applied potentials (probably less destructive for the material) reveals an average yield of 25–30% with less reduced stages. It is possible that CO_2 reacts with charged graphite both as a nucleophile (in carboxylation) and as an acceptor (in primary electron transfer leading to the CO_2 anion radical with subsequent reactions taking place outside of graphite). However, preliminary reduction of CO_2 and the addition of its anion radical on graphene could be also considered. Under the conditions of carboxylation described here, the reduction of carbon dioxide occurs in competition with an S_N2-type carboxylation. This fact is undoubtedly in agreement with the value of the standard potential for the monoelectronic reduction of CO_2 (−2.2 V vs. SCE under similar conditions [21]).

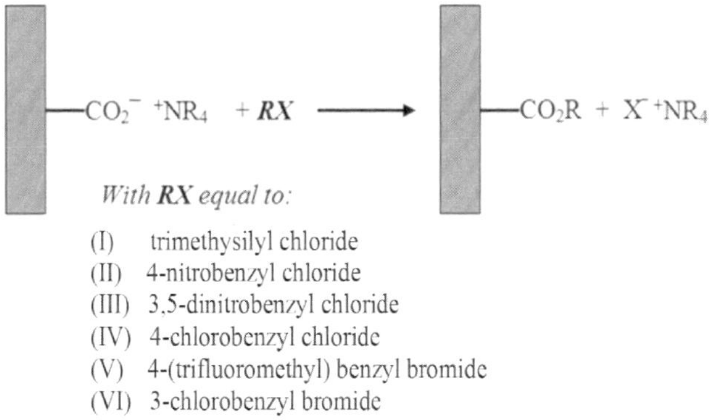

Scheme 11.3. Modifications of carboxylated carbons with various electrophiles [41].

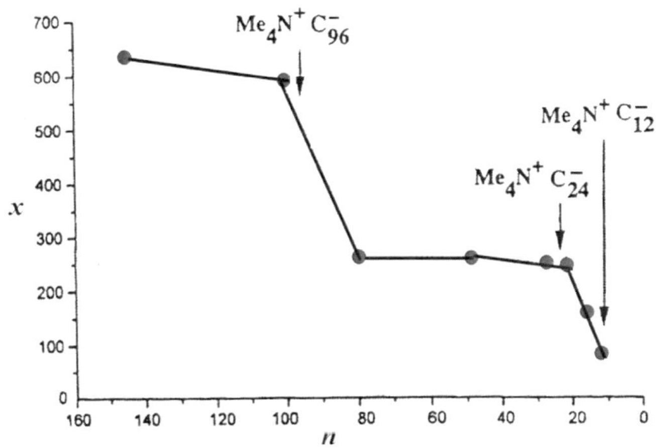

Figure 11.6. Cathodic carboxylation of natural graphite by carbon dioxide. The level of carboxylation x (C_xCO_2H) vs. the amount of charge n (R_4NC^-n). The n values were obtained by gravimetry. Reduction and carboxylation were achieved on 200 mg of natural graphite as reported in Ref. [41].

Nevertheless, two other factors should certainly be taken into account:

(i) The hindrance factor since the negative charge located on the graphene surface is replaced with a carboxylate function.
(ii) The structure factor because of the C–CO_2 bond formation (corresponding to the carboxylation of graphene planes) changes

the structure of carbon atoms from sp^2 to sp^3. For some reasons, incorporation of scattered functions onto graphene should require maximum energy (the presence of single sp^3 carbons among a large number of sp^2 ones). Consequently, one may expect then that carboxylation could principally occur first at the edges or at the defects of graphene planes. In the case of a total (but improbable) per-carboxylation where all carbons become sp^3-hybridized could be proposed for giving an analogous structure to graphite fluoride [22, 23]. Anyhow, acting of a strong base (concentrated NaOH) on carboxylated material provokes an instantaneous release of carbon dioxide (Scheme 11.4), implying a relative instability of the grafted forms as soon as strongly basic zones are growing. The reversibility of the carboxylation process could be envisaged for a temporary storage of CO_2.

The data obtained when using different hetero-atom electrophiles appeared to be poorly reliable. Thus, large discrepancies were observed in most of experiments with the data from acid–base and gravimetric methods. Such differences could be explained by the fact that RX reagents, supposed to quench the nucleophilic centers, were too bulky to reach the target fast enough before the by-reactions intervene. Only two results appear in agreement with other analytic estimations: one is that obtained with trimethylsilyl chloride, obviously of compact size and very reactive; the other corresponds to the electrophile V (Scheme 11.3) possessing a strongly withdrawing CF_3 substituent that powerfully increases the rate of carboxylation, despite a much higher bulkiness.

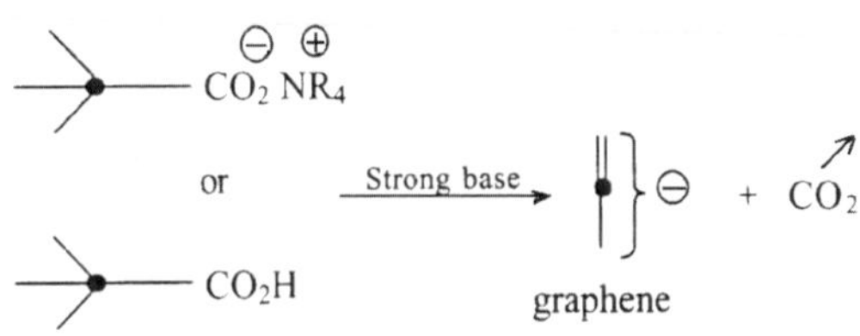

Scheme 11.4. Decomposition of carboxylated graphite by a strong base [41].

11.6. Cathodic Carboxylation of Graphenes: A New Versatile Material

11.6.1. *Introduction*

Carbon dioxide was already mentioned as a poorly reactive under cathodic conditions [17]. When covered by graphene, GC electrodes are cathodically charged at the potentials < −1.7 V vs. Ag/AgCl in aprotic solvents in the presence of CO_2. Specific signals attributed to the charge of GC rapidly vanish and a sudden inhibition takes place during further recurrent scans. This process suggests that the principal electrode reaction is now cathodic charge of graphene resulting in polynucleophilic species. Within a potential range where CO_2 is not reduced ($E > -2.2$ V), its reaction with the charged carbon may occur. A quite dense polycarboxylation of the material is chemically confirmed. The grafting of a large panel of reactive electrophiles (eventually carrying a redox group) on these graphene–carboxylate surfaces was achieved and then allowed the easy building of a wide variety of modified electrodes. Moreover, the enhanced hydrophilicity of those new modified surfaces makes possible their use in aqueous media.

11.6.2. *GC electrode modified with graphene*

It was already stressed in the present chapter that GC is a common electrode material for analyses and electrolyses, especially in the cathodic domain. Preliminary experiments on the modification of GC, owing to the presence of graphite, fullerenes, and possibly graphene as electro-active impurities, were recently reported (see Chapter 10) and may support that carbons (whatever their structure) can react, on aging, with oxygen leading to their degradation easily visualized around the edges or defects by the presence of keto and carboxy groups. On the other hand, the growing importance of graphene in electro-chemistry (possibly functionalized for catalytic purposes) allows one to build a new kind of modified electrodes that permits developing multiple applications [12]. All this may prompt electro-chemists to consider that its charged forms could present

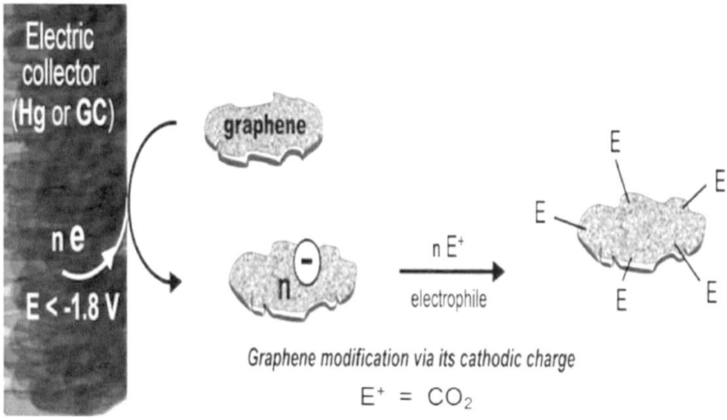

Scheme 11.5. Generation of a graphene polynucleophile at GC or Hg electrodes that allows the poly-carboxylation of graphene flakes in a fluidized bed inside the cathodic compartment of the cell shown in Figure 11.10 [44].

of addition of different groups (redox, hydrophilic, lipophilic, organo-metallic, etc.) to this material is of great importance. Here, we focused on electro-chemically induced chemical changes of graphene.

As underlined above and put forward recently [12], charged graphene can be seen both as a pro-nucleophilic material and as a radical scavenger, depending on the applied potential and on the nature of reagents present in the solution (Scheme 11.5).

Profound chemical functionalizations of graphene, and in quite large amounts, can be achieved by fixed-potential macro electrolyses using quite simple electrolytic cells. Under the experimental conditions described in the previous section, the process can further be improved by follow-up reactions, leading to graphenes largely modified with redox entities like Fc and AQ, or else with aromatic moieties (anthracene, pyrene, naphthalene). Further easily modifiable functions like allyl and propargyl could be also attached. Obviously, this scope is far from being fully explored. The innovative capability of graphene to be electro-chemically charged in polar organic solvents allows the building of macro-nucleophilic transients active toward carbon dioxide and easily enabling the formation of [graphene–carboxylate] materials.

A typical procedure could be generalized as follows: about 150 mg of industrial graphene is introduced to the cathodic compartment of an H-type electrolytic cell with the available volume of at least 10 mL (Figure 11.10), which was then filled with 6 mL of DMF containing $TBABF_4$. An efficient stirring permits to form and to maintain graphene suspension similar to a fluidized bed system. The main current collector can be either a mercury pool (convenient when quite negative potentials, $E \cong -2$ V, have to be reached for inducing electro-chemical charge of graphene) or a GC plate (about 3–4 cm^2) placed in the center of the cathodic compartment. The anodic compartment simply contains an anode (carbon, graphite, or platinum grid) immersed in the same electrolyte. Basically, a potentiostatic system is required (in order to control the potential of the collector electrode). However, with small amounts of electricity injected in this process, a galvanostatic procedure could easily be used as well.

11.7.3. *What is the average level of grafting by the proposed procedure?*

This is a question difficult to solve in the absence of a marker. We opted for the use of redox markers such as AQ, Fc, 4-nitrophenyl, 4-cyanophenyl, 4-iodophenyl, and other groups providing a distinct redox response. Especially, in the case of carboxylated graphene, additional post-grafting transformations of the $-CO_2H$ group were achieved (e.g., easy addition of AQ).

The working principle is the same in all cases: the deposition of modified graphene (typically from 0.15 to 0.5 mg) in fine layers onto a GC plate (average area from 0.5 to 1 cm^2) followed by the integration of the global redox response of the GC coverage used as an electrode. Several measurements (with different amounts of modified graphene) allowed to check that the electro-chemical signals remain quite well proportional to the deposited mass. Modification by CO_2 appears quite efficient and corresponds to $\approx 10^{-2}$

carboxylate groups per carbon atom. Experiments with much larger amounts of electricity remain to be done to produce more extensively modified graphenes. For comparison, under the experimental conditions summarized above, the attachment of Fc corresponds to $(5 \pm 2) \times 10^{-4}$ Fc per graphene carbon atom; that of AQ (using AQ–CH_2–Br) similarly exhibits levels of $(5 \pm 1) \times 10^{-4}$ AQ/carbon.

11.8. Conclusion

The cathodic charge of graphene, deposited as a thin layer onto a conducting support such as GC, achieved in the presence of carbon dioxide (in saturated solution in a non-aqueous polar solvent like DMF) makes possible an easy carboxylation of this material to a quite large extent. The present study therefore reveals the capacity of graphene deposited onto GC to react with CO_2. Under the conditions proposed in this presentation, the fixation of CO_2 onto glassy carbon might reach very large amounts equivalent to up to 100 layers (compared to the compact carboxy-monolayer at a flat GC surface). The reactivity of CO_2 was confirmed by FTIR spectroscopy and by the ensuing functionalization of graphene-incorporated carboxylate groups with suitable redox markers. Chemical versatility of this poly-carboxylic graphene material and its technological applications are obvious: even preliminary macro-electrolyses of graphene under CO_2 atmosphere in a conventional electro-chemical cell enables for the moment to prepare a graphene–carboxylate on a quite large scale (700 mg). Also let us recall that the present approach makes possible the use of these three-dimensional electrodes in aqueous solvents owing to their pronounced wettability.

Macro-electrolyses performed with graphene in fluidized bed mode lead to chemical modifications of graphene in a large amount. These functionalizations are exemplified with extensive carboxylation followed with grafting of redox moieties. Experimental conditions of this process are very simple and could be applied, after deposition of modified graphenes onto conducting solid supports

(GC, metals like gold, palladium, platinum and silver) to a wide variety of modified electrodes usable in catalysis and analysis.

References

[1] D.G. Peters, in *Organic Electrochemistry*, eds. H. Lund and O. Hammerich, M.M. Dekker, New York, Basel, p. 341 (2001).

[2] R.L. McCreery, *Chem. Rev.* **108** (2008) 2646.

[3] H.J. Schäfer, *Chem. Ing. Tech.* **42** (19701) 164.

[4] D. Bélanger, J. Pinson, *Chem. Soc. Rev.* **40** (2011) 3995.

[5] K.H. Vase, A.H. Holm, S.U. Pedersen, K. Daasbjerg, *Langmuir* **21** (2005) 8085.

[6] K.H. Vase, A.H. Holm, K. Norman, S.U. Pedersen, K. Daasbjerg, *Langmuir* **24** (2008) 182.

[7] C. Combellas, K. Kanoufi, Z. Osman, J. Pinson, A. Adenier, G. Hallais, *Electrochim. Acta* **56** (2011) 1476.

[8] V. Jouikov, J. Simonet, *Electrochem. Commun.* **15** (2012) 93.

[9] G.M. Jenkins, K. Kawamura, *Nature* **231** (1971) 175.

[10] P. Bernier, S. Lefrant, *Le carbone dans tous ses états*, Gordon and Breach Science Publishers, Amsterdam (1997).

[11] J.O. Besenhard, *Carbon* **14** (1976) 111.

[12] J. Simonet, *Electrochem. Commun.* **21** (2012) 22.

[13] S. Wawzonek, R. Berkey, E.W. Blaha, E.M. Runner, *J. Electrochem. Soc.* **102** (1955) 235.

[14] S. Wawzonek, D. Wearring, *J. Am. Chem. Soc.* **81** (1959) 2067.

[15] C. Amatore, J.-M. Savéant, *J. Am. Chem. Soc.* **103** (1981) 5021.

[16] B.R. Eggins, C. Ennis, R. McConnell, M. Spence, *J. Applied Electrochem.* **27** (1997) 706.

[17] D. Pletcher, *Electrochem. Commun.* **61** (2015) 97.

[18] Q. Jianping, T. Juntao, S. Jie, W. Cuiwei, Q. Mengqian, H. Zhiqiao, J. Chen, S. Shuang, *Electrochim. Acta* **203** (2016) 99.

[19] V. Jouikov, J. Simonet, *Langmuir* **28** (2012) 931.

[20] D. Lorcy, K.-S. Shin, M. Guerro, J. Simonet, *Electrochim. Acta* **89** (2013) 784.

[21] J. Mering, in *Les Carbones*, ed. Vol. I (Masson, Paris, 1965), p. 129.

[22] N. Platzer, in *Les Carbones*, ed. Vol. II (Masson, Paris, 1965), p. 465.

[23] R. Setton, in *Le Carbone Dans Tous Ses Etats* eds. P. Bernier and S. Lefrant, Gordon and Breach, Amsterdam (1997).

[24] J.O. Besenhard, H.P. Fritz, *Angew. Chem. Int. Ed. Engl.* **22** (1983) 950.

[25] M.J. Bottomley, G.S. Parry, A.R. Ubbelohde, D.A. Young, *J. Chem. Soc.* (1963) 5674.

[26] J.O. Besenhard, H.P. Ftitz, *Z. Naturforsch* **26** (1971) 1225 and *Z. Naturforsch* **27** (1972) 1294.

[27] Z. Zhang, M.M. Lerner, *J. Electrochem. Soc.* **140** (1993) 742.

[28] N. Iwashita, H. Shioyama, M. Inagaki, *Synth. Met.* **73** (1995) 31.

[29] T. Ohzuku, Y. Iwakoshi, K. Sawai, *J. Electrochem. Soc.* **140** (1993) 2490.

[30] D. Billaud, F.X. Henry, M. Delaurain, P. Willmann, *J. Phys. Chem. Solids* **57** (1996) 775.

[31] J.O. Besenhard, *Carbon* **14** (1976) lll.

[32] J. Simonet, H. Lund, *J. Electroanal. Chem.* **75** (1977) 719.

[33] J. Berthelot, M. Jubault, J. Simonet, *J. Chem. Soc. Chem. Commun.* (1982) 759.

[34] W. Roman, E. Steckhan, *Electrochim. Acta* **42** (1997) 2027.

[35] E. Simon, E. Sablé, H. Handel, M. L'Her, *Electrochim. Acta* **45** (1999) 855.

[36] A. Deronzier, D. Eloy, P. Jardon, A. Martre, J.C. Moutet, *J. Electronal Chem.* **453** (1998) 179.

[37] F. Geneste C. Moinet, G. Jezequel, *New J. Chem.* **26** (2002) 1539.

[38] P. Jego, C. Moinet, in *Novel Trends in Electroorganic Synthesis,* ed. S. Torii, Springer, New York, p. 161, (1997).

[39] F. Geneste, M. Cadoret, C. Moinet, G. Jezequel, *New J. Chem.* **26** (2002) 1261.

[40] M.F. Nielsen, J.H.P. Utley, in *Organic Electrochemistry,* eds. H. Lund and O. Hammerich, Marcel Dekker, New York, Fourth Edition, p. 836 (2001).

[41] C. Dano, J. Simonet, *J. Electroanal. Chem.* **564** (2004) 115.

[42] L. Dai, *Acc. Chem. Rev.* **46** (2013) 31.

[43] J. Simonet, *Electrochem. Commun.* **34** (2013) 223.

[44] J. Simonet, Unpublished results.

[45] J. Simonet, H. Lund, *J. Electroanal. Chem.* **75** (1977) 719.

[46] V. Jouikov, J. Simonet, *Electrochem. Commun.* **43** (2014) 67.

Chapter 12

Cathodic Immobilization of Functionalized Alkyl Chains at Polarized Glassy Carbon

12.1. Introduction

Functionalized alkyl halides (RX), in contact with different kinds of carbon polarized at reducing potentials more negative than −1.7 V vs. Ag/AgCl, react with these cathodically charged surfaces in aprotic polar solvents. This form of a pre-doping of carbons allows alkyl chain immobilization on these materials. The process is expectedly enabled by the cathodic charge of graphitized and fullerenized zones present in carbon (and especially at the surface), followed by a heterogeneous nucleophilic displacement reaction (analogous to a nucleophilic attack and proposed to be denoted as HS_N2) toward alkyl iodides and bromides. The new way of grafting is well supported by the fact that carbon charge (when achieved separately) permits to create a nucleophilic material quite stable in non-reactive atmospheres (argon and nitrogen). Grafting experiments taking place under insert atmosphere (in particular the transfer of the charged carbon to a solution of RX) allow to obtain quite similar results. Such charge processes also applied to graphite and graphene (and developed all along this book) allow to extend (using various organic electrophilic substances) the possibilities of surface modification of carbon surfaces. In this chapter, the grafting of

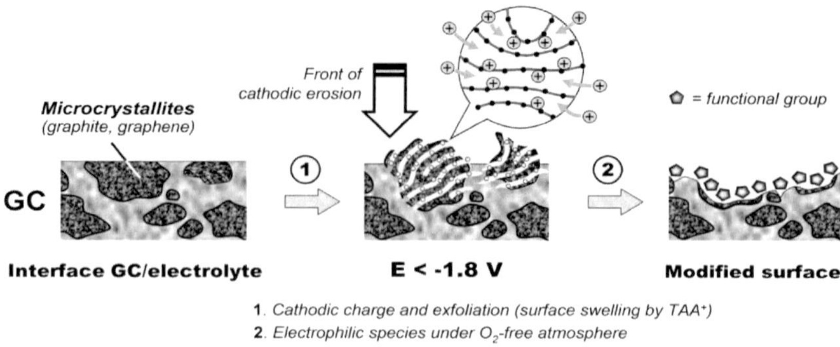

Scheme 12.1. Charge of GC and swelling brought by ion insertion [17].

ferrocene (Fc), tetrathiafulvalenes (TTF), alkenes, alkynes, long chain aliphatic acids, nitriles, is described. Coverage levels were found to be quite high, currently reaching apparent surface concentrations $> 5\times10^{-9}$ mol cm^{-2}. These high values are explained by swelling of the carbon interface (Scheme 12.1) provoked through its charge by insertion of TAA^+ cations that occurs concomitantly with the substitution process.

12.2. What is the Mechanism of the Immobilization of Alkyl Halides at Carbon Cathodes?

Glassy carbon (GC) is generally seen as a convenient electrode material to carry out analyses and electrolyses within cathodic and anodic ranges. For long, GC has been considered as an "inert" material made of unreactive carbon atoms [1]. However, this ideal structure does not seem compatible with the quite recent development of surfaces modified by addition of aryl radicals since the reaction obviously implies the presence of unsaturated systems in the material matrix [2]. This approach mainly concerned the chemical or electrochemical immobilization of aryl radicals through the reduction of diazonium salts [3, 4]. However, the structure of glassy carbon prepared from carbonization of phenolic resins [5] at temperatures often higher than 1500°C is complex and certainly depends on the industrial processing mode. Actually, GC contains sheets of graphite-like discontinuous structures in the carbon matrix with boundaries expected to correspond to all types of carbonaceous materials prepared this way. Therefore, GC should be considered as complex patchwork of conducting carbon reticulated zones in mixture with graphitized areas of different sizes that appear to strongly depend on temperatures (up to 2500°C) reached during pyrolysis. Apparently, a tight mixture of sp^2 and sp^3 carbon zones coexists both in the bulk and at the polished surface [6, 7]. Additionally, those graphite particles are concomitantly linked together with "polyaromatic materials" [8] expected to be progressively transformed by under aging complex quinones and ketones [9].

When the above-mentioned heterogeneity of GC is admitted to resemble a pudding riddled with graphitized and fullerenerized zones [7], one may reconsider this material (Scheme 12.1) as an electronic conductor candidate to be electro-chemically tailored at least at surface by means of cation insertion that occurs during its cathodic charging. As a matter of fact, highly oriented pyrolytic graphite (HOPG) provides, under cathodic charge [10], well-defined insertion stages [C_n^-,TAA^+] in contact with tetraalkylammonium (TAAX) salts. More concretely, in aprotic organic solvents, in the presence of electrolytes M^+X^- (with M^+ chosen to be not electroactive within the considered potential range), the polarization of GC at < −1.7 V vs. Ag/AgCl yields nucleophilic species enough reactive to induce to a large extent interfacial reac tions and exfoliation of graphite material [11]. Importantly in matter of reactivity, the concomitant insertion of electrons and cations into HOPG produces a nucleophilic material capable of reacting with a large panel of electrophilic species, for example, CO_2 that permits the formation of a graphite polycarboxylate [12]. Thus, these "electrified" materials (GC or/and graphite) put in contact with alkyl halides were first reported to react similarly since, resulting in carbon surface alkylation or graphite alkylation [13, 14]. While for alkyl halides the occurrence of a global two-electron scission appeared well established [15, 16], there emerged a basic idea that two concomitant processes (two-electron vs. one-electron process) implying either a charge (in mass) or a reduction via a radical transient (at surface) for RX present at GC should be taken into account. Therefore, the exact nature of interface reactions (electron transfer(s) versus nucleophilic displacements (SN)) deserved being reconsidered and the phenomenon of a real electro-chemical charge of GC and graphite established. This point is discussed with graphite, graphenes (industrial or those obtained via exfoliation), and GC. For better featuring this point, a reduction scheme of organic derivatives that may present electrophilic properties is summarized in Scheme 12.2.

In general, organic halides may exhibit electrophilic properties. Many studies, in the field of electro-chemical synthesis and analysis, concern these species [1] and their electro-chemical behavior is well documented [13]. Especially, it is clearly reported that 1-iodoalkane (RIs) are reduced at GC according to a two-electron cleavage of the C–I bond, mainly leading to the corresponding alkane (RH). Free radical $R^{\bullet}$ (expected to be the main transient) was shown to be reduced at a potential less negative than the reduction potential observed — in the total absence of catalysis — for the two-electron bond scission. However, in aprotic organic solvents, some amounts of homo-dimers could be also obtained possibly due to the Wurtz reaction that implies organic carbanions as transients. Additionally, in polar aprotic solvents, there is no evidence of the existence of free radical intermediates

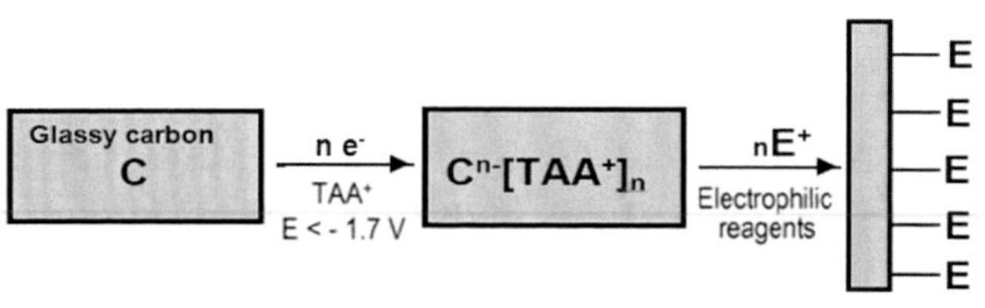

Scheme 12.2. Glassy carbon considered as chargeable material (in the presence of TAAX salts, aprotic conditions, the potential < –1.7 V) and transformed under electron transfer into a polynucleophilic material capable of reacting with organic electrophilic reagents (e.g., iodoalkanes) [25].

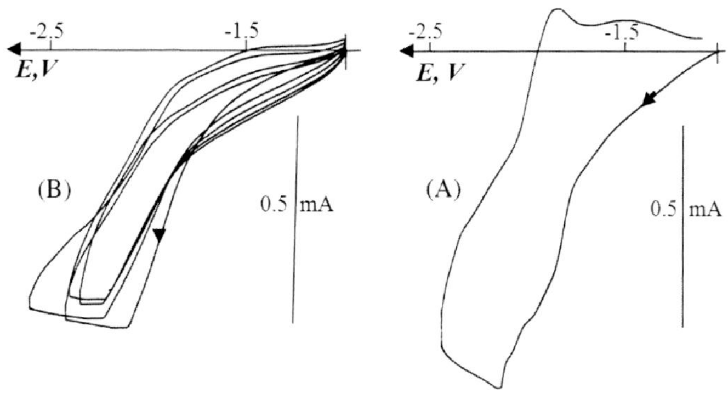

Figure 12.1. Voltammetry of a micro-piece (9.8 mg) of broken (non-polished) piece of GC in dimethylformamide (DMF)/0.1 M $TBABF_4$. The cycling (A) permits to exhibit a reversible step assigned to the superficial charge of the material. Scan rate: 100 mV s^{-1}. (B) Multi-scans of 1-iodo-hexadecane (11 mmol L^{-1}) at the same electrode. The anodic step vanishes and a significant shift of the reduction step is seen. Sample of GC from Carbone Lorraine CV-1500 [25].

in electrode reactions and of the grafting via this transient onto "standard" carbon surfaces [14]. Interestingly, quite recently, cathodic voltammetric experiments performed especially with 1-iodooctadecane showed a progressive inhibition of the electrode (until a total suppression of current), assigned to an insoluble organic deposit at the GC surface possibly provoked by radical intermediates [11].

If the cathodic voltammetry of a small piece of GC is carried out now under aprotic polar conditions (Figure 12.1), one may observe a large reversible step ($E_0 \approx -1.91$ V) assigned to superficial electrochemical reduction of graphite inclusions (via the intercalation of TBA^+ cations) with a charging threshold at about −1.75 V. The same experiment with HOPG leads to a spectacular change in the morphology of the material. Quite similarly, when much larger amounts of electricity are concerned, HOPG was reported to lead (with TAAX salts) to a huge exfoliation of the material [21]. Accordingly, SEM images of GC surfaces submitted to short electric pulses at −2.0 V provokes many minor modifications of the GC interface that could support local swelling of the material presumably due to the insertion of TBA^+ cations into graphitized microzones present at the surface. The quasi-reversible system observed with GC (Figure 12.1) would only concern graphite-like external planes. Under these conditions, a total charge (the insertion of TBA^+ in the whole mass of defects contained in GC samples) is therefore totally excluded.

It is proposed to reconsider the reduction of RIs (seen under a widest scope) in aprotic polar solvents with a special focus on researching evidences of surface modifications at GC and graphite. The building of compact films from primary RIs is described and discussed as a function of the chain length. Furthermore, quantification of surface deposits prompts us to consider alkyl chains marked with a ferrocenyl group, with an ethylene double bond or with a carboxyl group. In particular, ω-iodoalkylferrocenes are used as a convenient probe. For the first time, alkylferrocenes are directly fixed in large amounts onto carbon interface [14]; moreover, alkyl chains in C_{12} may lead to Fc-containing SAMs. Actually, those probes were chosen because they are electro-chemically active within the anodic range (reversible formation of ferrocenium cation) and then may bring valuable information beyond the reduction zone of remnant organic iodides. This way, a better knowledge on electroactive layers at GC surface could be gained while obtaining high-density coverages allow considering the method as very promising for layer-wise deposition of chemical groups and specific functions.

Before discussing and presenting different possibilities of immobilizing functionalized alkyl chains at carbon surfaces, we wish to underline the exact nature of the process by a simple experiment: the carbon sample is ***first*** charged (at the potential < -1.7 V vs. Ag/AgCl electrode) in DMF or acetonitrile (ACN) containing different TAAX salts (for example, tetra *n*-butylammonium tetrafluoroborate ($TBABF_4$) and then transferred (***still*** under inert argon atmosphere) to a solution of functionalized RI compound. See Scheme 12.3.

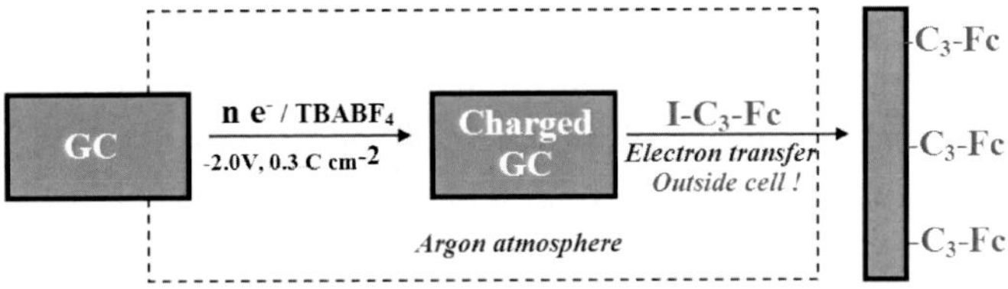

Scheme 12.3. Schematic representation of a two-step grafting via (i) charging GC carbon in a separate vessel, independently of the reduction of the RI and (ii) transfer of the charged material to the solution of Fc-marked iodoalkane still under argon atmosphere [25].

12.3. Results

12.3.1. *Reduction of alkyl iodides and bromides*

Experimentally, the cathodic reduction of primary alkyl iodides was shown to lead to structural changes of GC surfaces when recurrent scans are run within a quite negative potential range (say between −1.5 and −3 V vs. Ag/AgCl). Accordingly, in ACN containing $TBABF_4$, progressive modifications of GC surface appear upon repetitive voltammetric scans. Thus, a gradual potential shift toward more negative potentials is observed. The shift is often quite large (up to −0.4 V); additionally, a decay of peak current (especially in the case of long alkyl chains) is noticed. An ultrasonic rinsing shows that the reduction step of the alkyl iodide is definitively shifted. This behavior can be exemplified with 1-iodooctane and 1-iodododecane. After several scans, surface deposits lead to a progressive inhibition of the electrode surface that is well seen when comparing the anodic oxidation of potassium ferrocyanide at this electrode and at a freshly polished carbon surface. All primary alkyl iodides $C_nH_{2n+1}I$ (with $n > 3$) exhibit a similar behavior (sometimes, very spectacular blocking of the GC surface). Moreover, secondary iodides (2-iodopropane, 2-iodobutane, iodo-cyclohexane) do not exhibit large potential shifts in the course of multi-scan voltammetries under the above conditions (solvents like ACN and DMF). At the same time, any strong blocking of GC surface could be evidenced. This rules out the involvement of a free radical in the grafting process since free secondary alkyl radicals are known [17] to be more stable than the primary ones. If now these secondary alkyl iodides are reduced in solvents of higher dielectric constant (such as propylene or ethylene carbonates, pure or in mixtures), the deposits onto carbon can be observed. This solvent effect is even stronger with alkyl bromides (primary, secondary, and also tertiary) and then confirms the role of the solvent in the polarization of the C–X bond [18] in the substitution process for organic halides. These experimental evidences would greatly support the occurrence of a "nucleophilic grafting" at the cathode interface, in competition with the regular electron transfer. Under those conditions, the proximity (if not the adsorption) of the RI substrate to carbon appears quite indispensable. Such reactivity should be favored at the threshold of reduction steps. This point receives some support when quantified by using of 1-iodoalkylferrocenes at the step threshold with the injection of very small amounts of electricity.

Therefore, it looks consistent that the immobilization of alkyl groups at carbon is mainly favored by the polarization of C–I bonds. On the contrary, primary alkyl bromides — less reactive in ACN and DMF — appeared to be cathodically boosted by the use of much stronger polar solvents (PC eventually with the addition of ethylene carbonate, EC (Figure 12.2)).

Additionally, secondary and tertiary alkyl bromides, using these experimental conditions, may lead to the immobilization of 1-adamantyl and 2-adamantyl moieties [19, 20]. The immobilization, to a large extent, onto carbon is of interest for surface modification by further transformations of primarily grafted functionalities (Scheme 12.4).

Lastly, low reactivity of C–Cl bonds forbids (in the absence of an appropriate catalyst) any kind of grafting. This evidence is illustrated by the electro-chemistry of α,ω-chlorobromoalkanes (Figure 12.3). The cathodic reactivity does concern only the C–Br bond (large

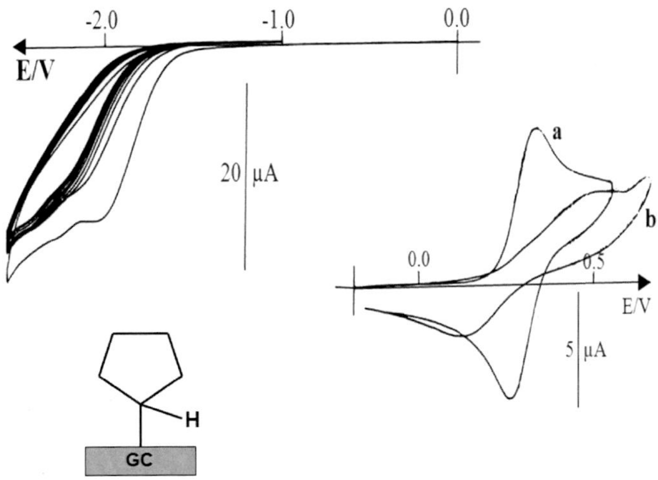

Figure 12.2. Multi-scan voltammetry of bromo-cyclopentane (18 mmol L^{-1}) in PC–EC (50/50) mixture containing $TBABF_4$ at GC electrodes (0.8 mm^2). Scan rate: 50 mV s^{-1}. The efficient coverage of the carbon surface is underlined by the anodic voltammetry of potassium ferrocyanide (curve b) at the modified surface to be compared to the curve (a) obtained at a smooth carbon electrode [13].

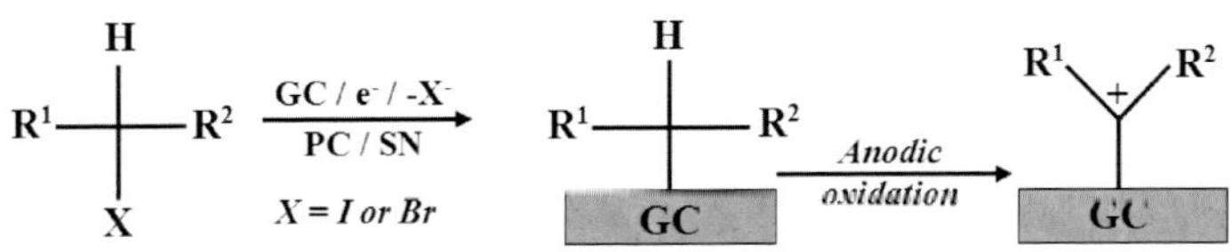

Scheme 12.4. Coverage of GC by secondary alkyl groups. The anodic oxidation may provide carbocations (oxidation that allows to estimate the grafting level of the alkyl group) and also permits functionalization if the reaction is carried out *in situ* in the presence of nucleophiles [13].

potential shift upon scans) while the reduction potential of step II of the chloride group (located at very negative potentials, $E < -2.8$ V vs. Ag/AgCl) remains invariant upon sweeps.

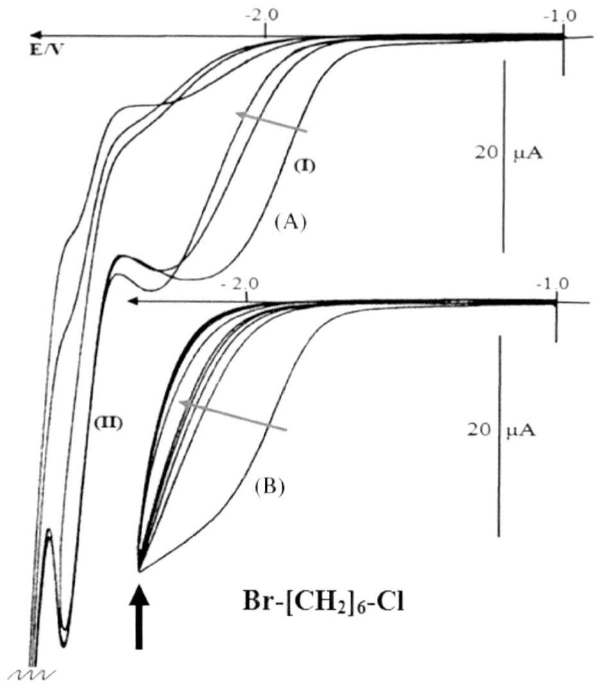

Figure 12.3. Reduction of 1-bromo-6-chlorohexane (13 mmol L^{-1}) in DMF/0.1 M $TBABF_4$. Scan rate: 50 mV s^{-1}. GC micro-electrode (0.8 mm^2). The C–Br bond is cleaved at wave (I). The multi-scan voltammetry (both in A and B) reveals a shift of the reduction step supporting the continuous grafting along the scans. A multi-scan limited to the arrow (only focused on step I) permits an efficient coverage of GC by 1-chlorohexyl groups [25].

12.3.2. *Functionalization of GC*

Several functional groups ***F*** can be efficiently grafted onto GC at the conditions to belong to the general family I–C_n–***F***. The iodide group is preferred to ensure the stability of several functions that could be electro-active at $E > -2$V and therefore destroyed at the cathode. Functions branched on iodoalkyl chains such as trimethyl silane (TMS), acethylenic and nitrile groups, as well as sulfonyl group (–SO_2–R) were shown [25] to be immobilized onto GC according to the process mainly developed with primary RIs (Scheme 12.5) based on a certain selectivity of reaction (3) toward a reduction via electron transfer (according to 2).

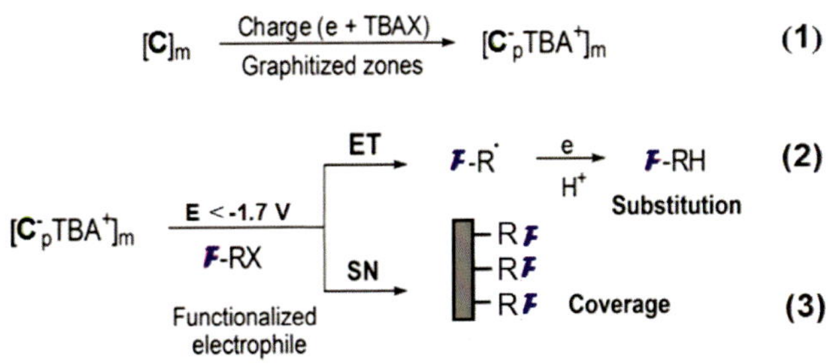

Scheme 12.5. A general scheme for the modification of carbon surfaces at quite reducing potentials through the cathodic charge of carbon and the cathodic reactivity of functionalized electrophiles ***F***–RXs (ET vs. SN).

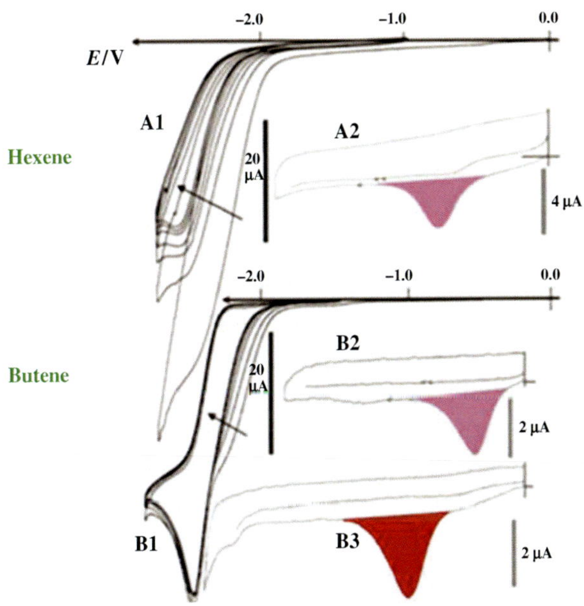

Figure 12.4. Voltammetric responses of ω-bromo-1-alkenes in DMF/0.1 M $TBABF_4$ at GC (all steps located at potentials < −2 V, curve in black: scission of the C–Br bond after six recurrent scans in A1 and B1). Curves A2 and B2 display the responses of electrode deposits (in ACN + $TMABF_4$) after marking them with halogens (I_2 in purple and Br_2 in red-brown). These peaks attest to the reduction of vicinal di-halides to regenerate the corresponding double bonds [19]. Refer also to Figure 12.5 showing the EDX mapping.

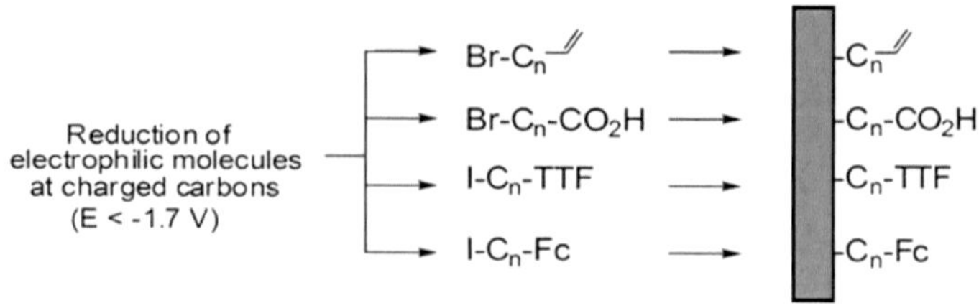

Scheme 12.6. Four examples of decoration of carbon surfaces as detailed in the text.

Furthermore, it is emphasis is placed (Scheme 12.6) on the decoration of GC using different functional groups via the grafting by means of long alkyl chains.

12.3.3. *Vinylation of carbon surfaces*

In continuity of works described in the present chapter, the immobilization of *n*-alkyl chains ending with a vinyl group has been developed [19] with the aim of constructing the surface layers where carbon atoms of the cathode material are electronically disconnected from the terminal vinyl moieties. The coverages of that kind appear to be highly convenient for preparing chemically enriched carbon surfaces. In particular, vinyl groups are known to be rather good traps for free radicals. Additionally, nucleophilic species (carboxylates, primary amines) are efficient for transforming such vinylized carbon interfaces. Also, epoxidation reactions (along with 1,2-hydroxylations) could also be envisaged for subsequent specific modifications. Lastly, [2 + 2] and [2 + 4] additions, under specific conditions, could as well be considered for bringing a large palette of additional functions and redox groups to the carbon surface. Long carbon chains in C_{10} and C_{12} can be considered for building SAMs with appropriate functions.

Practically, such grafting operations were realized in two complementary modes: (i) one being based on electro-chemical charging of GC at < −2.0 V in an aprotic polar solvent containing TAAX salts in the presence of ω-bromo-1-alcenes and the other (ii) by taking advantage of the formation of free vinyl-alkyl radicals at carbon surfaces lightly doped with Pd nanodeposits that are easy to

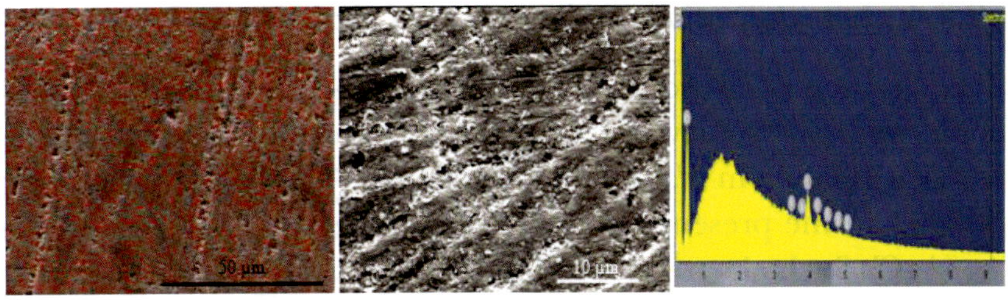

Figure 12.5. The SEM image pictures the very large surface modification provoked by reaction of charged GC with bromobutene. Additionally, the EDX mapping (obtaining of red spots) after reaction with iodine attests the iodination of the layer. The integration of the reduction peaks of di-iodoalkanes reveals alkene coverages of $(1.3 \pm 0.2) \times 10^{-9}$ mol cm^{-2} [19].

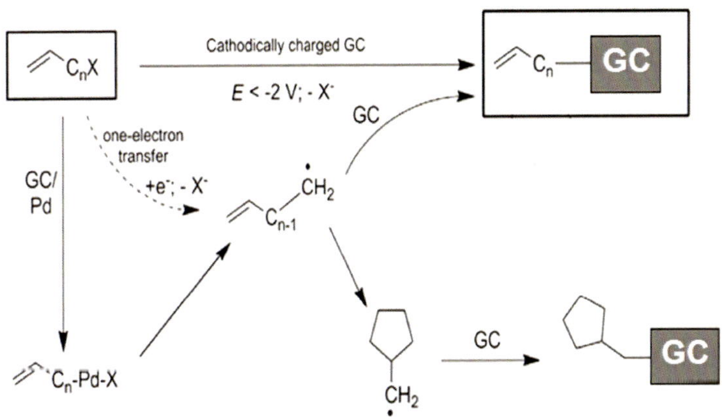

Scheme 12.7. A general way of the immobilization of a pure vinyl layer is depicted by the first line of the scheme. The possibility of creating a radical transient (with or without Pd catalyst) is also mentioned. Lastly, the use of a halogen (Br_2 or I_2) for saturating the vinyl group is also mentioned [19]; it is practical for assessing the number of double bonds grafted at the GC surface [19].

eliminate by sonication and rinsing after the grafting. The use of Pd (route (ii)) is only given as a second possibility of grafting and is not described in much detail in this chapter. In fact, Pd catalyst would create the risk of producing free alkene radicals prone to cyclization distorting the homogeneity of the deposited organic layer (see Scheme 12.7).

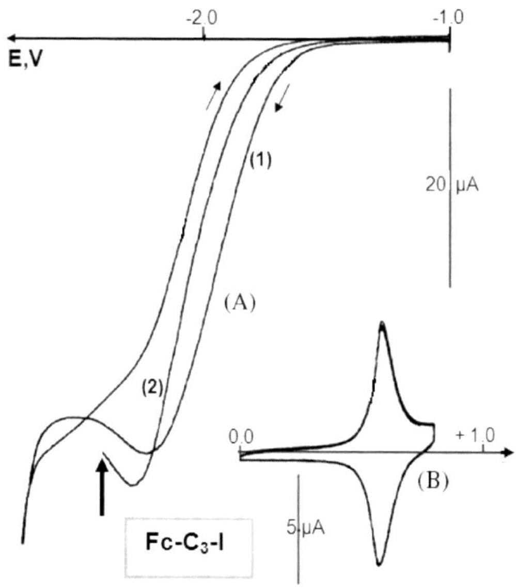

Figure 12.6. Voltammetry of 3-iodopropylferrocene (17 mmol L^{-1}) in ACN containing 0.1 M $TBABF_4$) and immobilization of the propylferrocene moiety at GC (surface area: 0.8 mm^2). (A) Reduction. Scan rate 50 mV s^{-1}. Hold of 30 s at the second scan (arrow at –2.4 V). (B) The response of the deposit after sonication and rinsing in water and acetone. Ten successive scans (100 mV s^{-1}) between 0 and +0.8 V show the high stability of the layer under oxidation. The density of ferrocenyl units at the GC interface is 8.2×10^{-9} mol cm^{-2} [21].

the formation of electrically charged micro or nanozones, they could behave as nucleophilic reagents dispersed in the polarized carbon material and initiate the polarization of C–I bonds. The macro-electrolyses carried out at GC to charge the material in order to render it sufficiently nucleophilic toward RIs carrying the Fc moieties were particularly spectacular when the charged GC was transferred (under inert atmosphere) to another vessel only containing the solution of an RI substrate.

12.3.6. *About the immobilization of tetrathiafulvalenes*

The grafting of π-donors obtained onto carbon substrates by means of alkyl links was particularly original [22] and successful. Here, it is

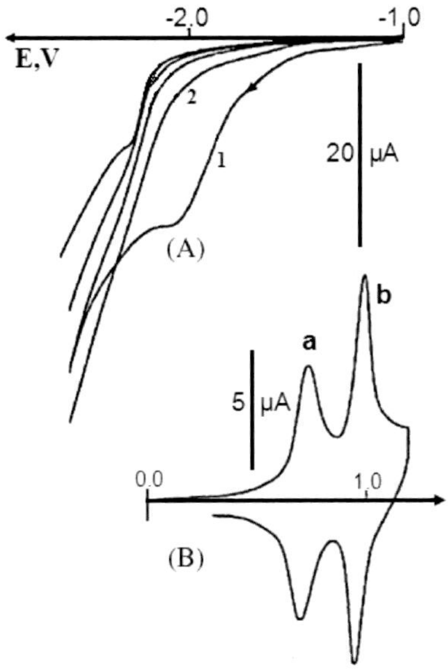

Figure 12.7. (A) Repetitive voltammetric (cathodic) responses of TTF–C_5–I (7.5 mmol L^{-1}) at a GC electrode (0.8 mm^2). Electrolyte: CH_3CN/0.1 M $TBABF_4$. Scans from –1.0 to –2.5 V. Scan rate 50 mV s^{-1}. (B) Anodic response of the deposit on the GC electrode observed after the fifth cathodic scan (after a 2 min ultrasonic rinsing) [22].

illustrated by the immobilization of TTF known to be prone to π-stacking and to exhibit two reversible oxidative electron transfers (Figure 12.7).

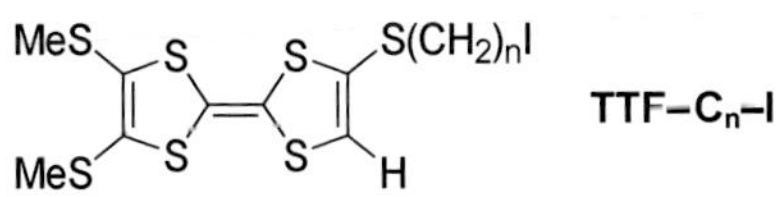

$$TTF - e^- \rightleftharpoons TTF^{\bullet +} - e^- \rightleftharpoons TTF^{++}$$

This method presents an original protocol permitting the robust immobilization of large amounts of TTF moieties onto GC using the TTFs with ω-iodo-alkylsubstituents. This procedure is also valid at charged graphite. The process at GC needs very negative potentials [18, 20]. Specifically with TTFs, the deposition efficiency was determined: TTF coverage levels were found to be large whatever the chain length (the linker with 3–10 carbon atoms). Such TTF layers are stably attached to carbon surfaces, suggesting that covalent linkages were formed with the graphene units activated in the course of bursting the graphite micro-inclusions by the insertion of bulky TAA^+ cations during the charging process. In contrast, the attempts of grafting conducted under the same conditions on gold and platinum did not lead to any depositions: this strongly reveals something specific to the complex structure of GC. Integration of the first oxidation step (corresponding to the TTF cation radical) permitted estimating the maximum amount of TTF species as equal to about $\Gamma_{TTF} = 10\times10^{-9}$ mol cm^{-2}. These values are quite high when compared to those already reported for the compact films obtained when attaching TTF to smooth gold surfaces via adsorbed alkanethiol chains [23, 24]. Such excess of TTF concentration at the interface was expected to be provoked by preliminary cathodic charge of aromatic species (like graphite, fullerenes as well as polyaromatic inclusions) contained in GC that lead to the swelling of the material accounting both for a high porosity and an abnormally large roughness factor. The present method is certainly valid for immobilizing different chemical functions via polymethylene links of various lengths for producing many kinds of modified carbon electrodes.

In order to compare, the current method assigned to an SN process, and usually achieved at very negative potentials, was replaced by an alternative process based on a Pd-catalyzed immobilization of TTF groups. This latter allows the direct generation of free alkyl radicals at the solid surface. Preliminary experiments were carried out with catalytic amounts of palladium deposited onto carbon ($\Gamma_{Pd} \approx 10^{-9}$ atom cm^{-2}). This way, the amounts of grafted TTF were found to be significantly smaller than from the SN procedure. However, they remain quite large presumably due to the radical ablation of carbon as was already noticed before with alkyl radicals; it brings a large roughness to the solution-exposed carbon interface. Lastly, it is worth underlining that grafting of TTF may be achieved under very simple conditions with many varieties of carbons (graphite, industrial cokes, porous carbons and felts, etc.).

References

[1] D.G. Peters, in *Organic Electrochemistry*, eds. H. Lund and O. Hammerich, M.M. Dekker, New York, Basel, p. 341 (2001).

[2] R.L. McCreery, *Chem. Rev.* **108** (7) (2008) 2646.

[3] J. Pinson, F. Podvorica, *Chem. Soc. Rev.* **34** (2005) 429.

[4] D. Belanger, J. Pinson, *Chem. Soc. Rev.* **40** (2011) 3995.

[5] G.M. Jenkins, K. Kawamura, *Nature* **231** (1971) 175.

[6] P.J.F. Harris, *Philos. Mag.* **84** (2004) 3159.

[7] Y. Gogotsi, J.A. Litbera, N. Kalashnikov, M. Yoshimura, *Science* **290** (2000) 317.

[8] P. Bernier, S. Lefrant, *Le carbone dans tous ses états*, Gordon and Breach Science Publishers, Amsterdam (1997).

[9] A. Gennaro, A. Isse, C. L. Bianchi, P. R. Musssini, R. Rossi, *Electrochem. Commun.* **11** (2009) 1932.

[10] J.O. Besenhard, *Carbon* **14** (1976) 111.

[11] J. Simonet, H. Lund, *J. Electroanal. Chem.* **75** (1977) 719.

[12] C. Dano, J. Simonet, *J. Electroanal. Chem.* **564** (2004) 115.

[13] M.M. Chehimi, G. Hallais, T. Matrab, J. Pinson, F.I. Podvorica, *J. Phys. Chem.* C **112** (2008) 18559.

[14] J. Simonet, *Electrochem. Commun.* **13** (2011) 107.

[15] D.G. Peters, in *Organic Electrochemistry*, eds. H. Lund and O. Hammerich, M.M. Dekker, New York, Basel, p. 341 (2001).

[16] H.S. Sorensen, K. Daasbjerg, *Acta Chem. Scand.* **52** (1998) 51.

[17] V. Jouikov, J. Simonet, *Electrochem. Commun.* **45** (2014) 32. Mini-review.

[18] J. Simonet, *Electrochem. Commun.* **13** (2011) 661.

[19] P. Poizot, J. Simonet, *Electrochem. Commun.* **23** (2012) 137.

[20] V. Jouikov, J. Simonet, *Electrochem. Commun.* **13** (2011) 1296.

[21] V. Jouikov, J. Simonet, *Langmuir* **28** (2012) 931.

[22] D. Lorcy, K.-S. Shin, M. Guerro, J. Simonet, *Electrochim. Acta* **89** (2013) 784.

[23] Y. Yokota, A. Miyazaki, F. Fukui, T. Enoki, *J. Phys. Chem. B* **110** (2006) 20401.

[24] C.M. Yip, M.D. Ward, *Langmuir* **10** (1994) 549.

[25] J. Simonet, Unpublished results.

Chapter 13

Grafting at Solid Surfaces via Trimethylsilyl Derivatives

13.1. An Overview

The capability of a large allotropic variety of carbons (e.g., glassy carbon (GC) known to possess nanoinclusions of graphite and fullerene-like) to be, under well-defined conditions [1], anodically charged at potentials over +1.5 V vs. $Ag/AgCl_{sat}$ accounts for the formation of electrophilic centers on carbon able to react with appropriate nucleophiles, in particular with trimethylsilyl-substituted derivatives $R–Si(Me)_3$. The method, first developed for allylation and benzylation of different carbons under specific conditions, is extended to a larger variety of R groups. This chapter aims to expose new anodic modes of grafting on solids.

Nowadays, GC remains being considered as an efficient electrode material for carrying out electro-chemical analyses and electrolyses, mainly within the cathodic domain [2]. Several times in this book, we insisted that this material contains, and in large amounts, ribbons of graphite-like structures and therefore should be seen as a complex solid possessing conducting zones (possibly reticulated) in tight admixture with micro-graphitized and micro-fullerenized areas that endows this material with a specific reactivity due to the large presence of sp^2 carbon structures [3, 4].

Usually, the electrolytic transients formed in the course of electro-organic reactions at carbon interfaces both in cathodic and anodic ranges may act as reactive intermediates in the heterogeneous reactions with the electrode. Accordingly, most of organic deposits at the GC interface are anchored to the surface by addition of free radicals. The reactivity of such transients toward the solid surface underlines the high significance of unsaturated "impurities" present in the used carbons. Reactions of this kind were reported to occur via radical intermediates formed from the oxidation of alcohols [5], amines [6], and carboxylates [7] or from the reduction of electro-active salts like aryl diazonium [8], iodonium [9], and sulfonium [10]. Cathodic scission of the activated carbon–halogen bonds (allylic, benzylic, and propargylic halides) revealed the efficiency of this process for C–C grafting onto various carbons via the corresponding C-radicals, together with the reaction of the charged carbon if rather negative potentials are reached [11–14]. Additionally, the one-electron cleavage of C–X bonds can be achieved using catalysis by ad-atoms of transition metals (deposition of transition metals in subnanomolar amounts at the GC surface, reduction of species resulting from oxidative addition of RX to the metal to give C-radicals, grafting of the latter, and elimination of remaining ad-atoms by sonication during the workup), allowing one to build dense organic deposits at the carbon interface.

For innovating in the domain of surface chemical modification, it may obviously be of interest to explore and to develop further the electro-chemical charging of carbon itself, developing an "inert" substrate for radical additions and an active and efficient reactive interface with specific nucleophilic or electrophilic reactivity. As shown in Chapter 12, this approach allowed to introduce and to enlarge considerably the possibilities of nucleophilic surface modification within the cathodic range at the condition that appropriate electrophilic species are present in the vicinity of cathodically doped electrode surface.

Still in the field of electro-chemical charging of different carbons, let us recall that these materials are also anodically chargeable in the presence of suitable and poorly reactive anions. It should also be remembered that graphite can efficiently incorporate bromine that corresponds to the insertion of Br^- into an anodically charged carbon. To come back to GC, such possibility allows the creation of electrophilic centers at the carbon surface. This phenomenon is worth being carefully considered and exploited. A first contribution in this field [15] has recently underlined the interest of oxidizing different types of carbon (GC, graphite, deposited graphene), and prompted an overview of the capability of several trialkylorganosilanes to react with electro-generated carbocations, leading to the scission of carbon–silicon bonds in the molecular precursor and giving rise to the grafting of organic free radicals at electrophilic surface-localized carbon atoms.

The elevated kinetic instability of silicenium cations, in particular of TMS^+ group, is provoked by their extremely high electrophilicity. Very efficient reactions with a large variety of both π- and σ-electron donors, exploiting this feature, are reported in organic chemistry. The field of the oxidative cleavage of C–Si bonds, enhanced by the excellent leaving group ability of TMS^+ cations, has been recently reviewed [15]. Therefore, many useful synthetic reactions based on such chemistry can be expected. Many examples of electrophilic displacement of trimethylsilyl group are well documented [16]. Just to name a few, the oxidative coupling of benzyltrimethylsilane with aromatic compounds has been reported to lead to benzylated products in high yields (up to 88%) while oxidative desilylation of allylic silanes or of benzyl trimethylsilane gave 1,5-hexadienes and aromatic products, respectively. Let us mention that allyl and benzyl nitrates were formed under oxidative conditions from allyl and benzyl trimethylsilanes [17–19]. Electrophilic substitution of TMS group of benzyltrimethylsilane with IBr in acetonitrile (ACN) leads to benzyl bromide in 60% yield [20]. Still in the same spirit, the electrophilic substitution of TMS in the reaction of substituted benzyl and heteroaryl trimethylsilanes with electrophiles such as PhCHO and $PhCH_2Br$ could also be mentioned [21]. Now under the conditions of electro-chemical redox catalysis, electro-generated thianthrene cation radical (E^+) may react (1) with

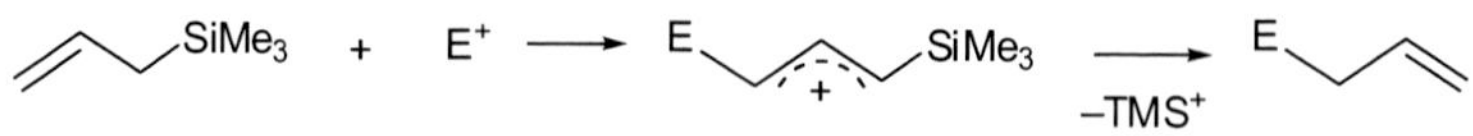

Scheme 13.1. Reactivity of allyl trimethylsilane with electrophiles.

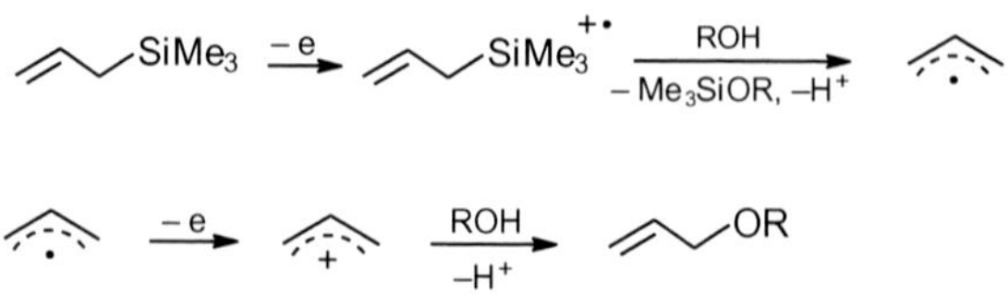

Scheme 13.2. Anodic oxidation of trimethylsilyl derivatives in nucleophilic media.

allyl trimethylsilane, permitting (Scheme 13.1) the condensation of the latter with the mediator [22].

More specifically, in an electro-chemical context, TMS group placed in the β-position to double bonds raises the HOMO level of allyl and benzyl silanes owing to $\sigma_{(C\text{–}Si)}$-π conjugation [11, 12] and consequently lowers their E_{ox}. Upon electron withdrawal, the β-effect renders Si more electrophilic, strongly favoring Si–C bond cleavage (Me_3Si^+ then acts as a "super-proton" [24]) and nucleophilic reactions at the concerned carbon atom. For allyltrimethylsilane, the value of $E_{p/2}$ = 1.70 V vs. Ag/AgCl [16] was reported. Benzyltrimethylsilane is oxidizable at a rather close potential [1, 15, 18]. Electron withdrawal from these trimethylsilylmethyl derivatives leads, through the elimination of the TMS^+ fragment (scavenged by an appropriate nucleophile or solvent) and the instantaneous oxidation of the intermediate C-radicals, to allyl or benzyl cations. In the presence of nucleophiles (Scheme 13.2), these cations form the corresponding products of anodic substitution [15, 23–28].

With the focus on modification of solid surfaces, such reactions based on the electron transfer initiated cleavage of the Si–C bonds — either by direct oxidation at the electrode or via anodic charging of carbon creating electrophilic centers at the interfaces — were developed with a series of trimethylsilylated precursors. The superficial

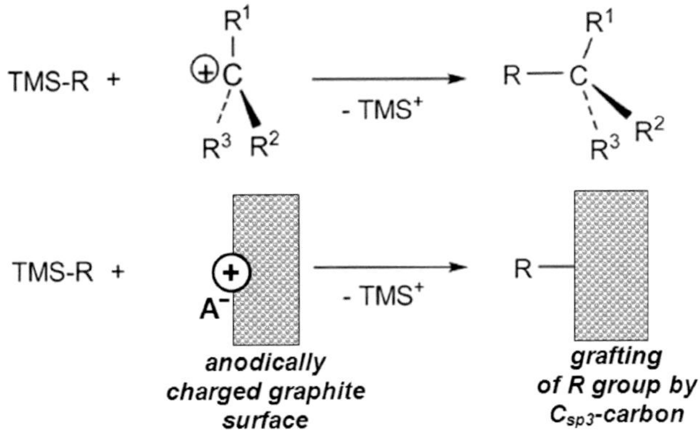

Scheme 13.3. The principle of a novel grafting process at graphite and carbon interfaces via TMS derivatives by taking advantage of the anodic charge of the material [29].

concentration of allylic groups was calculated through the marking of the ethylenic bond by iodine and bromine [1] with an estimated $\Gamma_{All} \approx 9 \pm 1\times10^{-8}$ mol cm^{-2}. The focus of this chapter developed hereafter is summarized by Scheme 13.3.

13.2. Oxidation of Allyltrimethylsilane (ATMS) at Carbon Anodes

Carbon samples used for the experiments in this chapter were purchased from Tokai carbon C° (GC rods for electrodes) and from Carbone Lorraine (large plates, VD 1500 and 2500, thickness 3 mm). For polishing these GC electrodes, the Strucrs P1200 papers (waterproof silicon carbide) as well as Emery papers 02 and 03 were used. After polishing, the electrodes were rinsed in an ultrasonic bath in order to eliminate both grease and carbon dust. The electro-chemical anodic charge of a graphite sample shown in Figure 13.1(A) produces the formation of electrophilic centers able to react with R-TMS species.

The anodic oxidation of ATMS at GC electrodes in ACN containing $TBABF_4$ permits to observe a diffusional oxidation peak at E_p = +1.8 V (Figure 13.1, curves B). Also, one may check that platinum,

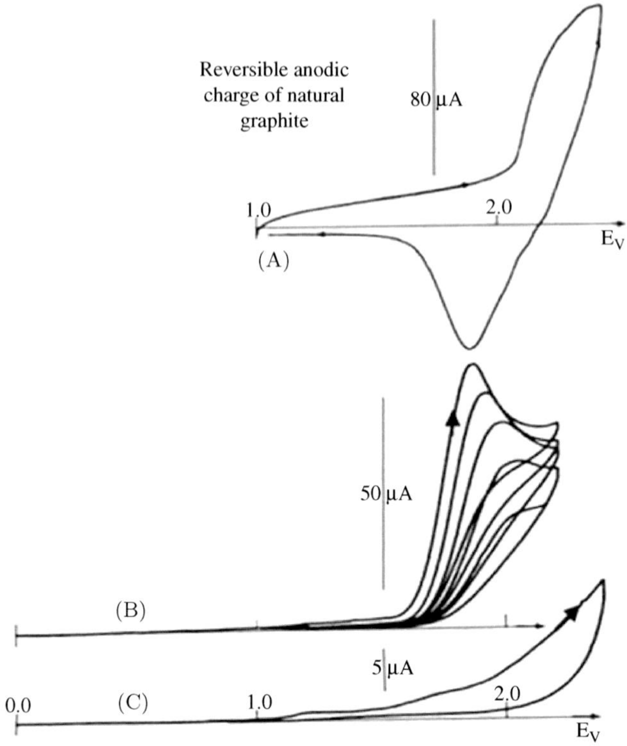

Figure 13.1. Compared to the oxidation of natural graphite (A), voltammetries in (B) display the oxidation of ATMS (19 mmol L^{-1}) in ACN + tetra *n*-butylammonium tetrafluoroborate ($TMABF_4$). Five recurrent scans up to +2.1 V at GC are compared with the response of electrode material alone in A. Scan rate: 50 mV s^{-1}. Surface area of the electrodes: 0.8 mm^2. (C) Smooth Pt electrode does not permit any oxidation within the same potential range [29].

when used as an anode material, does not allow any significant oxidation step (curve C).

Thus, employing a pure graphite anode (as displayed in Figure 13.2, curve A) allows one to observe the reversible charging of graphite concomitantly with the oxidation of ATMS. The latter occurs precisely at the threshold of graphite oxidation. We proposed (as summarized in Scheme 13.3, with allyl group being denoted as R) that carbon plays a catalytic role in the oxidation

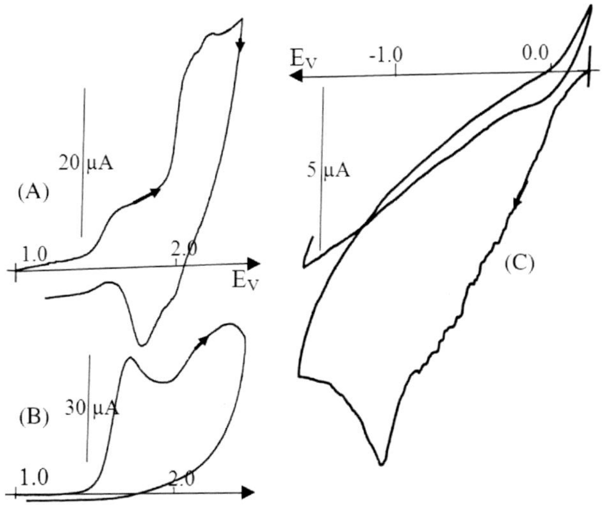

Figure 13.2. Voltammetry of ATMS (19 mmol L^{-1}) in ACN + $TMABF_4$. Scan rate: 50 mV s^{-1}. Surface area of the electrodes: 0.8 mm^2. (A) Response at graphite electrode. The reversible charge of graphite corresponds to the oxidative insertion of tetrafluoroborate in galleries. (B) Response at a GC electrode. (C) Oxidation of ATMS at +1.8 V. The charge of 1.2×10^{-3} C was followed by iodination of the polyallylic layer deposited onto the GC surface. Reduction of the iodinated layer in ACN + $TBABF_4$. The superficial concentration of immobilized allyl units was found to be $\Gamma_{All} = 7.4 \times 10^{-8}$ cm^{-2} [1].

process of ATMS, quite certainly via the electrophilic centers formed at the anode surface.

Thus, substitution of all these electro-generated reactive centers by allyl groups should lead to a GC surface coverage progressively decreasing the activity of the anode interface. This is exactly what is depicted seen in Figure 13.1(B): upon recurrent scans, the oxidation peak of ATMS progressively shifts toward more anodic potentials concomitantly with the total current decay. The inhibition of electrode turns being quasi-total would reveal a large coverage of the active anode surface.

In order to quantify the amount of allyl groups immobilized at the GC interface, a simple method of surface halogenation (bromination and iodination of the unsaturated units) was employed,

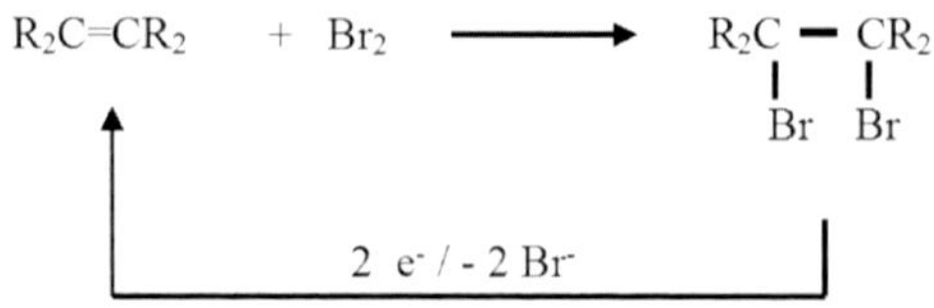

Scheme 13.4. Marking of grafted ethylenic functions (as-formed according to Scheme 13.5) by addition of bromine followed by reductive de-bromination.

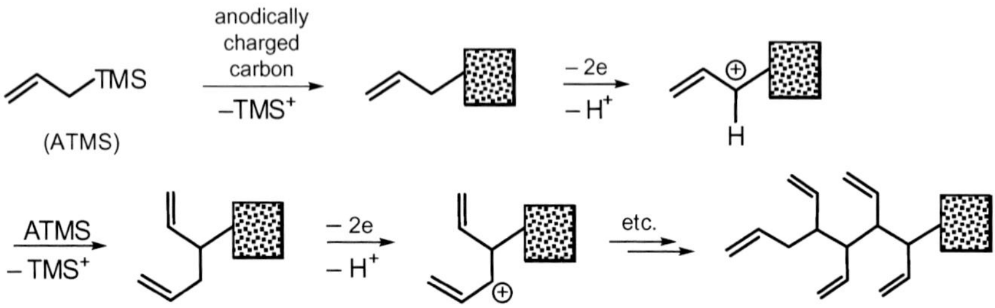

Scheme 13.5. Anodic deposition of allylic groups at carbon via ATMS [1].

leading to the corresponding vicinal dihalo derivatives that are readily de-brominated (or de-iodinated) by a two-electron reduction as depicted in Scheme 13.4.

The halogenation of the grafted layer was achieved by their contact (30 s) with the solutions of I_2 (5% in ethanol) or of Br_2 (ca. 3% in cyclohexane). Afterward, the electrodes were twice rinsed in acetone and dried in a hot air flow. The electro-chemical reduction of thus formed dihalo-derivatives (Figure 13.2(C)) permitted assessing the number of allyl units at the carbon surface. Under the conditions specific to Figure 13.2, current integration provided the values of 7.4×10^{-8} mol cm^{-2} (via iodination) and 9.0×10^{-8} mol cm^{-2} (via bromination). These rather close values are certainly much larger than those expected for allyl mono-layers (about 5×10^{-10} mol cm^{-2} for an ideally smooth solid surface). Therefore, the formation of a multi-layer coverage should be envisaged. This is quite certainly related to the ability of C–H and C=CH centers in the primarily grafted allylic layer to form carbocations upon (partial) oxidation, these latter being themselves good candidates for the reaction with the ATMS substrates. Scheme 13.5 takes into account the elimination of allylic proton under oxidation, favoring the formation of a multi-allylated deposit.

Scanning Electron Microscopy (SEM) images point to the formation of a dense, thick, and uniform layer at the surface. Its Fourrier Transform Infra Red (FTIR) spectra show characteristic C–H and C=C vibrations. IES also confirmed the presence of a large deposit at the interface exhibiting two very different resistances, one at the carbon surface (R_{CT} = 317 ohm) and the other, much larger, associated with the deposited organic layer, R_{CT} = 470 kohm.

13.3. Benzyltrimethylsilane (BTMS): Oxidation at Carbon Anodes

The oxidation of BTMS occurs at quite positive potentials with first well-shaped peak at $E_{p/2}$ = +1.44 V followed by a very broad one ($E_{p/2}$ = +2.04 V). Recurrent scans lead to a fast decay of their currents suggesting the coverage of the carbon electrode via a polymerization process underlining the specific catalytic character of the oxidation and the concomitant grafting process at a GC electrode. The almost total inhibition of the anode and the resulting dense coverage of the electrode were explained by the succession of anodic oxidation steps of benzyl groups following the initial grafting to the surface (first grafted layer) as displayed in Scheme 13.6. SEM images of this surface confirm the formation of a dense and uniform organic layer. The inhibition of the carbon electrode was fully coherent with Electro-chemical Impedance Spectroscopy (EIS) experiments.

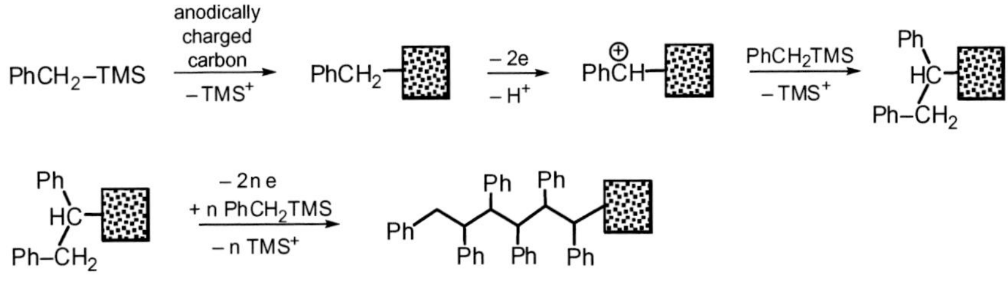

Scheme 13.6. Immobilization of benzyl groups at carbon anodes [1].

13.4. Oxidation of Phenyltrimethylsilane at GC Electrodes

The oxidation of PhTMS at GC electrodes under similar conditions was also undertaken. The anodic reaction appears to be extremely effective permitting a very fast and dense coverage of the carbon surface with a compact layer that blocks (almost definitively) the electrode surface. This could be due to an efficient polymerization involving the oxidation of phenyl rings (through an electrophilic aromatic addition) that takes place not only at the surface, but presumably also at fractals and flaws in the GC. This kind of deposit makes difficult a fine electro-chemical and chemical tuning of the surface for taking advantage of this particular multilayer system.

13.5. TMS as a Vector for the Immobilization of RX at Solid Surfaces

13.5.1. *Grafting of alkyl halides at carbons*

The immobilization of $-(CH_2)_n-X$ moieties (with X = I or Br) is of tremendous interest due to their electrophilic properties that provide new possibilities for filling carbon interfaces with various functional groups (Scheme 13.7). For example, the expected superficial reactivity of C–I bonds toward anion radicals (electro-generated or chemically prepared), nucleophilic bases, and in electron transfer reactions could enable many new processes. This original approach is shortly presented hereafter.

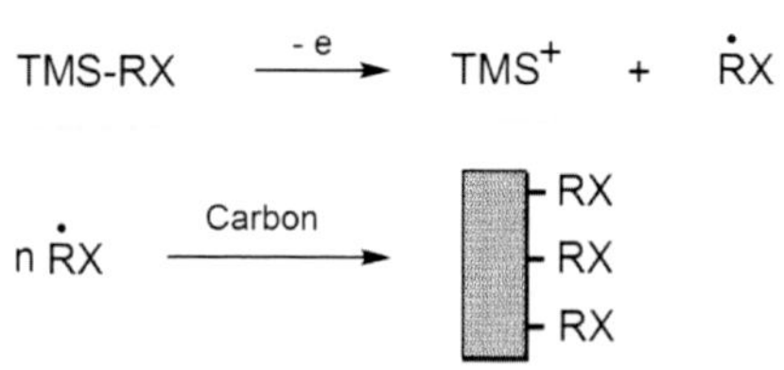

Scheme 13.7. A general process for the immobilization of RXs at carbon via TMS^+ as a leaving group [29].

Concerning the immobilization of RIs, this preliminary presentation is focused only on the use of the following precursors:

$$TMS\text{–}(CH_2)_6\text{–}I$$
$$TMS\text{–}(CH_2)\text{–}I$$

Upon oxidation at a smooth carbon electrode with the potential scans up to +2.6 V vs. Ag/AgCl, the first substrate with a rather long alkyl chain exhibits a slow but steady current decay associated with a progressive grafting of $-(CH_2)_6I$ moieties. Therefore, voltammetry of the produced electrode surface reveals a reduction step (E_p = −1.54 V) seen only in the course of the first scan, which accounts for the cathodic cleavage of C–I bonds belonging to the grafted layer (Figure 13.3(B)). EDS performed after the ultrasonic cleansing of

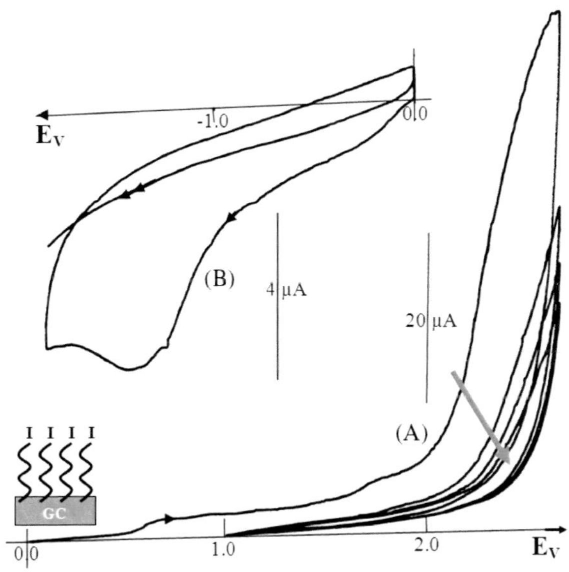

Figure 13.3. Grafting of $-(CH_2)_6$–I onto smooth GC using TMS–$(CH_2)_6$–I (8.7 mmol L^{-1}) in ACN containing $TMABF_4$. Scan rate: 50 mV s^{-1}. (A) Recurrent oxidation scans between 0 and +2.7 V. (B) The reduction step of the grafted iodoalkyl moieties onto carbon [29].

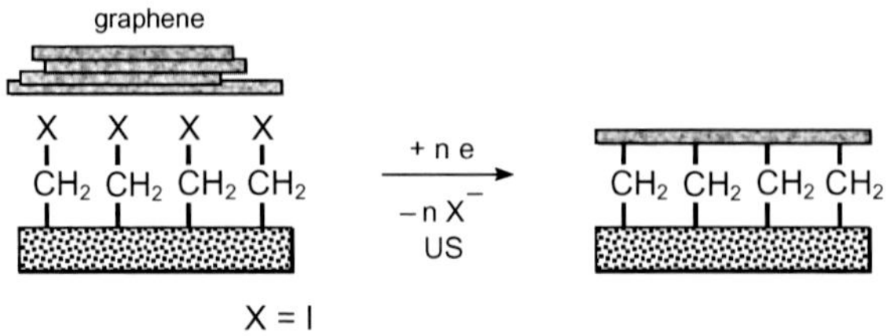

Scheme 13.9. Process for an attachment of graphene flakes using CH_2I anchors dispersed onto a solid conductor (here carbon) and activated through electron transfer to give free radicals that add to graphene. Using this system, one may operate the building of a "multi-floor" graphene electrode highly reacting on the edges and presumably presenting a possible absence of π,π-interactions between graphene planes that could allow diffusion of ions during charge and discharge of this "multilayer" electrode. The attachment of graphene monolayer could be supported by a preliminary sonication [30] together with a solid attachment of graphene flakes [29].

able to add to graphene layer and further immobilize it, leading to a robust modification (good resistance to sonication) of GC that may be helped by the presence of graphene on its external surface (Scheme 13.9).

13.5.2. *Scission of organo-trimethylsilanes at gold and platinum*

Additionally, it could be shown [29] that other activated organo-trimethylsilanes are anodically and cathodically reactive at gold and platinum in ACN containing $TMABF_4$. Thus, fixation of trimethylsilyl-1-propyne onto gold visible through a large self-inhibiting peak (E_p = +1.54 V) produces a very stable and compact surface coverage. The triple bond in the immobilized layer reacts with iodine (the reduction of the corresponding diiodo form observed at −0.49 V). Figures 13.5 and 13.6 illustrate the FTIR and EIS responses of the grafted layer attesting the presence of –C≡C–Me units on the interface. The usefulness of the easy modification of gold using this commercially available reagent is noticeable for building versatile surfaces (Scheme 13.10).

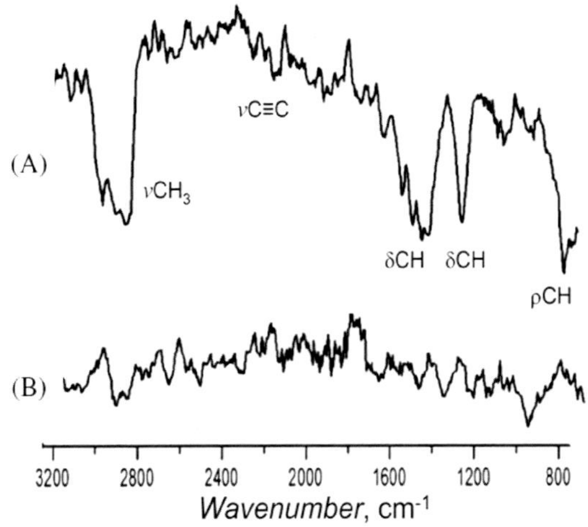

Figure 13.5. Addition of TMS–propyne at gold: (A) FTIR spectrum of the C≡C–Me layer grafted to the electrode surface by oxidizing TMSC≡CMe (12 mmol L^{-1}) in ACN + $TMABF_4$ at $E = +1.54$ V. (B) Bare electrode [31].

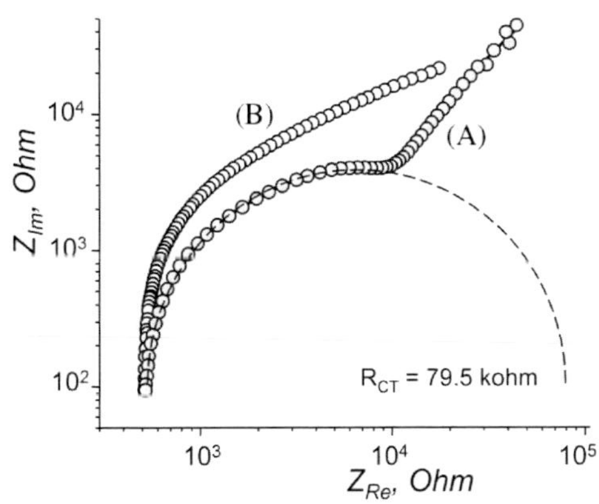

Figure 13.6. Nyquist plot for the reduction of chloranil (3 mmol L^{-1}) in AN/0.1 M $TBAPF_6$. (A) Freshly immobilized C≡C–Me layer on gold. (B) Same electrode after the removal of the grafted layer. $E = -0.16$ V, $\Delta E = 10$ mV, frequency range from 0.306 to 43 mHz [29].

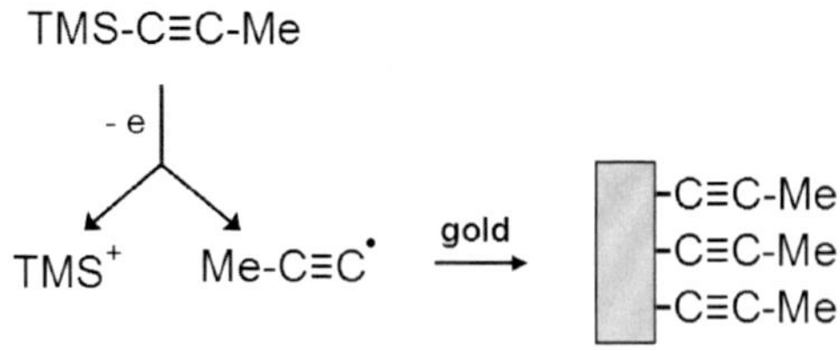

Scheme 13.10. A novel process for the attachment of acethylenic moieties at gold [29].

13.5.3. *The versatile use of TMS–CH$_2$–I under cathodic or anodic electron transfer*

The interest in this reagent is underlined by Scheme 13.11. The use of two kinds of differently polarized electrodes (anode vs. cathode) may lead to two distinctly very different scission patterns, differentiating the valuable functionalizations of gold (and platinum as well). Quite surprisingly, for the chemist (and for the reader), two different sequences (changing the polarity of the working electrode, which can either first act as a cathode and then as an anode, or first act as an anode and then as a cathode) produce similar layer of methylene-centered transient radicals whose fate depends on their environment. Obviously, such a poly-radical interface (as well as both precursor interfaces) could be considered as a valuable platform for further gold modification. On the other hand, the facile use of this precursor and the clarity of the anodic and cathodic results should encourage supplementary works in this field [29].

Additionally, certain possibilities with TMS–C≡C–CH$_2$Br (commercial) to build [30] different layers associated to gold were disclosed. Beside graphene flakes, elaborations of double layers are possible. Notice that the unsaturated adduct form is always of interest after immobilization (Scheme 13.12) as exemplified in the case of C_{60}.

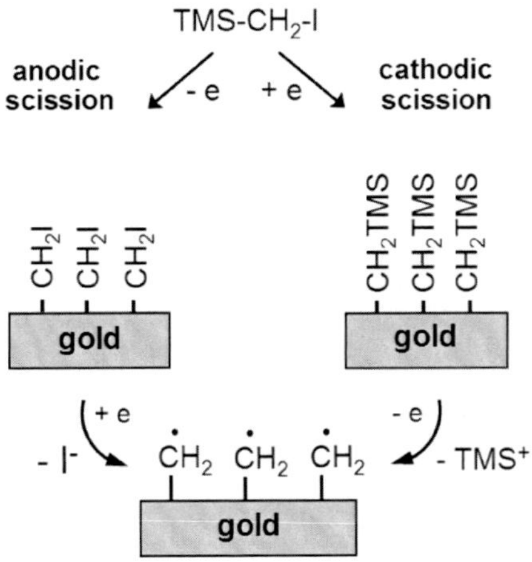

Scheme 13.11. Immobilization at gold (through a one-electron bond scission) of versatile reagents such as TMS–CH_2–I by mean of electron transfer(s) [29].

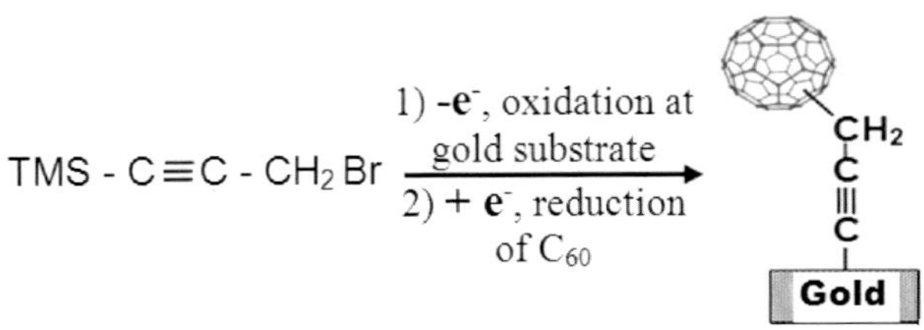

Scheme 13.12. An example of a double cleavage/immobilization with a TMS–R–X compound. The experimental conditions are: use of DMF containing $TMABF_4$. The oxidation is achieved at smooth gold, but smooth GC could be used as well. Sonication and rinsing after immobilization. The reduction of an acceptor (or any radicalophilic system) can be achieved and plays the role of homogenous reducer toward the C–Br bond. Therefore, the *reduced* form of C_{60} can be immobilized. Notice that the process essentially lies on the grafted propargylic bromide moiety (obviously catalytically increased by the vicinity of gold). The propargylic link (or pillar) could be seen as an expander, and similar to a chemical "caryatid" between two different solid (or compact) layers. The extension [30] to multi-layers (possibly considered as host structures) has not been carried out so far. A chemical interpretation of Greek architecture is attempted in Figure 13.7.

Figure 13.7. A representation of a "propargylic pillar" considered as a caryatid to separate two very different solid layers thanks to the electro-chemical use of several TMS–R–X compounds (see text, Ref. [30]).

13.6. Conclusion

Anodic electro-chemistry of organo-trimethylsilanes (R–TMS) either at carbon (taking advantage of its positive doping) or at gold and platinum, offers a wide range of new possibilities for modifying solid interfaces. The high versatility of the TMS units as a potential leaving group in anodic oxidation has to be underlined. On the other hand, the use of organic halides (activated or not) associated with TMS may provide highly valuable openings in the field of surface substitutions and free radical chemistry. This contribution does not cover all the results in hand as it is clearly intended to be essentially an introduction to this promising field.

References

[1] V. Jouikov, J. Simonet, *Langmuir* **28/1** (2012) 931.

[2] R.L. McCreery, *Chem. Rev.* **108**(7) (2008) 2646.

[3] G.M. Jenkins, K. Kawamura, *Nature* **231** (1971) 175.

[4] P. Bernier, S. Lefrant, *Le carbone dans tous ses états*, Gordon and Breach Science Publishers, Amsterdam (1997).

[5] H. Maeda, Y. Yamauchi, M. Hosoe, T.-X. Li, E. Kasamatsu, H. Ohmori, *Chem. Pharm. Bull.* **42** (1994) 1870.

[6] B. Barbier, J. Pinson, G. Desarmot, M. Sanchez, *J. Electrochem. Soc.* **137** (1990) 1757.

[7] C.P. Andrieux, F. Gonzalez, J.-M. Saveant, *J. Am. Chem. Soc.* **119** (1997) 4292.

[8] J. Pinson, F. Podvorica, *Chem. Soc. Rev.* **34** (2005) 429.

[9] K.H. Vase, A.H. Holm, S.U. Pedersen, K. Daasbjerg, *Langmuir* **21** (2005) 8085.

[10] K.H. Vase, A.H. Holm, K. Norrman, S.U. Pedersen, K. Daasbjerg, *Langmuir* **24** (2008) 182.

[11] V. Jouikov, J. Simonet, *Electrochem. Commun.* **12/3** (2010) 331.

[12] V. Jouikov, J. Simonet, *J. Electrochem. Commun.* **15/1** (2012) 93.

[13] V. Jouikov, J. Simonet, *ChemPlusChem* **78** (2013) 70.

[14] P. Poizot, J. Simonet, *Electrochem. Commun.* **23** (2012) 137.

[15] P.D. Lickiss, in *The Chemistry of Organic Silicon Compounds*, eds. Z. Rappoport and Y. Apeloig, Vol. 2, John Wiley and Sons, Chichester, p. 575 (1998).

[16] I.S. Arkhrem, N.M. Chistovalova, D.V. Avetisyan, V.F. Sizoi, M.E. Vol'pin, *Izv. Akad. Nauk SSR, Ser. Khim.* **8** (1987) 1868.

[17] F. Takashi, H. Toshikazu, O. Yoshiki, *Tetrahedron Lett.* **34/35** (1993) 5601.

[18] M. Ochiai, E. Fujita, M. Arimoto, H. Yamaguchi, *Chem. Pharm. Bull.* **32** (1984) 887.

[19] M. Bordeau, P. Villeneuve, B. Bennetau, J. Dunogues, *J. Organomet. Chem.* **331/2** (1987) 169.

[20] A. Ricci, M. Fiorenza, M.A. Grifagni, G. Bartolini, G. Seconi, *Tetrahedron Lett.* **23/48** (1982) 5079.

particular, GC should be considered as a cheap electrode material to perform analyses and build modified interfaces. During the last three decades, despite GC being often considered by electro-chemists as an "inert" substrate, various developments regarding the addition of free radicals (chemically or electro-chemically generated) offered new opportunities, leading to reconsider the specific reactivity of this type of carbonaceous substrate. That was, in particular, the case of aryl radicals generated by reduction of related diazonium cations. Thus, the one-electron reduction of diazonium cations remains an efficient tool for forming chemically modified surfaces [8–15]. Unfortunately, this use is more or less limited to aryl radicals as an almost obligatory pathway.

Nowadays, GC can be considered as a non-graphitizing carbon with a structure that has been for long a subject of controversy. However, it now seems accepted that different forms of GCs are partially made of pure sp^2 bonded carbon atoms possibly organized into sp^2 clusters/ribbons of graphite-like sheets interspersed with other regions of amorphous carbon [16] or into a fullerene-related microstructure [17]. Consequently, GC could be perceived as a pudding-like conducting material gathering graphene-like structures and more or less stable toward oxidation, especially on aging, which leads to slow superficial transformations into aromatic ketones and acids [18]. Consequently, almost all of the electro-chemistry specific to graphite (in particular on HOPG [19, 20]) could be applied into a certain extent to GC.

Recent studies have shown that the electro-chemical cleavage of activated organic halides like benzylic bromides [21–23], 9-bromofluorene [24], or haloacetonitriles [25] lead to free radical grafting onto graphite or GC. On the catalytic point of view, a very simple approach can be developed for producing free alkyl radicals $R^\bullet$ by means of the electro-chemical reduction of RXs. A simple and efficient method uses silver-based electrodes at moderate negative potentials [26] or silver nanoparticles [27] that allow to open up new possibilities in the genesis of free alkyl radicals within a potential range where they are not reduced [28]. The possible addition of such alkyl radicals toward reactive sp^2 carbons as impurities belonging to GC substrates deserves to be more deeply studied. Grafting mechanisms of simple primary alkyl chains on macro GC cathodes is investigated here and compared to the electro-chemical reactivity of flexible graphite used under similar conditions.

14.2. Generation of Free Alkyl Radicals at GC Surfaces

14.2.1. *Methodology*

The general approach to modify silver using palladium(II) and the corresponding characterizations has been previously detailed [29]. Briefly, the procedure is based on the contact of metallic silver powder with a fresh acidic bath of palladium sulfate, making the displacement reaction of Ag^0 by Pd^{II} possible with alloy formation.

The GC samples used were purchased from Tokai Carbon Co. (code: GC Rod, for building GC stationary electrodes) and from Carbone Lorraine (large plates VD 1500 and 2500, thickness: 3 mm). All GC electrodes, when used as substrate in voltammetry, had an apparent surface area of 0.8 mm^2 whereas GC substrates used for surface modifications had an apparent surface area ranging from 0.5 to 2 cm^2. Prior to being used, the electrodes were twice rinsed with water, alcohol, and then acetone before drying.

The general strategy focused in this report for generating free alkyl radicals at different kinds of carbonaceous substrates derives from recent reports [29–32]. Silver-based layers (doped with Pd or made of electro-generated Ag nanoparticles) were essentially used and applied owing to their efficient catalytic effect, especially toward the cleavage of primary RXs. Alkyl iodides (RIs) and alkyl bromides (RBrs) are readily reduced at moderate cathodic potentials according to an one-electron step, for example, with formation of the corresponding free radicals $R^\bullet$. For instance, the main reduction step of 1-iodobutane at the Ag–Pd interface illustrates the catalytic effect by a large potential shift in comparison with a pristine GC electrode ($\Delta E > 0.9$ V). The appearance of a one-electron peak in voltammetry supports the transient formation of the free alkyl radical. A further reduction step is assigned to the reduction of this radical (Figure 14.1). The peak potential observed for this second step,

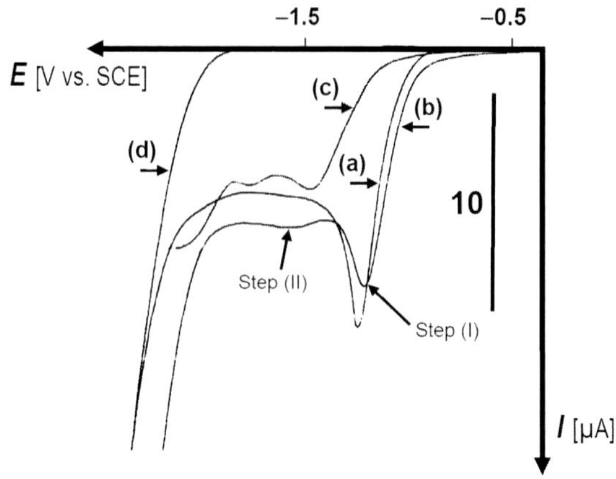

Figure 14.1. Voltammetries of 1-iodobutane (concentration: 12 mmol L^{-1}) in 0.1 M $TBAPF_6$/propylene carbonate recorded at 50 mV s^{-1} (surface area of used electrodes: 0.8 cm^2). (a) Reduction of 1-iodobutane at Ag–Pd; (b) at GC covered with Ag–Pd (thickness: 0.1 µm); (c) at GC covered with galvanostatically deposited Ag (thickness: 0.1 µm); (d) Reduction of the electrolyte at an Ag–Pd electrode [36].

which is lower than −1.5 V, makes it possible to estimate that such radicals [26] are cathodically stable at the potential of their formation. It should be underlined, however, that the step corresponding to the reduction of the radical is, in general, relatively small since consuming reactions such as radical–radical coupling reactions may take place.

Thus, the presence of palladium was found to dope the catalysis process although the electro-active form appears to be mainly R-Ag^+, X^- species [26]. Scheme 14.1 traces the proposed reduction mechanism of RX compounds (oxidative addition of Ag° inside the carbon–halogen bond) at a thin layer of silver (or modified silver with palladium) [26, 27]. This underlines the possible pathway for $R^\bullet$ radicals to react with sp^2 carbon at GC surface (dashed line). To generate such free alkyl radicals at a pristine carbonaceous electrode (free of silver-based coating) and then study their possible reactivity with GC surface that depends on their electro-chemical stability [28], it appeared suitable to use a kind of "fluidized

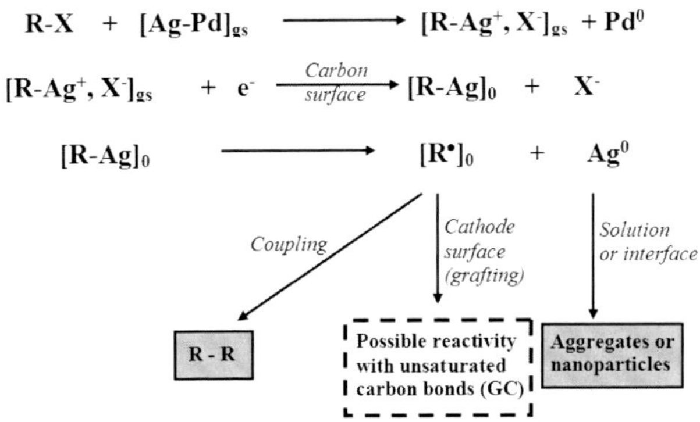

Scheme 14.1. Proposal for the formation of free alkyl radicals at a carbon surface via a process using a suspension of grains of Ag–Pd (noted *gs*) possibly discharged by contact at an electrified solid conductor (here GC). Produced Ag_0 could possibly lead to nanoaggregates (i.e., NP) staying at the GC surface and then eliminated by ultrasonic rinsing [36].

electrode" based on an Ag/Ag–Pd powder suspension placed in the cathodic compartment and polarized through an inactive central cathode (see experimental section for details). It is worth noting that this technique has been recently developed to form silver nanoparticles (denoted AgNPs) at a wide series of polarized substrates in the course of RX reduction [27].

This particular process, which can be viewed as a large three-dimensional (3D) metallic electrode, is practically able to generate free radicals in the vicinity of a pristine carbonaceous electrode concomitantly with the formation of AgNP as schematically shown in Scheme 14.2.

In such an extent, the as-produced radicals may react with unsaturated sp^2 carbon bonds. A large series of primary RXs was electro-chemically cleaved (one-electron process) when the method described above was applied. For that, many RIs and a few bromides were used for their possible capability to readily form radicals together with concomitant by-reactions such as grafting, polymerization, or ablation when in contact with carbon substrates (Scheme 14.2). Hereafter, the reported results corresponding to

Table 14.1. Evidence for the unexpected side-products obtained from RXs in 0.1 M $TBABF_4$/DMF. Electrolyses performed at $E_c = -1.3$ V vs. SCE of GC coated with an Ag/Ag–Pd layer. Identification of reduction products by GC–MS. Experimental details given in Ref. [26]).

RX	Main side-products (alkanes)	Other products
1-Bromooctane	C_{17} (12%)	—
1-Bromodecane	C_{14} (2%), C_{16} (15%)	Olefin in C_{16}
1-Bromohexadecane	C_{18} (4%)	—
1-Bromopentane/ 1-Bromooctane (1:1)	C_{14} (2%), C_{24} (3%), C_{25} (4%) and C_{26} (4%)	—

0.1 mg cm^{-2}. The apparent pitting of the carbon surface by free alkyl radicals is expected to be assisted by the concomitant deposit of AgNP onto GC. At this stage, let us recall that previous analyses [26] of reaction products formed during reduction of RXs at GC (covered with a thin layer of Ag/Ag–Pd) well support a real reactivity of carbon atoms of the cathode toward electro-generated free alkyl radicals.

Actually, it is worth revealing that the catalytic reduction process of RBr and RI compounds (where free radicals as transients are clearly involved) led, besides coupling products obtained in high yield, to unexpected by-products characterized by extra carbon particles added to the starting carbon chain R. Table 14.1 (dealing with electrolyses of several RBr compounds in dimethylformamide (DMF) as solvent [26]) underlines evidence of the participation of electrode material in a complex electrode process (it could be that the catalytic process has something in common with a form of microcracking implying both alkyl radicals, solvent, and some carbon atoms of the material) for leading to a rather complex pathway).

Similar data were also obtained with RIs. However, the formation of numerous side-products remains in limited amounts. When reduction of RIs was performed in the presence of silver catalyst in PC or DMF as a solvent, spit-like morphology is observed in certain locations (Figure 14.3), although these cavities are not well defined

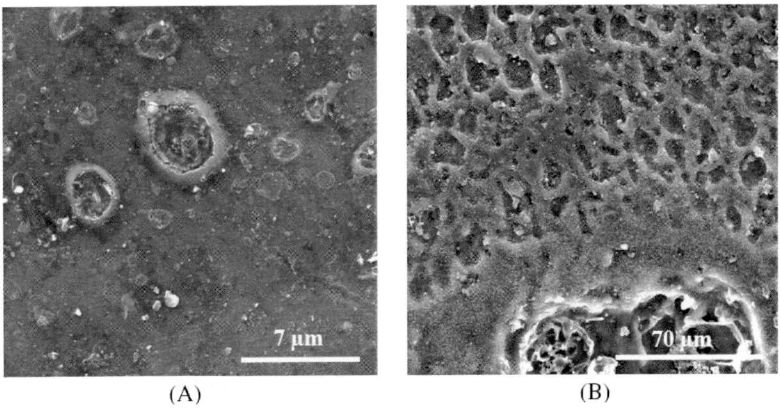

Figure 14.3. SEM images of pristine GC used as electrode material after electrochemical reduction at $E_c = -1.65$ V/SCE ($Q = 5$ C cm^{-2}) in contact with a fluidized electrode system based on Ag particles only. (A) 1-iodooctane (12 mmol L^{-1}) in 0.1 M $TBABF_4$/PC. (B) 1,3-Diiodopropane (9.5 mmol L^{-1}) in 0.1 M $TBABF_4$/DMF [36].

compared to those already exhibited in Figure 14.2. Thus, SEM images could reveal a kind of surface swelling, being possibly a phase that just precedes cavities formation. This behavior is particularly visible in the case of the 1-iodooctane reduction when making a comparison between Figures 14.2 and 14.3. One may also notice the presence of bumps spread over the surface after electrolysis. 1,3-Diodopropane in PC (Figure 14.3(B)) behaves similarly. It is worth noting that the use of silver powder modified with palladium appears to act by reinforcing the catalytic action and obviously efficiently favors the C–Br bond cleavage (when appropriate potentials are applied to the surface).

SEM characterizations provide information on the transformations of GC substrates after electrolysis, with the appearance of diggings and large modified zones looking like canyons, apparently in association again with a primary swelling effect of the carbon surface. For example, Figure 14.4 exhibits well-patterned zones after 1-bromopentane electrolysis ($E_c = -1.35$ V/SCE in 0.1 M $TBABF_4$/DMF) at a potential that corresponds to that of radical formation. The process could start with a surface swelling (grafting/addition

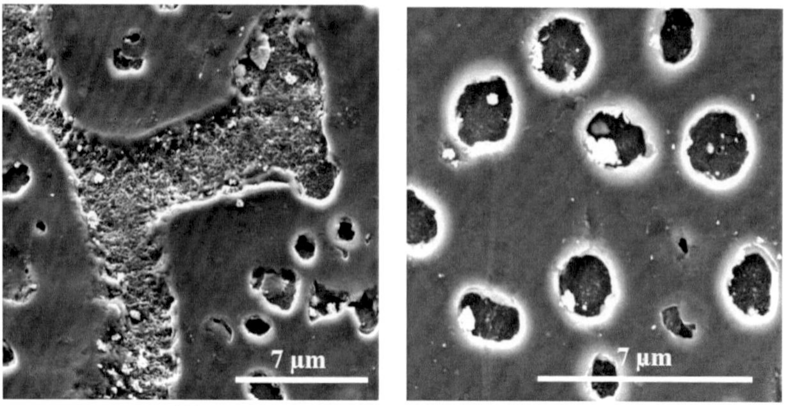

Figure 14.4. Typical SEM (A) and (B) images focused on a modified zone of a GC pristine substrate, focusing on points of initial corrosion extended afterward on much larger canyons. Case of the electro-chemical reduction of 1-bromopentane (concentration: 9.5 mmol L^{-1}) in 0.1 M $TBABF_4$/DMF at $E_c = -1.35$ V/SCE ($Q = 2$ C cm^{-2}). Catalyst: fluidized electrode system based on Ag/Ag–Pd particles. Mass loss of the electrode ~0.12 mg cm^{-2} per passed coulomb [38].

reactions) while further radical addition in large quantities induces subsequent stresses concerning the surface to yield to mechanical degradations and diggings that disaggregate the electrode material progressively at random locations (Scheme 14.3). Although the nature of surface reactivity still remains uncertain, it seems to depend on the amount of charge passed through the cell and the nature of the RX. It is also quite obvious that the presence of Ag nanoparticles present at the GC surface may induce radical addition at certain locations followed by a digging effect in the vicinity of AgNP deposits, as shown in Figure 14.4.

These widespread surface degradations can be interestingly exemplified by means of some relevant SEM images. Thus, Figures 14.5 and 14.6 illustrate the fact that the used catalyst and the nature of the RX compound may greatly (but specifically) modify the GC surface. Thus, long-chain RBr compounds, both with

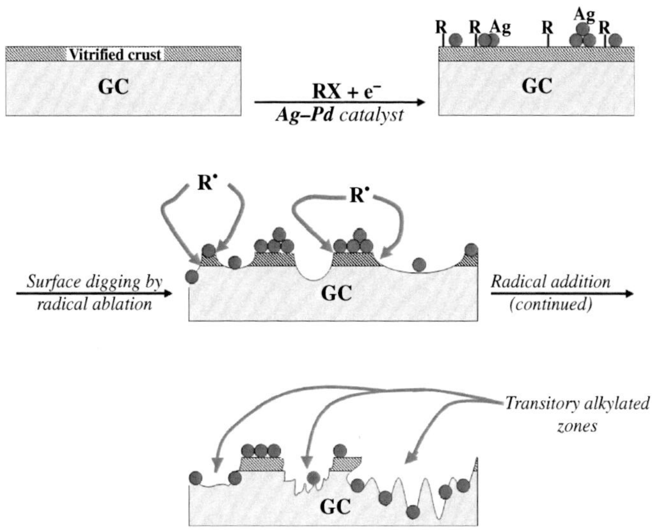

Scheme 14.3. Corrosion of GC vitrified crust by electro-generated alkyl radicals. This proposal would concern GC pristine samples from Carbone Lorraine [38].

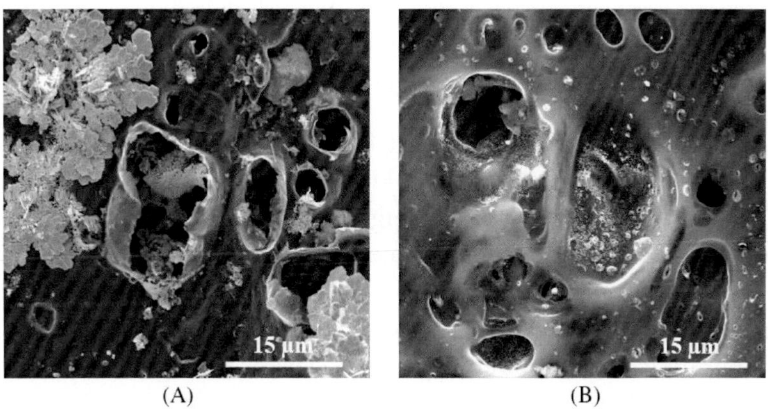

Figure 14.5. Reduction of long-chain RBrs in the presence of silver catalysts at GC surfaces. (A) Reduction of 1-bromodecane in propylene carbonate containing $TBABF_4$. Catalyst: silver NPs. Electricity: 1.2 C cm^{-2}. (B) Reduction of 1-bromododecane at GC under similar conditions that in (A) but in the presence of Ag–Pd. Applied potential: −1.3 V. Amount of electricity: 6 C cm^{-2} [38].

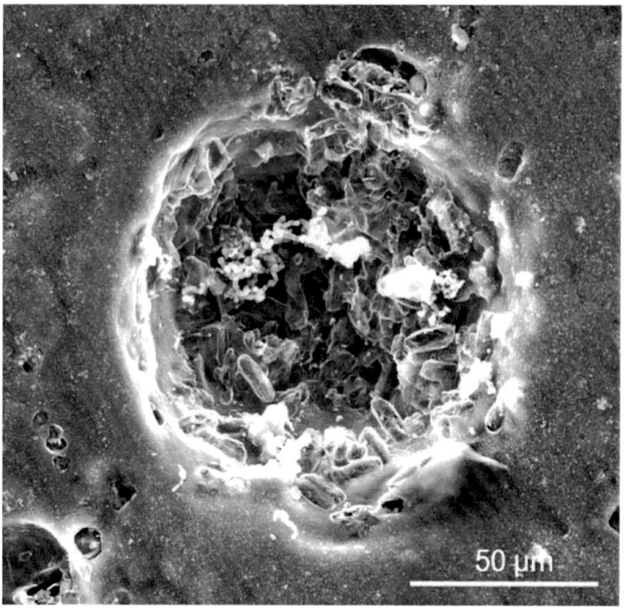

Figure 14.6. Impact and digging of GC surface via the reduction of 1-iodooctane in the presence of Ag–Pd catalyst. Applied potential: −1.4 V. The picture allows to see the real fragmentation of GC inside the crater which leads to the formation of vitreous sticks of almost equal shape and size [38].

Ag (slide (a)) and Ag–Pd (slide (b)), surprisingly lead to a plasticity of the carbonaceous material (a kind of "electro-chemical" melting) for giving, in the GC crust, strange bubbles or excrescences associated with deep digging remanence in all cases of silver at the state of flakes or nanoparticles (Scheme 14.4). When these data are compared to those obtained with short-chain RBrs, it is expected that the pre-eminence of self-assembled mono-layers (SAMs) relative to the adsorption of {alkyl chain–Ag^+} moieties at GC surface and their role in the early stage of the reaction characterized by the presence of AgNPs surrounded by grafted organic matter (white dots). This organic layer should be considered as a source of free alkyl radicals ready to further react with the surface, leading to an apparent and a very profound pitting of carbon.

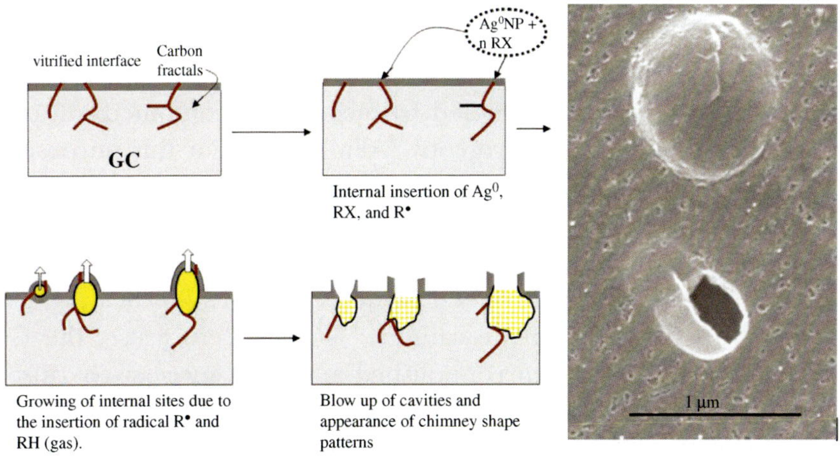

Scheme 14.4. Modification of GC via fractals and concomitant insertion of Ag–Pd particles and RX. Generation of cavities and protuberances explained by Figure 14.5. On the right side is seen a picture of two cavities, with one in formation [38].

Very curiously, the use of certain 1-iodoalkanes and α,ω-diiodoakanes (such as 1,3-diiodopropane) offers quite different actions with the formation of spectacular cavities, as illustrated in Figure 14.6. The presence of such circular cauldrons, showing some sorts of chain at its center, quite certainly implies that silver is very reactive. With time, the digging of GC spreads out while one can still observe the presence of AgNPs in the cavities concomitantly with other carbonaceous debris. However, it remains risky to connect the charge passed to an alteration degree because radicals also couple very quickly and, for instance, some amount of reduction product such as dimers often insoluble in the cathodic bulk (in the case of long aliphatic chains > 8 carbons) stays at the carbon surface. Such surface reactivity being uncommon and subtle, a series of specific experiments were carried out in an attempt to support this suggested reactivity mode. It is quite probable, however, that the generation of bubbles is due to gaseous materials generated in the course of the electrolysis, inside the GC (as shown in Scheme 14.4).

14.2.3. *Preliminary step: A radical grafting at GC?*

The incorporation of radicals at the GC surface during the early stages of the reaction is studied by means of a combined infrared (IR) and atomic force microscopy (AFM) study. For this purpose, a specific RBr containing a well-defined IR-active group (–CN) such as 4-bromobutyronitrile was used (Figure 14.7). The reduction was performed in 0.1 M $TBABF_4$/DMF in the presence of a fluidized electrode system based on Ag/Ag–Pd particles ($E_c = -1.35$ V/SCE, $Q = 0.3$ F mol^{-1}, surface area of the cathode: 4 cm^2). Before use, the GC substrate was carefully mirror-polished and then successively rinsed with water, ethanol and acetone before drying prior to being used. IR spectra of the GC substrate before and after electrolysis of 4-bromobutyronitrile in the 800–1800 cm^{-1} and 2700–3400 cm^{-1} range were obtained: Peaks or large bands centered around 1050, 1500, and 1650 cm^{-1} are usually attributed to carbon–nitrogen and carbon–carbon bonds in sp^2 and sp^3 configuration [37]. The major peak centered around 2855 and 2915 cm^{-1} can be attributed to the symmetric and asymmetric stretching mode of the CH_2 bond with the C

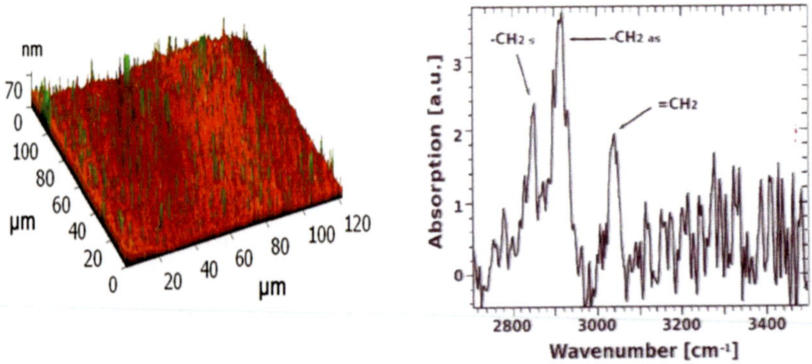

(A) (B)

Figure 14.7. AFM image of GC surface after electrolysis of 4-bromobutyronitrile in 0.1 M $TBABF_4$/DMF with a limited charge passed ($E_c = -1.35$ V/SCE, $Q = 0.3$ F mol^{-1}, surface area of the cathode: 4 cm^2). (A) AFM of the deposit. (B) FTIR spectrum of the surface showing the deposition of alkyl chains [36].

atom in the sp^3 hybridized state. A peak around 3040 cm^{-1} is due to the presence of CH_2 with the C atom in the sp^2 hybridized state. These peaks indicate the presence of carbon chains on the surface of the GC after electrolysis of 4-bromobutyronitrile.

Regarding the AFM study of the substrate, Figure 14.7 reveals a new material with a growth on most of the surface until heights up to 50 nm, which are quite certainly due to the addition of free radicals, not only onto the surface, but also onto the nitrile moieties themselves. This addition reaction is quite similar to the formation of multi-layers already described with aromatic radicals obtained by reduction of aryl-diazonium salts [32]. We may also stress that the amount of electricity involved is too small to start digging the GC, as mentioned above. The background of AFM image colored in brown/orange corresponding to polishing scratches are also observed on the pristine surface, having an average depth around 15 nm. These observations do confirm a constant reactivity of free radicals with carbon during the electrolysis process in the presence of Ag/Ag–Pd catalysts.

14.3. Polarized Graphite as a Trapping Material for Alkyl Radicals

The results reported above aim to support the reactivity of present sp^2 carbons quite certainly in abundance on GC surfaces. Additionally, it appears important to explore the activity of $R^•$ toward other sp^2 carbon-based substrates known to be, in essence, more reactive. The use of materials derived from graphite was also tested using a Grafoil®-based electrode within a potential range where the suspected insertion phenomenon of graphite with tetraalkylammonium salts is much disregarded (i.e., $E_c > -1.7$ V vs. SCE). Cathodic degradation of TAA^+ cations at very reducing E_c with possible liberation of alkyl radicals is an important reaction, but not discussed in this chapter. Radical reactivity has been recently observed with benzyl and fluorenyl radicals at HOPG [22, 24] obtained from electron transfer onto corresponding bromides (both commercially available). Under the experimental conditions used for the process developed hereafter (in the presence of Ag–Pd catalyst), weakly activated organic halides (such as RBr compounds) can react with graphite via transient free radicals. This reaction can be considered as a kind of radical trapping by graphite. The reaction is easy and applicable to natural graphite and under the conditions of an electro-chemical exfoliation to supported graphene deposited onto GC, silver, copper and gold (see, in particular, Chapter 15).

In this presentation, the reduction of several RXs is achieved in DMF and PC containing $TBABF_4$ or $TBAPF_6$ in the presence of Ag–Pd as a catalyst. For this, organic halides such as 1-iodophenylpropane, 1-iodooctane, 1,3-diodopropane, 4-butyronitrile, benzylbromide and 1,8-dibromooctane were reduced at $E > -1.6$ V. With relatively low amounts of electricity, impressive formations of localized hole in graphite were observed (Figure 14.8). These profound and quite unexpected morphology changes that especially concern virgin Grafoil® surface could be provoked by the diffusion of free radicals inside the graphite with a kind of "radical exfoliation" (Figure 14.8). A digging process based on an efficient polyalkylation of graphite

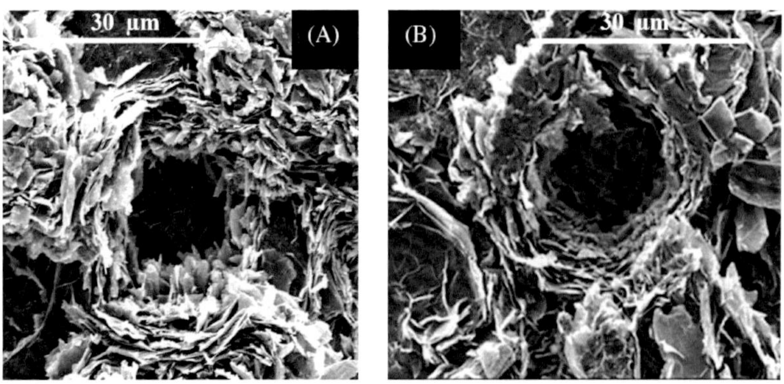

Figure 14.8. Two examples of localized radical ablation on graphite characterized by SEM images. (A) Electro-chemical reduction of 1,3-diiodopropane (concentration: 15 mmol L^{-1}) in 0.1 M $TBABF_4$/DMF at $E_c = -1.05$ V/SCE ($Q = 8$ C cm^{-2}) until total completion using a fluidized electrode system based on Ag/Ag–Pd particles. (B) Electro-chemical reduction in similar conditions of 4-bromobutyronitrile (concentration: 13 mmol L^{-1}) at $E_c = -1.15$ V/SCE ($Q = 5$ C cm^{-2}). In both experiments, samples were shortly sonicated after electrolysis to eliminate a large amount of AgNPs and graphite fragments [36].

planes leads to boost a boosting of graphite around initial points of radicalar corrosion. The digging process is, in fact, very similar to the one shown in Scheme 14.3 with GC (but at the nanometric scale). The piercing effect throughout graphite planes happens around deposited AgNPs according to a catalytic chain organized in the region of a catalyst NP, which is constantly regenerated. Furthermore, it is worth to imagine that the presence of silver nanoparticles staying in cavities also contributes to extend progressively not only their size (lateral action) but also their piercing contribution. Thus, a spectacular exfoliation process around cavities may take place owing to the lateral grafting of bulky radicals (butyronitrile or benzyl radicals) or via a bi-radical reaction with of α,ω-dihaloalkanes (case of 1,3-diiodopropane and 1,8-dibromooctane), which may induce huge deformations of concerned materials. Possibly, demolition of the graphite structure around holes supports the possible role of AgNPs by diffusion inside the layers with more profound devastations (until digging of tunnels).

Finally, it has been found that polyaromatic hydrocarbons such as pyrene, anthracene, naphthalene or acenaphthylene could also react in the same way as solid carbon under the same experimental conditions (same medium, presence of the silver-based catalyst, and so on). The difference between redox catalysis (anion radical considered as reducing species of RIs) and catalysis by Ag–Pd or AgNPs toward RIs (then leading to free alkyl radicals able to add to π-acceptors) has been discussed in recent contributions [33–36]. Importantly, it was observed here that free alkyl radicals can add not only onto graphite, but also onto fullerenes and other aromatic species. The implication of polyaromatic hydrocarbons as redox catalysts (reduced at potential significantly more negative ($E_c < -1.6$ V vs. SCE) than those for RIs) was achieved. Under such experimental conditions, reductions specifically based on redox catalysis processes were totally excluded. Additionally, analytical results have shown that alkyl radicals derived principally from iodomethane, 1-iodopropane, 1-iodododecane, 2-bromo-2-methylpropane, 1,4-diiodobutane, and 1,6-dibromohexane led in an interesting way to mono or polyadditions of RXs as well as to bridged forms and polymers (case of

Table 14.2. Effective reactivity of reduced π-acceptors (in order to mimic the reactivity of graphite) toward electro-generated free radicals obtained in the presence of Ag–Pd powders (grain size: 250 μm) in 0.1 M $TBABF_4$/DMF ($E_c < -1.6$ V vs. SCE Use of a GC cathode). The alkylation yield corresponds to an electricity amount of 0.5 F per mol of π-acceptor. Products were identified by GC–MS [36].

π-acceptor	RX	Catalyst	Products (ratio yield)
Pyrene	2-bromo-2-methylpropane	Ag–Pd	1-*t*Bu-pyrene 40%/1-*t*Bu-dihydropyrene 2.4%/Poly*t*Bu-pyrenes ~3%
Phenanthrene	1-iodopropane	Ag–Pd	Ethylphenanthrene 2%/diethylphenanthrene 1.3%
Anthracene	1-iodopropane	Ag–Pd	9-propylanthracene 0.1%/dipropylanthracene 1.1%/tetrapropylanthracene 0.9 %
Acenaphthylene	Iodomethane	Ag–Pd	Mono-methylacenaphtylene 32%/dimethylacenaphtylenes 2.1%/trimethylacenaphtylenes 4.2%

α,ω-dihaloalkanes). A few significant examples are reported in Table 14.2. Thus, these experiments well confirm the reactivity of unsaturated impurities inserted in carbon and therefore support (even with low electricity amounts) carbon digging by a progressive and large alkylation of the polyaromatic material until a spectacular ablation via swelling of graphite.

References

[1] Design and control of structure of advanced carbon materials for enhanced performance, *Proceedings of the NATO Advanced Study Institute*, eds. B. Rand, S.P. Appleyard and M.F. Yardim, NATO Science Series E, Kluwer Academic Publishers, Berlin (2001).

[2] H.W. Kroto, J.R. Heath, S.C.O'Brien, R.F. Curl, R.E. Smalley, *Nature* **318** (1985) 162.

[3] S. Iijima, *Nature* **354** (1991) 56.

[4] O.V. Prezhdo, P.V. Kamat, G.C. Schatz. *J Phys Chem. C.* **115**(8) (2011) 3195.
[5] K. Kinoshita, K. Zaghib, *J. Power Sources* **110**(2) (2002) 416, and references cited therein.
[6] P. Bernier, S. Lefrant, *Le carbone dans tous ses etats*, Gordon and Breach Science Publishers, Amsterdam (1997), and references cited therein.
[7] R.L. McCreery, *Chem. Rev.* **108**(7) (2008) 2646.
[8] M. Delamar, R. Hitmi, J. Pinson, J.M. Savéant, *J. Am. Chem. Soc.* **114**(14) (1992) 5883.
[9] J. Pinson, F. Podvorica, *Chem. Soc. Rev.* **34**(5) (2005) 429.
[10] A.J. Downard, D.J. Garrett, E.S.Q. Tan, *Langmuir* **22**(25) (2006) 10736.
[11] J. Lyskawa, D. Bélanger, *Chem. Mater.* **18**(20) (2006) 55.
[12] L.T. Nielsen, K.H. Vase, M. Dong, F. Besenbacher, S.U. Petersen, K. Daasbjerg, *J. Am. Chem. Soc.* **129**(7) (2007) 1888.
[13] K. Malmos, M. Dong, S. Pillai, P. Kingshott, F. Besenbacher, S.U. Pedersen, *et al. J. Am. Chem. Soc.* **131**(13) (2009) 4928.
[14] J.M. Noël, D. Zigah, J. Simonet, P. Hapiot, *Langmuir,* **26**(10) (2010) 7638.
[15] Y.R. Leroux, H. Fei, J.M. Noël, C. Roux, P. Hapiot, *J. Am. Chem. Soc.* **132**(40) (2010) 14039.
[16] N.M.D. Brown, N. Cui, A. McKinley, *Appl. Surf. Sci.* **133**(3) (1998) 157, and references cited therein.
[17] P.J.F. Harris, *Philos. Mag.* **84**(29) (2004) 3159.
[18] A. Gennaro, A.A. Isse, C. L. Bianchi, P.R. Mussini, M. Rossi, *Electrochem. Commun.* **11**(10) (2009) 1932.
[19] J.O. Besenhard, H. Möhwald, J.J. Nickl, *Carbon* **18**(6) (1980) 399.
[20] J.O. Besenhard, *Carbon* **14**(2) (1976) 111.
[21] V. Jouikov, J. Simonet, *Electrochem. Commun.* **12**(3) (2010) 331.
[22] F. Hui, J.M. Noël, P. Poizot, P. Hapiot, J. Simonet, *Langmuir* **27**(8) (2011) 5119.
[23] V. Jouikov, J. Simonet, *Electrochem. Commun.* **12**(10) (2010) 1262.
[24] J. Simonet, V. Jouikov, *Electrochem. Commun.* **13**(3) (2011) 254.
[25] C. Combellas, F. Kanoufi, Z. Osman, J. Pinson, A. Adenier, G. Hallais, *Electrochim. Acta* **56**(3) (2011) 1476.

[26] P. Poizot, J. Simonet, *Electrochim. Acta* **56**(1) (2010) 15.
[27] J. Simonet, *J. Appl. Electrochem.* **39**(10) (2009) 1625.
[28] K. Daasbjerg, S.U. Pedersen, H. Lund, *General Aspects of the Chemistry of Radicals*, ed. Z. B. Alfassi, p. 385 (1999).
[29] P. Poizot, L. Laffont-Dantras, J. Simonet, *J. Electroanal. Chem.* **622**(2) (2008) 204.
[30] J. Simonet, *Electrochem Commun.* **7**(1) (2005) 74.
[31] M.S. Yazici, D. Krassowski, J. Prakash, *J. Power Sources* **141**(1) (2005) 171.
[32] D. Bélanger, J. Pinson, *Chem. Soc. Rev.* **34** (2005) 429.
[33] P. Poizot, L. Laffont-Dantras, J. Simonet, *J. Electroanal. Chem.* **624**(1-2) (2008) 52.
[34] J. Simonet, *Electrochem. Comm.* **13**(2) (2011) 107.
[35] V. Jouikov, J. Simonet, *Electrochem. Commun.* **11**(9) (2009) 1785.
[36] P. Poizot, O. Durand-Drouhin, M. Lejeune, J. Simonet, *Carbon* **50** (2012) 73.
[37] M. Benlahsen, M. Therasse, *Carbon* **42**(11) (2004) 2255.
[38] J. Simonet, P. Poizot, Unpublished works.

Chapter 15

Silver–Graphene and Gold–Graphene Electrodes: Two New Catalytic Materials

15.1. Introduction

The cathodic polarization of smooth metal surfaces (especially silver and gold) sheltered with graphene, which is easily built by means of the exfoliation process displayed in Chapter 11, is described. In solvents such as dimethylformamide (DMF) or acetonitrile (ACN) containing tetraalkylammonium (TAAX) salts, a spectacular exfoliation process takes place, which leads to the almost total release of graphite particles. This cathodic ablation provoked by insertion of bulky TAA^+ cations in the material goes practically to the last (few) layer(s) of graphene in immediate contact with the metal. Other efficient methods to form such metal–graphene interfaces may exist: let us look at the deposit of graphite micro-flakes (beforehand pre-exfoliated at a mercury pool at −2 V) followed by rubbing them onto microelectrodes. These as-prepared metal/graphite surfaces are astonishingly stable and allow quite reproducible voltammetric responses linked to the electro-chemical doping of a very few remaining graphene layers. Their nature and associated charge currents depend essentially on the size of the employed TAA^+ cations. The amounts of electricity involved in global reduction/oxidation processes are found to be stable and obviously assigned to the doping of graphene layers attached to Ag or Au. The average coverage based on the hypothetical TAA^+ mono-layer formation was estimated to be $<10^{-9}$ mol cm^{-2}. Electro-chemical doped Ag–graphene and Au–graphene then appear to behave as new catalytic materials both due to the charged graphene (as reducing species) and metals whose catalytic properties were already discussed in this book. Additionally, chemical modifications of carbon layers by grafting of free radicals or by means of nucleophilic substitutions permit to assign these materials with catalytic properties and allow the formation of new modified surfaces.

15.2. Ag–Graphene Surfaces

Glassy carbon (GC) is typically used as a convenient and popular electrode material appropriate to the cathodic domain [1], especially at $E < -1.8$ V in polar organic aprotic solvents. Quite recently, its structure was discussed [2, 3]. On the other hand, smooth silver electrodes displayed interesting possibilities in the field of catalytic scission of carbon–halogen bonds [4–6]. Silver cathodes were applied for de-pollution even in aqueous media [7, 8]. The findings and uses of new heterogeneous electrodes based on silver, copper, and palladium alloys were also recently published [9–11]. Herein, we report a new model of electrode that allies silver and graphene both with their specific catalytic capabilities. As recalled above, silver presents a general capacity for activating the scission of carbon–halogen bonds while graphene — as was recently shown within a wide interval of potentials — combines the possibilities of trapping free radicals by its large unsaturated network and enabling S_N2-type reactions with electrophilic organic species. The preparation of this composite electrode, its characteristics, and its reacting capacities are described together with preliminary experiments on catalysis and grafting of redox groups to highlight potentialities of this material.

Electro-chemical analyses were performed in 0.1 M solutions of TAAX salts dissolved in upper grade DMF. Measurements were performed in three-electrode cells separated with a fritted glass. All potentials are referred to an aqueous Ag/AgCl/$KCl_{(sat)}$. The electrochemical instrumentation has been previously reported [9]. Silver electrodes made of Ag wire (∅ 1 mm), sealed in a Pyrex tube, were carefully polished first with coarse (Struers 500) and then with finer (Struers 1200) silicon carbide polishing paper. Prior to being deposited onto smooth Ag surface, natural graphite (from Ceylon) was partly exfoliated by the electrolytic method [12] as recently used for exfoliation processes at carbon electrodes [13]. Pre-conditioned graphite was wetted with acetone to form slurry prior to the deposition onto Ag. Alternatively, a micro-flake of graphite can be deposited on Ag using a small rubber eraser, which enables also a nice and reproducible deposit.

15.2.1. *Electro-chemical properties of Ag–graphene layers*

For best and reproducible results, natural graphite has been in all cases pre-electrolyzed in DMF (mixed with tetramethylammonium tetrafluoroborate or perchlorate) at a mercury pool at –2.1 V (with a large and efficient stirring) with a passed charge of 12 C per gram of carbon. Under these conditions, one may produce a soft material particularly convenient for an easy coverage of smooth silver interfaces. This process producing an efficient swelling of graphite can be confirmed by SEM analysis. This pre-conditioning procedure appears to be of high interest for getting an almost total exfoliation, finally obtaining of a very thin deposit of graphite at silver. Most conveniently, a slurry of this "expanded" material was deposited onto the polished metal surface. After drying with hot air flow ($T° < 50°C$), the electrode was first electrified by means of voltammetric scans or fixed potential electrolyses ($E < -1.8$ V). One may also achieve the deposit of graphite flakes through a precise contact of a rubber eraser and the electrode. Technically, these electro-chemical methods efficiently induce graphite exfoliation and a consequent release of almost all carbon deposits. Recurrent scans achieved in different TAAX salts in DMF/$TMAClO_4$ allowed visualizing the process already fully described in the case of GC–graphite exfoliation (see Chapter 11).

More specifically with the case of the Ag–graphite system, after a huge cathodic step in the course of the first scan that corresponds to progressive exfoliation, the decay of the current reaches the merging of sharp and reproducible reduction steps (as in Figure 15.1, obtained with $TEABF_4$). Basically, the deposits prepared according to these two procedures show quite reproducible voltammograms on a large number of recurrent scans. For example, TEA^+-based salts (triflate, BF_4^-, and ClO_4^-) exhibit specific

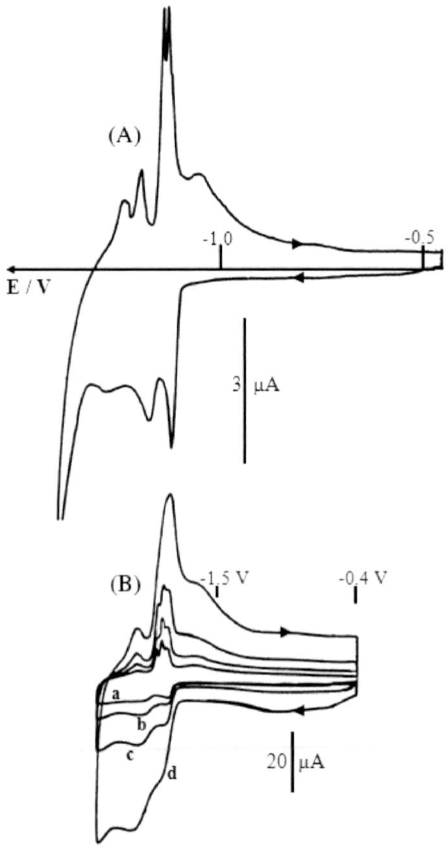

Figure 15.1. Voltammetries of a Ag–graphene electrode in a solution of 0.1 M $TEABF_4$ in DMF. Surface area of smooth silver substrate: 0.8 mm^2. (A) Scan rate: 50 mV s^{-1}. (B) The same electrode at different scan rates from 50 (a) to 500 mV s^{-1} (d) [32].

voltammetric responses for which no noticeable decay of current is observed during repetitive cycles. Below a potential boundary $E > -2.2$ V, three reversible charge processes are observed and each cathodic step is linked with a corresponding anodic discharge step. Currents are proportional to the scan rate (Figure 15.1(B)), which suggests that cathodic and related anodic currents are diffusion controlled (at least up to 1 V s^{-1}).

The current is assigned to the cathodic charge associated with the electron transfer onto the graphite plane that could imply the formation of a cationic layer at its surface. Especially, with TEA^+, the sharp first cathodic step corresponds to an amount of electricity that can be associated with the electron density on graphene layer and to a mono-layer of TEA^+ ions covering reduced graphene. Under these conditions, a simple calculation allows to propose a total amount of electrons (8×10^{-9} elemental charge cm^{-2}) dispersed in graphene. If one admits that the total negative charge is equally dispersed in the graphene plane, it should be equally compensated on both sides of the graphene layer by two symmetrical TEA^+ ion mono-layers; therefore a coverage of 4×10^{-9} ion·cm^{-2} fits well with a very thin coverage of graphene onto the smooth metallic substrate.

By the same way, the charge with $TMeABF_4$ and THexAI gave a coverage of 5×10^{-9} ion cm^{-2} through a similar approach. Figure 15.2 displays a different charge through voltammetric responses where the bulkiness of electrolyte cations play a prominent role in the building of a compact coating against the charged graphene.

When the potential is made more negative, several sharp cathodic peaks merge and could show a progressive filling of the graphene layer by additional electron transfers with total transformation of the TAA^+ layers (possibly on the form of bi- or tri-layers). With TEA^+, this could be supported by the fact that the two first cathode steps are by far the most significant (that also remains true at the end of the discharge). Similar features are observed with $THexA^+$ and $TOctA^+$ although only one step is seen. Very thin layers of graphite deposited on Ag were then formatted in DMF using different TAA^+ salts. As the cation radii increase, the charge intensities of the peaks observed between −1.7 and −2.2 V (at a given scan rate)

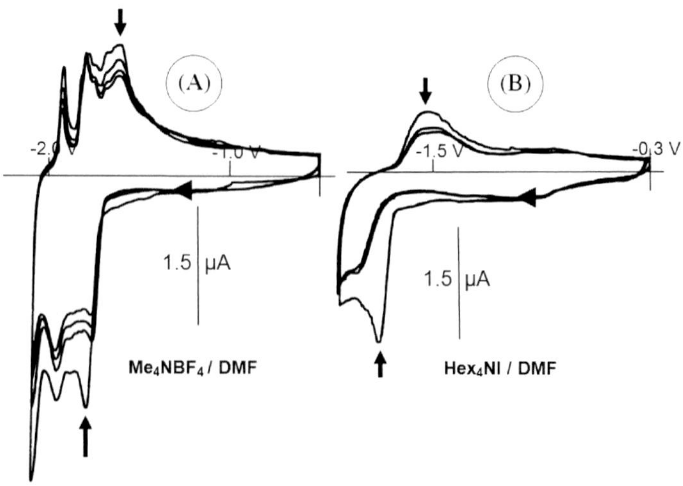

Figure 15.2. The ionic coverage as a function of the bulkiness of TAA^{+} cations associated with the charge–discharge processes in the case of an Ag–graphene electrode. Multi-scan experiments at 50 mV s^{-1}. Arrows underline the first charge–discharge processes [32].

slightly decay while cathodic charging steps (for $E < -2.2$ V) progressively drift to more negative potentials. In moving from TMA^{+} to TBA^{+}, total charges (calculated for apparent surface area 0.8 mm^{2}) decline from 2.5×10^{-3} to 2.0×10^{-3} C cm^{-2}. For very bulky salts like tetra-*n*-hexyl-ammonium and tetra-*n*-octyl-ammonium, the total charge value is even smaller, 1.6×10^{-3} C cm^{-2}. Also, it is significant to stress that with those bulky cations, a sole reversible charge process appears at $E < -2$V. This strongly supports the significance of TAA^{+} size, which is implied in the charge process. Lastly, the nature of the counter-anion was shown to have a minor influence. In particular, these rather high apparent coverage values could be due to high roughness of the graphene layer and edge effect (assumed by some authors to be the sole reactive) without omitting the possibility of insertion of cations through the still associated graphene planes. All preceding values should be compared to those of $TOctA^{+}$, $(1.8 \pm 0.5) \times 10^{-9}$ mol cm^{-2}, for which a sole reduction step was observed, supporting a mono-layer as most likely in this series.

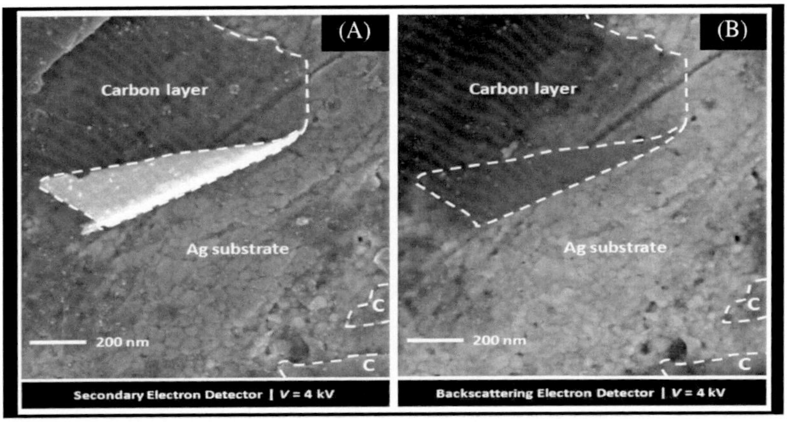

Figure 15.3. SEM images obtained for Ag–graphene partial surfaces [30, 32].

The characterization of composite silver–carbon electrodes after cathodic polarization was also checked by means of SEM technique. As expected, such electrodes exhibited only small amounts of carbon. Imaging based on the collect of both secondary electrons (SEs) and backscattered electrons (BSEs) obtained at low voltage revealed the attachment of mono-layers or very thin graphite layers (quite certainly a few graphene sheets as shown in Figure 15.3).

15.2.2. *Modification of the graphene layer by addition of in situ generated free radicals*

Both silver and graphene possess real capacity to activate the formation and the grafting of free radicals. Maybe, the best example of the complementarity of Ag and graphene could be established in the catalytic grafting process of alkyl iodides at Ag–graphene surfaces (Scheme 15.1).

Therefore, deposition of graphite onto smooth silver permits the easy building of new reactive interfaces. Electro-chemical and chemical stability of these composite materials is good in the presence of TAA^+ cations, irrespective of their size. The interest in such new modified electrodes that allies silver and graphene appear promising (Figures 15.4 and 15.5). Additionally, the grafting of free

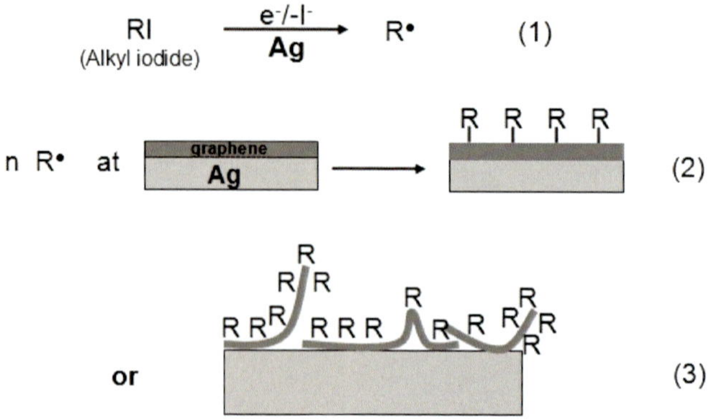

Scheme 15.1. A schematic representation of the graphene layer reactivity under a large addition of *in situ* generated radicals [32].

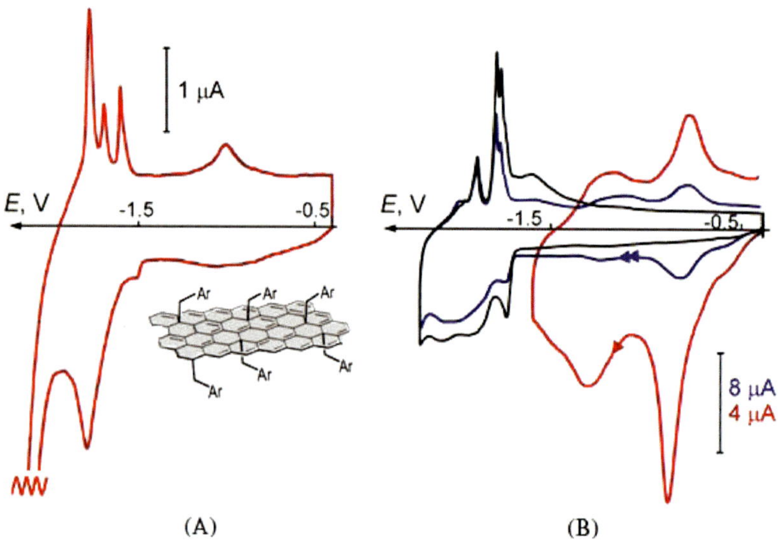

Figure 15.4. Addition of radicals onto the graphene network associated with silver. All the cyclings are achieved in $TEABF_4$/DMF solutions. (A) The Ag–graphene electrode is dipped in a solution of Br–CH_2–Py (10^{-2} mol·L^{-1}) for 2 min at –0.85 V). The resulting electrode appears totally modified (significant change in the cathodic response of the graphene network) together with a total current drop assigned to pyrene radical grafting. (B) The same kind of grafting is obtained now with 2-bromomethylanthraquinone by means of a scan in red until –1.7 V (scan rate: 100 mV·s^{-1}) followed (now cycling in a solution free of RBr) by the two steps (scan in blue) specific to anthraquinone. Black line represents the response of the electrode without a chemical modification [30].

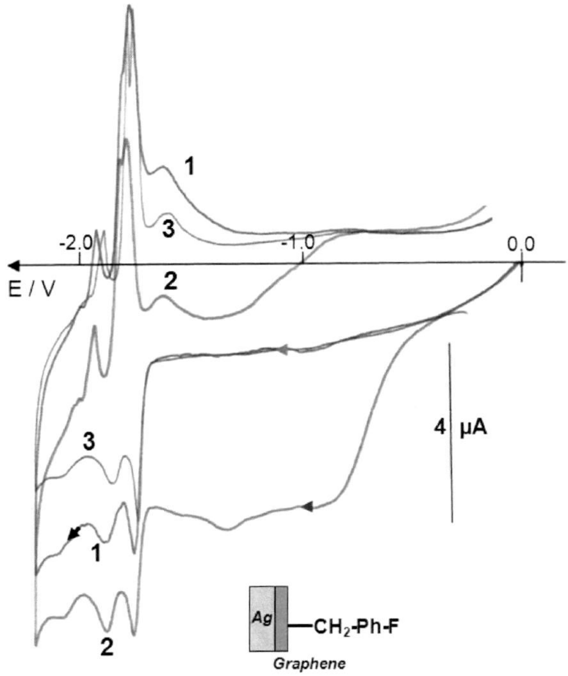

Figure 15.5. Addition of a marked benzyl radical (here 4-fluorobenzyl radical) onto the graphene network. Curve (1) displays the behavior of Ag–graphene electrode cycled in $TEABF_4$ (curve 1). The cycling of the electrode achieved in the presence of 4-fluorobenzylbromide (commercially available) is displayed in trace 2. After rinsing in water and then in acetone, the resulting response of the surface after immobilization of the benzyl radical is shown in (3). One may notice a significant decay between currents of (1) and (3) attributable to alteration of the graphene network by immobilization of a large amount of radicals [32].

radical generated by several redox systems (anthraquinone, fluorene, pyrene as well anthracene) to the Ag–graphene network was successfully achieved and allows to assume that the radical density present on the carbon layer is significant although a precise compactness might lead to a substantial degradation of the graphene layer. Moreover, the ability of graphene to form polynucleophilic structures (at $E < -1.8$ V) also allowed to simply attach alkyl chains (via marked 1-haloalkanes) by using simple technical conditions. The electro-chemical interest in the new concept of interface will certainly be taken into account for possible further development.

15.3. Electro-Chemical Exfoliation of Natural Graphite at a Gold Substrate

Continuing on the electro-chemical splitting of natural graphitic materials through intercalation of TAA^+ cations under specific conditions already reported almost four decades ago [14] by the author, this contribution aims to show the interest in building gold–graphene electrodes, especially for their stability through time. The present study also shortly focuses on a Raman study on Au interfaces covered with thin graphene layers, giving rise to quite steady graphene-modified gold material. Let us stress that the building of such interfaces is particularly easy for leading to Au–graphene micro-electrodes highly remarkable for being very reproducible in current and potential and then appearing helpful in electro-chemical analysis. Especially, diverse experiments can be accomplished with graphene layers (such as cathodic charge in the presence of different salts and the concomitant addition of functionalized free radicals). These new electrodes should quite certainly be considered as highly developed tools in the electro-chemistry of carbonaceous materials.

15.3.1. *How to prepare Au–graphene working electrodes*

Gold electrodes (S = 0.8 mm^2) were first carefully polished with rough (Struers 500) and then with finer (Struers 1200) silicon carbide polishing papers. Importantly, natural graphite (origin: Ceylon) was first partly exfoliated by a cathodic procedure. More concretely, some amount of graphite (say 80 mg) was cathodically doped (reduction at about $E_r = -2.1$ V vs. Ag/AgCl) in 0.1 M $TMABF_4$ using DMF as a solvent in a three-compartment cell. The cathodic compartment ($V \approx 8$ cm^3) was efficiently stirred. The working electrode was a mercury pool (surface area: 5 cm^2) or a large GC cathode polished beforehand to eliminate deposits. The graphite electro-chemical treatment was stopped after a passed charge of 7 C through the cell. The pre-intercalated graphite was filtered and then rinsed several times with water and acetone, and carefully dried for 20 min at 60°C. This expended "graphite" material was then finally transformed into a slurry (mixture with acetone) and immediately coated onto a gold disk. After drying, the electrode was firmly rubbed at a glassy cardboard for polishing until a shiny surface was obtained. If necessary, a very short sonication of a few seconds could permit eliminating residual graphite flakes.

15.3.2. *Au–graphene interface: Formation and structure*

As recalled above, the procedure starts by the use of a natural graphite pre-exfoliation procedure (reduction in 0.1 M $TMABF_4$ using DMF as a solvent). Then, the preliminarily "expanded" graphite is deposited onto a gold and reduced within a quite limited range of potentials in solutions of TAAX salts in DMF or AN. This ultimate process allows the obtaining of a very thin deposit of graphite onto the metal. Figure 15.6 depicts a typical cyclic voltammogram (CV) obtained in the presence of $TEABF_4$ (tetraethylammonium tetrafluoroborate) that triggers, via the exfoliation procedure, an almost total loss of graphite at the gold surface (that appears shiny). During the exfoliation, the cathodic current is assigned to an electrical charge associated with the insertion of TEA^+ cations between graphite layers. The main peak is followed by a sudden current decay that matches the graphite deposit dislocation assigned to the columbic repulsion of the charge layers. This form of "erosion" process is visible to the eye, while the working electrode turned from gray to yellow pristine gold. Different [TAA^+]-based salts could be used for achieving such exfoliation processes. However, fast and efficient exfoliations were obtained with TMA^+ (associated with BF_4^- or ClO_4^-). Electrolysis solutions could be reused several times after a simple filtration at the condition to avoid radical addition onto the graphene coverage.

Upon subsequent cycles (from the second scan as displayed below), reproducible voltammograms were systematically obtained

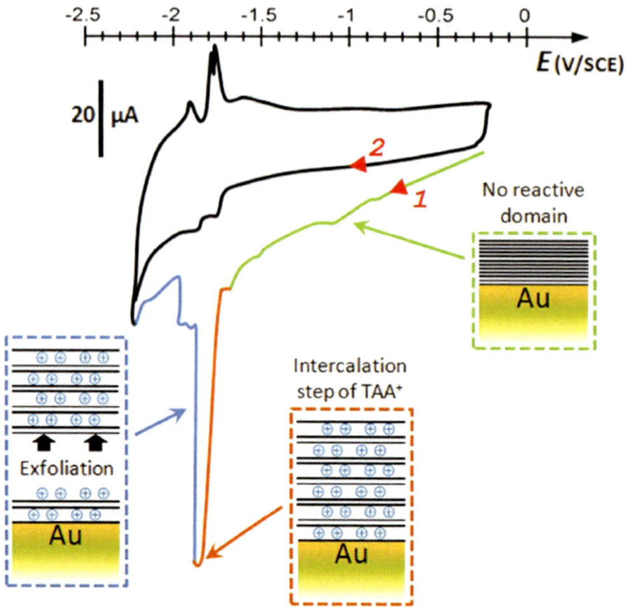

Figure 15.6. Large electro-chemical exfoliation process for a graphite deposit onto gold achieved in the course of two consecutive cathodic scans from −0.2 to −2.2 V/SCE. Scan rate: 50 mV s^{-1}; electrolyte: 0.1 M $TEtABF_4$ in DMF. Schematic exfoliation due to the lateral insertion of $TEtA^+$ ions inside the graphite until an almost total degradation of the layer. A strong affinity between gold and the first graphene layer is expected to render the process valid [32].

when TMA^+ salts are employed. This electro-chemical behavior means the thin and particularly stable electro-active interface corresponds to the charge of the layer assigned to a residual stacking of graphene planes well attached to the gold surface (Figure 15.7). From −1.5 to −2.2 V, several cathodic (during charge) and anodic ((during discharge) steps can be obtained: the shape of the CV specifically depends on TAA^+ salt solution in contact with the electrode.

The first cathodic step appears as the preliminary injection of electrons on the graphene plane surface sheltered previously by a TAA^+ layer (not necessarily a mono-layer) whereas complex cation

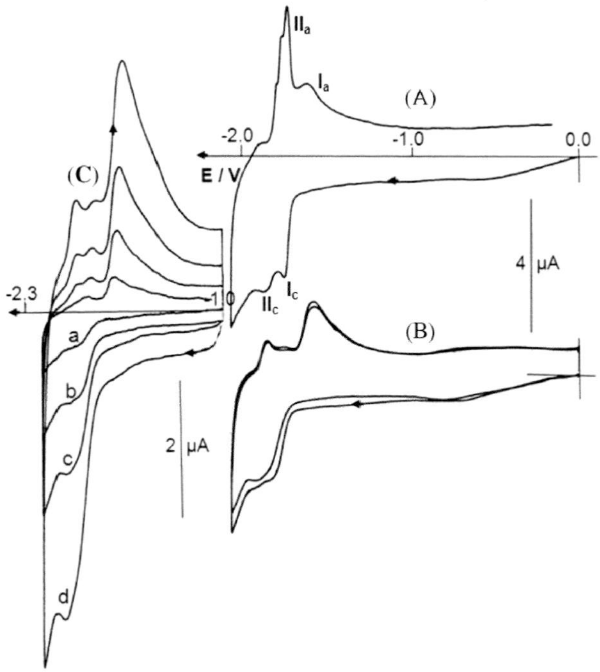

Figure 15.7. The resulting CV curves obtained after an electro-chemical exfoliation process for a thick graphite deposit onto a gold disk ($v = 50$ mV s^{-1}; electrolyte: 0.1 M $TEABF_4$ in DMF) as shown in (A). The two first charge processes are noted as **Ic** and **IIc**. Steps corresponding to associated discharge process are noticed as **Ia** and **IIa**. (B) and (C) display the responses of the same electrode now plunged in a solution of tetrahexylammonium iodide THexAI (0.1 M·L^{-1}). The corresponding influence of the scan rate on the CV curves (a) 20 mV s^{-1}; (b) 50 mV s^{-1}; (c) 100 mV s^{-1}; (d) 200 mV s^{-1}. The currents of main steps are proportional to the scan rate [31].

multi-layers are expected to be obtained at lower potentials. Of course, such assumption is related to the bulkiness of the TAA^+ cation and the level of dissociation of the salt TAAX in the considered solvent (in general DMF). A simple estimation based on the step current I_c by means of coulometry would suggest a TEA^+ coverage of 1.6×10^{-8} mol cm^{-2}. While this is quite large, we should take into account the highly disturbed geometry of the graphene deposit and the fact that both sides of the plane are involved with a coverage. A comparable analysis of the corresponding anodic step gives 9.9×10^{-9} mol cm^{-2}. Additional analytic estimations established with other TAA^+ of different radii check that coverage densities in the order $TMeA^+ > TEtA^+ > TBuA^+ > THexA^+ > TOctA^+$ were obtained. Furthermore, the cationic effect on charge–discharge processes is immediate and totally reversible: for example, a preliminary charge in the presence of $THexA^+$ (which does not give more than one main cathodic step due to its large size) turns to be at once specifically modified by contact of the electrode with $TMeA^+$. The procedure described above is a new contribution to the exfoliation process and to the building of composite materials containing graphite and/or graphene already abundantly documented [9–33].

Raman analysis was also performed [30] for drawing on both the orders of Raman spectra in a multi-wavelength approach, for identifying the remaining carbon phases present on the gold electrode after graphite exfoliation. Raman spectra obtained from Au-exfoliated graphene samples were recorded using a micro-confocal device as a function of laser and three excitation wavelengths (514.53 and 785 nm). First of all, Raman spectrum of the pristine natural graphite was drawn and was checked to correspond to the regular Raman signature of a perfect crystal without any measurable D and D′ bands.

Presently, a complete study on the structure of the gold–graphene interface is being discussed in a report presently in press [31]. One typical spectrum obtained after electro-chemical exfoliation is shown in Figure 15.8. It concludes on either a single Lorentzian profile centered at 2685 cm^{-1} with a FWHH of 42 cm^{-1} or a more complicate profile constituted by at least four components. On the basis of the literature [24, 25], it can be concluded that the mixture of these two profiles exists at the micron scale. The first is assigned to the spectrum of single-layer graphene (1LG), while the second is assigned to three-layer graphene (3LG). The 1LG spectrum is centered at 2685 cm^{-1}; this wavenumber is incompatible with an assignment to turbostratic graphite that displays a band at

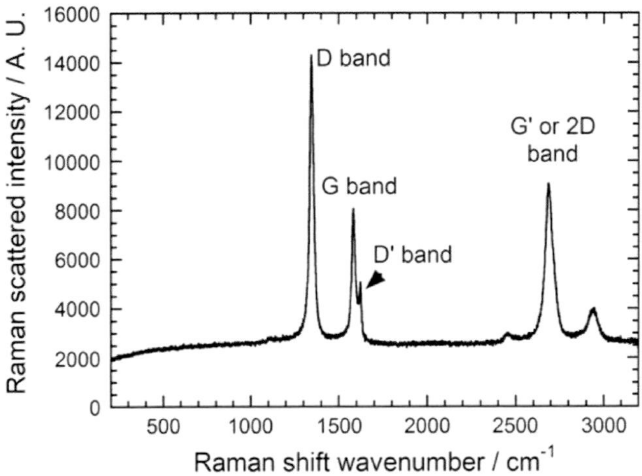

Figure 15.8. A typical spectrum obtained after electro-chemical exfoliation (2.41 eV excitation). The second order signals are labeled as D, D′, and G′ bands. The G′ band is clearly different from the Raman profile of a G' band characteristic of the pristine graphite phase [31].

around 2710 cm^{-1} with a green excitation. The 3LG profile is in agreement with profiles previously published in the literature. Finally, Figure 15.9 displays an example of Raman mapping obtained on a small spatial area by selecting the spectral range characteristic of graphene in the hyperspectral image (i.e., 2600 2800 cm^{-1} as shown in Figure 15.8).

The Raman spectra related to the blue surface area of the mapping correspond only to the graphene phases and never to the G′ profile, which is one characteristic of the graphite phase. The area to the right of the blue domain corresponds to bare gold substrate. In the top left corner, residual deposits of carbon are found, concomitantly with Raman signals of disordered carbon or graphite phases. The blue surface area was also probed by SEM imaging (Figure 15.9(B)). While the resolution of the corresponding picture is low due to this voltage, it shows a clear deposit spread onto the metallic substrate.

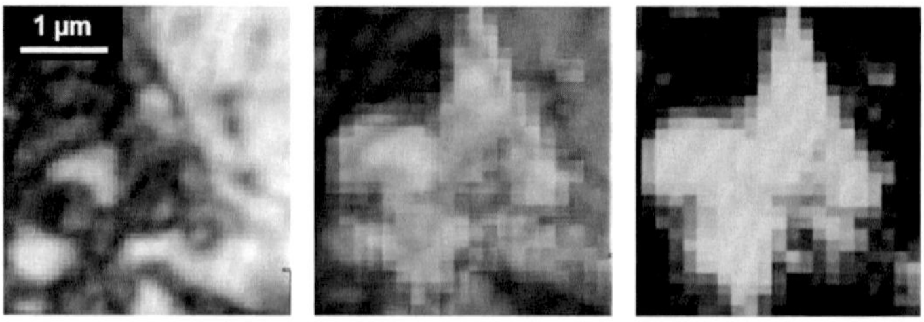

(A)

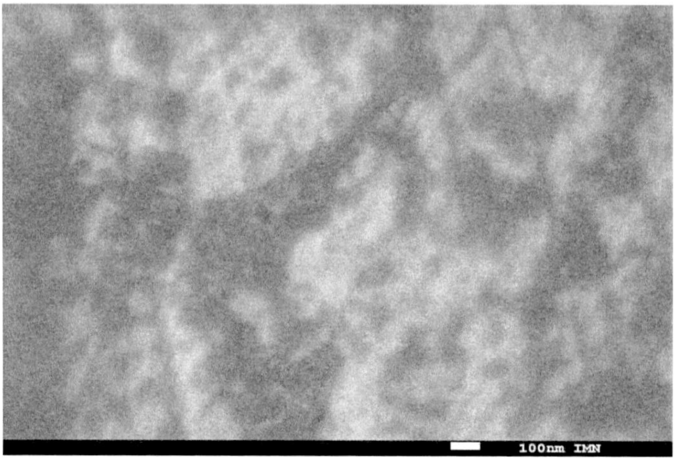

(B)

Figure 15.9. (A) Example of a representative surface area of the gold working electrode after electro-chemical exfoliation of the carbon coating: [left] optical image, [right] Raman map obtained by selection the spectral range between 2500 and 2700 cm^{-1} and [centre] the two images overlaid. (B) Corresponding SEM image recorded at very low accelerating voltage (500 V) for a selected area inside the blue region. The darker zones can be ascribed to the graphene deposit [31].

15.3.3. *Au–graphene interface as a spin-trap*

Many experiments were performed for testing the ability of the Au–graphene interface to fix different organometallic groups by employing the tailored interface as a catalyst. Figure 15.10 shows one graphene-modified surface where the immobilization of ferrocenyl groups is obtained, presumably mainly through a nucleophilic

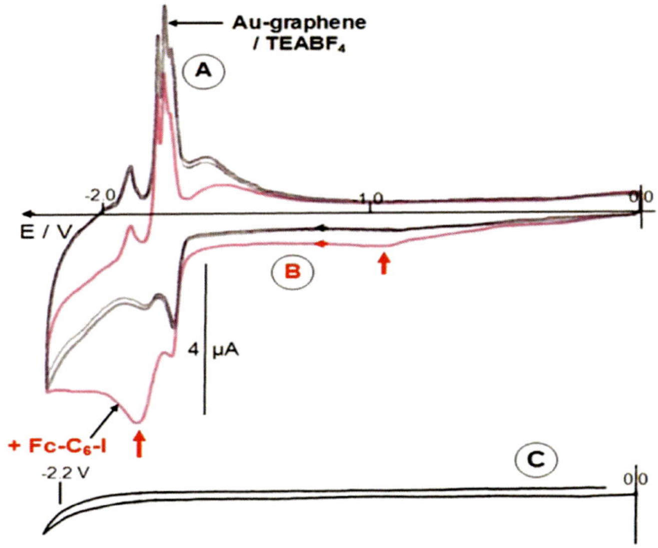

Figure 15.10. Voltammetries of a gold–graphene interface in TEA–DMF solution: (A) Response (in black) of the electrode. Scan rate: 50 mV s^{-1}. (B) With the same solution containing 1-iodohexyl-ferrocene (concentration: 11 mmol·L^{-1}), the two arrows display two kinds of catalysis: one located at about −0.95 V going through the alkyl-iodide adsorption onto gold and the second at −1.85 V corresponding to the S_N2 displacement occurring via the charged graphene. (C) Response of the bare gold substrate after polishing. Current scale is the same as (A) [32].

displacement, which takes into account the cathodic doping of graphene (the level of the coverage could be estimated to $\Gamma_{Fc} = 2.2 \times 10^{-9}$ mol cm^{-2}). Owing to the probable extreme roughness of the interface (exhibited by Figure 15.9), this value could appear as reasonable for a mono-layer. It is also probable that the entire surface is concerned and not only the edges of graphene flakes deposited onto gold (even if both sides of graphene could be, in principle, implied by the Fc addition).

References

[1] R.L. McCreery, *Chem Rev.* **108**(7) (2008) 2646.
[2] G.M. Jenkins, K. Kawamura, *Nature* **231** (1971) 175.

[3] P.J.F. Harris, *Philosophical Mag.* **84** (2004) 3159.
[4] G. Fiori, S. Rondinini, G. Sello, C. Alberto, M. Cirja, L. Conti, *J. Appl. Chem.* **35** (2005) 363.
[5] S.B. Rondinini, P.R. Mussini, F. Crippa, G. Sello, *Electrochem. Commun.* **2** (2000) 491.
[6] S. Rondinini, A. Vertova, *Electrochim. Acta* **49** (2004) 4035.
[7] A.A. Isse, B. Giacomo, L. Falciola, M. Rossi, P.R. Mussini, A. Gennaro. *J. Appl. Electrochem.* **39** (2009) 2217.
[8] C. Durante, A. A. Isse, A. Gennaro, *J. Appl. Electrochem.* **43**(2) (2013) 227.
[9] J. Simonet, *J. Appl. Electrochem.* **39** (2009) 1625.
[10] J. Simonet, *Electrochem. Commun.* **7** (2005) 619.
[11] V. Jouikov, J. Simonet, *Electrochem. Commun.* **12** (2010) 781.
[12] V. Jouikov, J. Simonet, *Electrochem. Commun.* **45** (2014) 32.
[13] V. Jouikov, J. Simonet, *Electrochim. Acta* **172** (2015) 28.
[14] J. Simonet, H. Lund, *J. Electroanal. Chem.* **75**(2) (1977) 19.
[15] C. Dano, J. Simonet, *J. Electroanal. Chem.* **564**(15) (2004) 115.
[16] V. Jouikov, J. Simonet, *Langmuir* **28**(1) (2012) 931.
[17] K.S. Novoselov, A.K. Geim, S.V. Morozov, D. Jiang, Y. Zhang, S.V. Dubonos, *et al.*, *Science* **306**(5696) (2004) 666.
[18] A.K. Geim, K.S. Novoselov, *Nat. Mater.* **6** (2007) 183.
[19] A.K. Geim, *Science* **24**(5934) (2009) 1530.
[20] X.Y. Yang, X. Dou, A. Rouhanipour, L.J. Zhi, H.J. Rader, K. Mullen, *J. Am. Chem. Soc.* **130**(13) (2008) 4216.
[21] A. Ambrosi, C.C. Chua, A. Bonanini, M. Pumera, *Chem Rev.* **114**(14) (2014) 7150.
[22] W. Sirisaksoontorn, A.A. Adenuga, V. T. Remcho, M.M. Lerner, *J. Am. Chem. Soc.* **133**(32) (2011) 12438.
[23] Y.L. Zhong, T.M. Swager, *J. Am. Chem. Soc.* **134**(43) (2012) 17896.
[24] W. Sirisaksoontorn, M.M. Lerner, *Inorg. Chem.* **52**(12) (2013) 7139.
[25] A.J. Cooper, N.R. Wilson, I.A. Kinloch, R.A.W. Dryfe, *Carbon* **66** (2014) 340.
[26] Y. Yang, F. Lu, Z. Zhou, W. Song, Q. Chen, X. Ji, *Electrochim. Acta* **113** (2013) 9.
[27] C.T.J. Low, F.C. Walsh, M.H. Chakrabarti, M.A. Hashim, M.A. Hussain, *Carbon.* **54** (2013) 1.
[28] J. Simonet, *Electrochem Commun.* **42** (2014) 34.

[29] V. Jouikov, J. Simonet, *Electrochem Commun.* **43** (2014) 67.
[30] V. Jouikov, N. Stéphant, P. Poizot, J. Simonet, *Electrochem Commun.* **51** (2015) 125.
[31] P. Poizot, B. Humbert, C. Ewels, J.-Y. Mevellec, N. Stéphant, J. Simonet, *Carbon* **107** (2016) 823.
[32] J. Simonet, P. Poizot, Unpublished works.
[33] L.M. Malard, M.A. Pimenta, G. Dresselhaus, M.S. Dresselhaus, *Phys. Rep.* **473** (2009) 51.

Chapter 16

Anodic Coupling of Aromatic Ethers: The Synthesis of Ionophoric Redox Materials Deposited at Solid Surfaces

16.1. Introduction

The anodic oxidation of aromatic ethers is a convenient source of cation radicals [1, 2] that may generate, under tricky experimental conditions, dimers, trimers, and polymers easily doped by further electron transfer(s). This class of π-donors offers wide possibilities [3] for generating structures that are difficult to access through classical organic chemistry. This chapter focuses on the organization of ethers as oxidation intermediates at solid anodes in weakly or non-nucleophilic media [4–6]. By this way, *in fine* complex ionophoric molecules becomes accessible and leads to new materials where both electrons and inorganic cations may have specific properties upon the application of potential and the liquid electrolyte.

The first noticeable reaction in the field is quite certainly the anodic coupling of anisole [7–10] achieved in a mixture dichloromethane–trifluoroacetic acid considered as a typical medium

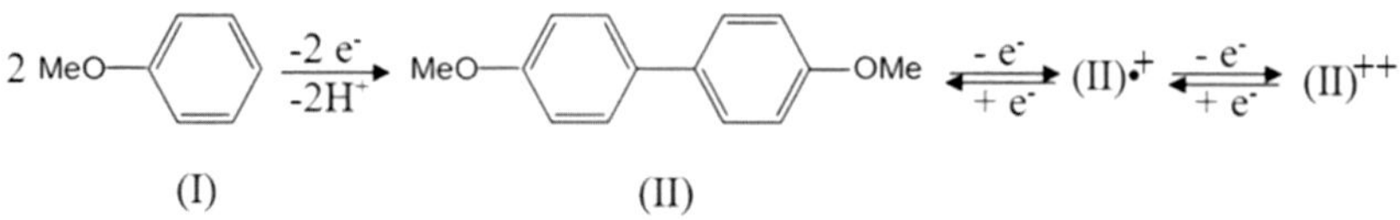

Scheme 16.1. Anodic coupling of anisole.

weakly nucleophilic for stabilizing dimeric cation radicals as shown in Scheme 16.1.

As observed with most of the dimeric forms obtained by anodic oxidation of aromatic ethers, the coupling product of alkoxybenzenes (4,4′-dialkoxybiphenyl (II)) is more easily oxidized than (I), leading to cation radical ($II^{\bullet+}$) and di-cation (II^{++}) (Scheme 16.1). The essential use of trifluoroacetic acid slows down the nucleophilic action of water contained in the solvent.

Surprisingly, when the oxidation of 1,2-dimethoxybenzene, under similar conditions, is carried out, a trimer is obtained (corresponding to three successive coupling reactions), yielding in one step hexamethyltriphenylene (HMT) as a major product [11, 12] as shown in Scheme 16.2.

Such structures of hexa-substituted triphenylene skeletons are of high interest:

— They are planar polyaromatic hydrocarbons.
— Most of these structures present a lone transition from the solid phase to the isotropic liquid phase and are classed among liquid crystals. The transition temperature is quite high (>400°C) and those structures are not stable under these conditions.
— They are produced (case of π-donors) in the form of charge transfer complexes (or radical-ions salts, lamellar structure): general formula: DX_x with $0 < x < 1$ [22].

We especially develop synthetic possibilities aimed at creating new redox ionophores as trimers, oligomeric and polymeric materials in the following (Scheme 16.2).

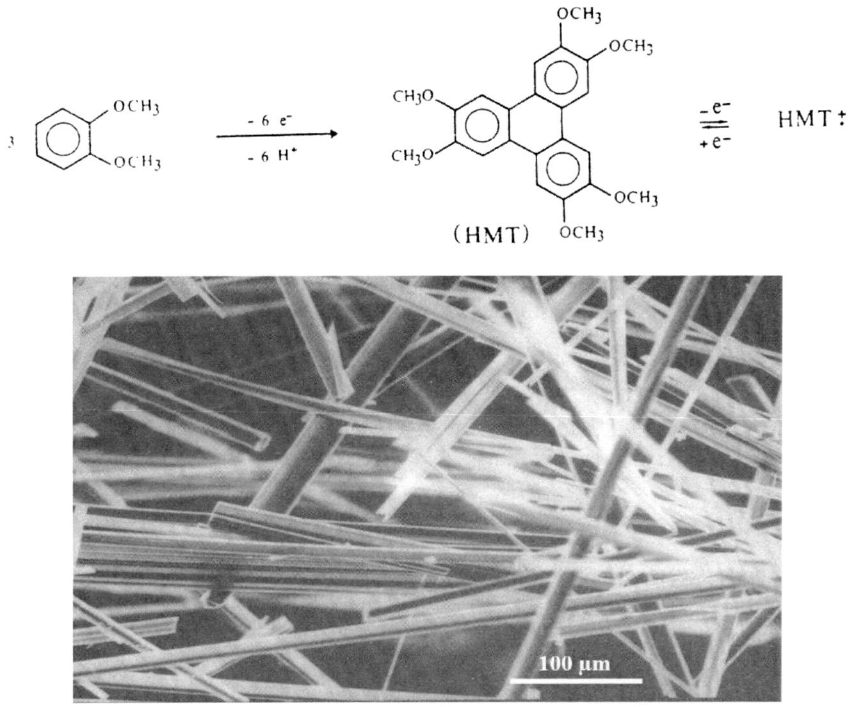

Scheme 16.2. Mechanism for the trimerization of 1,2-dimethoxybenzene together with the concomitant formation of micro-crystals at a polarized anode (a Pt wire of diameter of 0.5 mm) [22].

16.2. Anodic Trimerization of Orthodimethoxy Benzenes

16.2.1. *Oxidation of orthodialkoxybenzenzes*

Macro-electrolyses are based on the voltammetric results displayed in Figure 16.1, exemplified by veratrole (1,2-dimethoxybenzene). Thus, in the course of the first scan, only an anodic step $\mathbf{O_1}$ appears and is assigned to the formation of a trimer. In the course of the reverse scan at potential less anodic, a reduction step $\mathbf{R_2}$ is noticed and corresponds to the reduction of the trimer (II) deposited onto the platinum anode. At the second scan, a new oxidation peak is present and has almost the perfect stability of the trimer then

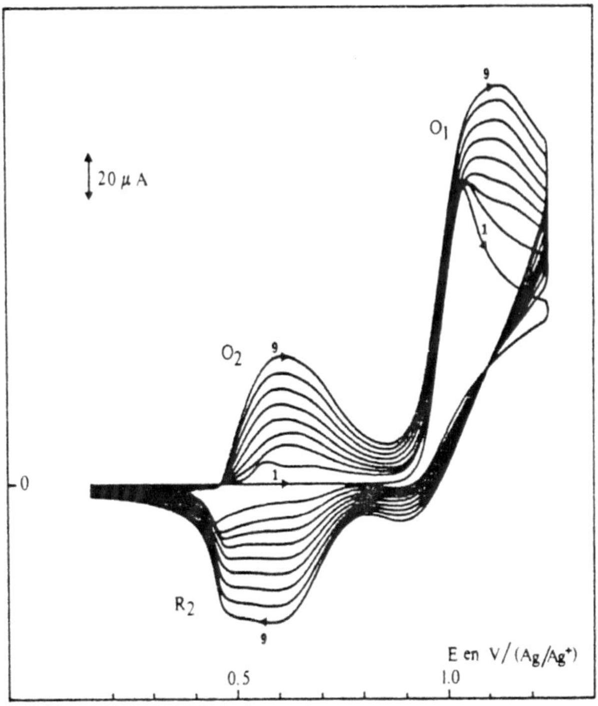

Figure 16.1. Recurrent voltammetries of veratrole (concentration: 1.5×10^{-2} M) in acetonitrile, Bu_4NBF_4 0.2 M. Scan rate: 0.1 V s^{-1}. Platinum disk electrode (from Ref. [13]).

deposited at the anode (the amounts of electricity relative to steps $\mathbf{O_2}$ and $\mathbf{R_2}$ irrespective of the number of scans are equal). In the course of successive anodic scans, the $\mathbf{O_2/R_2}$ system grows in intensity and mass of deposit at the electrode surface.

Preparative electro-depositions achieved at large-area anodes (example: a platinum foil of area >12 cm^2) at the level of the step $\mathbf{O_1}$ allowed to obtain with a good chemical yield (until 83% with $TBAPF_6$ used as salt in acetonitrile) a thick blue-green material. Elemental analyses confirmed that such deposits are salts. With veratrole, these salts correspond roughly to the formula: $(HMT)X_x$ with $X = BF_4^-$, ClO_4^- or PF_6^- ($x > 0$) (Figure 16.1). Stoichiometries are therefore relatively simple and permit to propose rough formula such as $(HMT)_2BF_4$, $(HMT)_3(ClO_4)_2$ and $(HMT)_3PF_6$. Access to neutral trimers is actually simple since they are achieved by the electro-chemical technique using electro-generated super-oxide as the reducing species (Scheme 16.3).

This mode of electro-chemical trimerization should be considered as useful for the synthesis of substituted triphenylenes. Thus, one may obtain, in one step, trimers, for example those of TMT, which could not be generated by organic syntheses, as shown in Scheme 16.4.

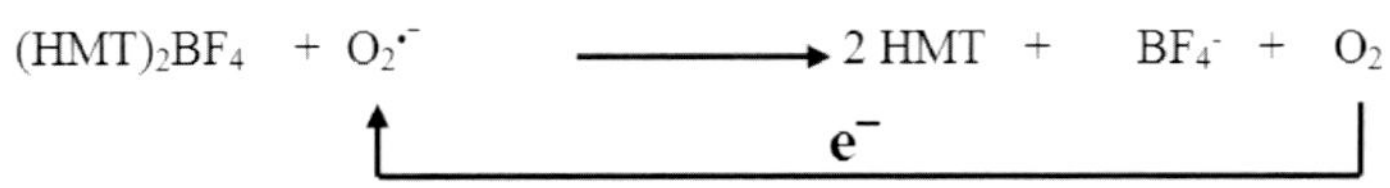

Scheme 16.3. Reduction of the mixed valence salt hexamethoxy triphenylene tetrafluoroborate by electro-generated superoxide.

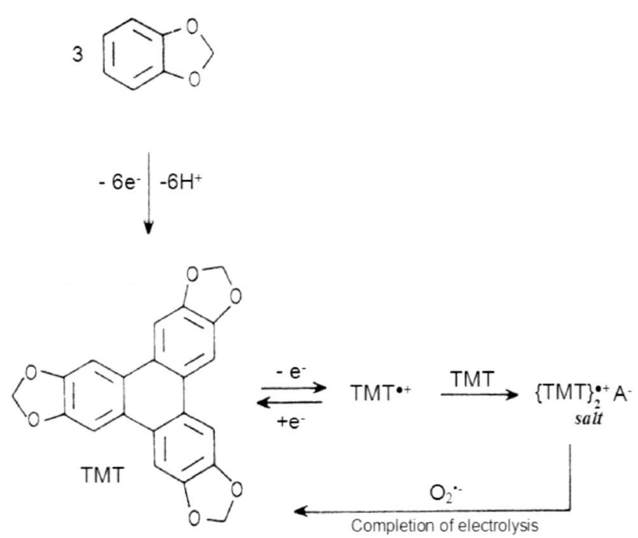

Scheme 16.4. Anodic trimerization and concomitant formation of a tris(methylene-dioxy)triphenylene via an insoluble mixed valence salt easily reduced at the end of the electrolysis by an electro-generated reducer (here super-oxide) [13, 23].

The easy access of such triphenylenes may afford multiple kinds of liquid crystals abundantly studied [14–20].

16.2.2. *A first approach toward the anodic synthesis of ionophores*

Trimerization of aromatic orthodiethers described above encouraged us [13] to undertake the synthesis of complex ionophoric molecules, which are of obvious interest in analytic chemistry (e.g., use for building indicative electrodes in electro-chemistry). The first objective was the trimerization of 2,3-benzo1,4,7,10,13-pentaoxacyclopentadeca-2-ene, more easily named by using Pedersen's nomenclature as benzo-15-crown-5 [21] and shown as a monomer in Scheme 16.5. The successful synthesis of the ionophoric trimer has been a difficult challenge owing the strong reactivity of the anodically charged electrolysis product and has been overcome through (a) the use of a poorly nucleophilic solvent (methylene chloride/trifluoroacetic acid containing tetraalkyl ammonium (TAAX) salts and (b) the building of an appropriate redox cell equipped with porous electrodes. Under those conditions, the highly reactive

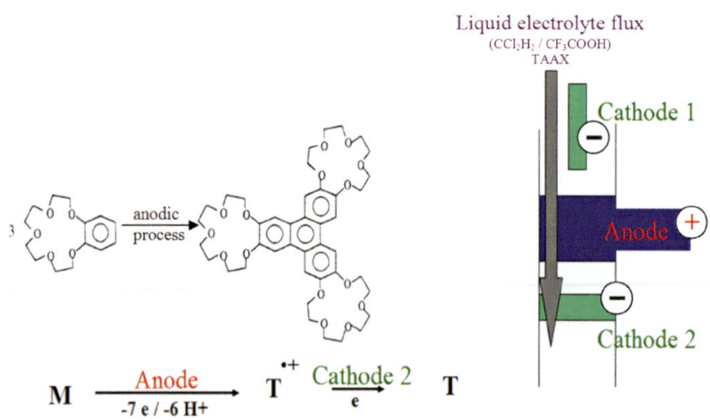

Scheme 16.5. Anodic trimerization of benzo 15-crown 5 at a redox cell equipped of porous solid electrodes [23, 24].

trimer cation radical is reduced fast (via electron transfer) at the cathode no. 2 made of carbon felt.

Other difficulties are not always minor: the regulation of the electrolyte flux through the cell, the concentration of the starting crown ether and the corrosion of rubber joints by CH_2Cl_2. This successful synthesis was really a result of Rennes' electro-chemistry group. Full details concerning the experiments and the glass cell plan have been reported [24].

This kind of synthesis has been further extended to mixtures of veratrole and different crown ethers as well as open chain complexants under quite simple conditions in two-compartment cells in acetonitrile containing 0.1 M $TBABF_4$. There is no deposit onto the Pt anode and the anolyte (presence of mixed valency salts is immediately reduced by triethylamine [17], Scheme 16.6). Triphenylenes equipped with a sole crown ($T_{1,1,2}$) were obtained with a yield of almost 50% in all cases and easily separated from the product mixture. Several preliminary attempts to introduce vinylic double bonds attached to the aromatic ether functions [18, 19], before trimerization processes were successful [26–28] and could be used for the attachment of trimers to solid conductors like glassy carbon or organic conducting polymers (Scheme 16.6).

16.3. The Concept of Ionophoric Polymers Based on Anodic Oxidation of Aromatic Ethers

Beyond obtaining of trimers as shortly presented above, the focus is on the oxidizing molecules that possess two dialkoxy-aryl units in order to build macromers that have poly-triphenylene structures. After considering trimerization of benzo 15-crown 5 as described above, the question of polymerization of dibenzocrown ethers shown in Scheme 16.6 deserves to be asked in order to build materials that possess a high degree of ionophoric sites. Also, and not the least, the concomitant formation of triphenylenes and their stacking in oxidized forms may build a class of

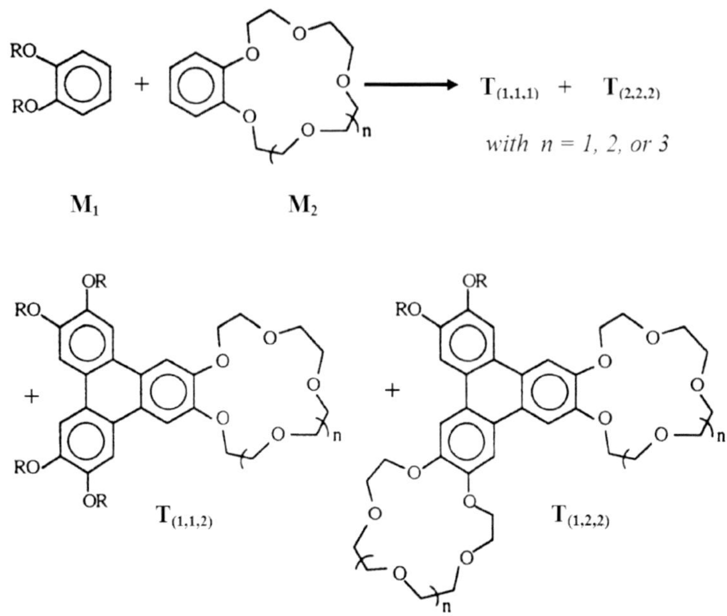

Scheme 16.6. Examples of mixed anodic trimerization reactions [25, 28] implying benzocrown ethers of different sizes (until 8 oxygen atoms participating to the crown). Conditions do not require highly sophisticated cells in the conditions defined by the reference [26]. Acetonitrile containing tetraalkylammonium salts can be used in classical two-compartment cells. Also, note that some OR groups can be replaced by $-O-(CH_2)_n-CH=CH_2$ for judiciously introducing a reactive double bond for facilitating further immobilizations at solid conductors.

uncommon architectures, which could depend on the size of the ring (values of *p* and *q*), the remnant charge in the polymer (conditions of electrolyses) and also the character of cations (organic or inorganic) possibly integrated — oxidation degree) in the mass of the material.

Therefore, one may imagine (Scheme 16.7) the formation by means of controlled potential electrolyses concomitant of structured polymers around triphenylene stacking (charge transfer complex depending on the total amount of electricity) considered as

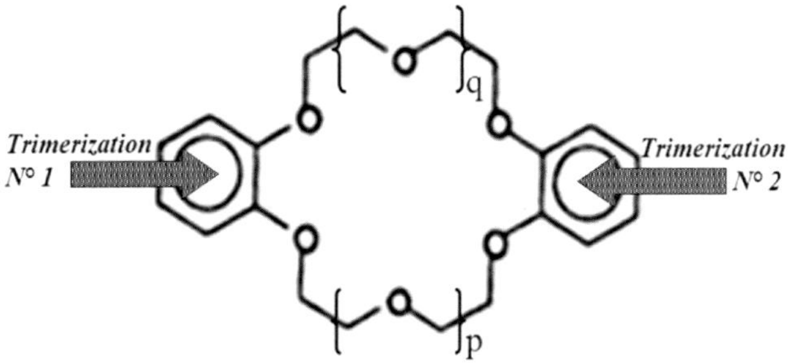

Scheme 16.7. Principle of the anodic polymerization of dibenzocrown ethers DBnCm [32].

vertical columns together with parallel piling of crown ethers organized as ion channels.

16.3.1. *Voltammetries and electro-chemical conversion of dibenzocrown ethers into ionophoric polymeric materials*

Voltammetric approaches (platinum electrode, solvent: methylene chloride containing 0.1 M Bu_4NBF_4, scan rate: 0.1 V s^{-1}) for DB15C5 (a central crown with 15 atoms including 5 oxygen atoms) is exhibited in Figure 16.2. As expected, the first scan (1) is assigned to the trimerization of activated aryl moieties, followed, from the second scan, by the redox response of electro-generated triphenylenes. The very regular growing of currents of both steps is quite remarkable and would support the organization of a redox deposit at the anode surface. The bending of steps underlines the electrical resistance of the growing deposit quite possibly through the electron hopping along triphenylene stackings together with that of Bu_4N^+ cations (Schemes 16.8 and 16.9).

Electrolyses of such ionophoric materials in methylene chloride or in acetonitrile at platinum foil anodes could be achieved. In most of the cases, salts such as tetraalkylammium tetrafluororate were, in principle, preferred to alkaline metal salts since they

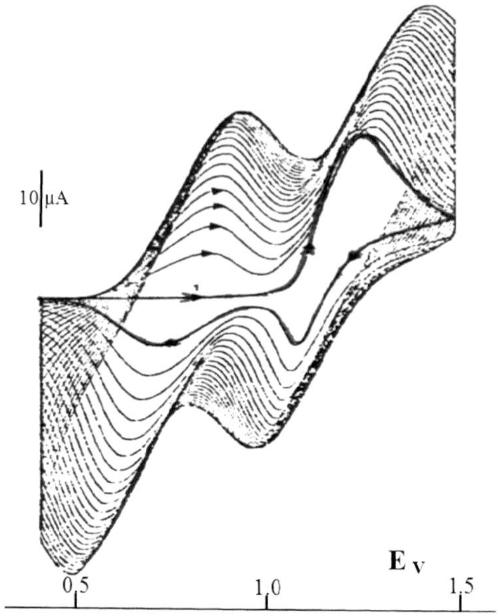

Figure 16.2. Anodic voltammetric behavior approaches (platinum electrode, solvent: methylene chloride containing 0.1 M Bu_4NBF_4, scan rate: 0.1 V s^{-1}) for DB15C5. Reference electrode: Ag/Ag^+ [13].

were reactive also with the starting monomer. Solvent–electrolyte mixtures were used under super dry conditions (preliminary treatment with freshly activated neutral alumina under vacuum at 340°C during several hours). Therefore, brown-red polymeric deposits could be pealed out the metal surfaces, rinsed by solvents such as acetone. They are quite stable and found to be paramagnetic. Different studies concerning their properties and stabilities, structure (solid NMR) and their reactivity toward aqueous or organic solution of different salts (essentially alkaline metal salts) were undertaken [29–35].

Here, electronic structures having qualitatively different oxidation states of thin rough materials were studied and discussed. Actually, five oxidation states (Scheme 16.10) were found and

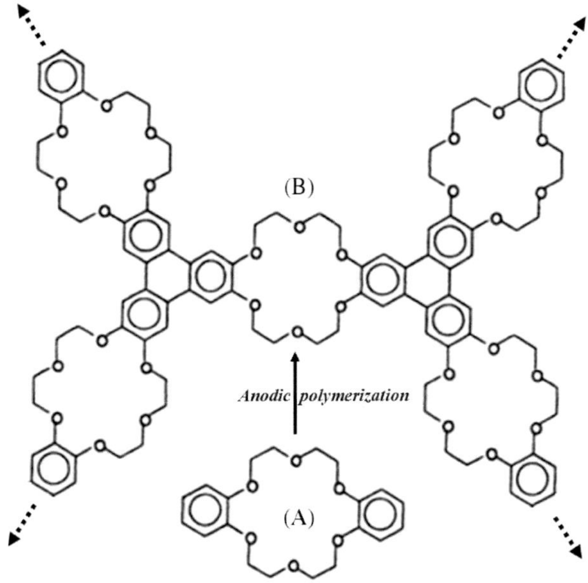

Scheme 16.8. Schematic proposal for a planar growing through anodic oxidation of dibenzocrown ether, DBC18C6 (A) that stresses the electro-chemical trimerization of each benzo moiety until first to a trimer, to a pentamer (B) and beyond to a planar polymer. The further oxidation of triphenylenes would permit columnar arrangements (via charge transfer complexes) and the extension in volume of a 3D polymeric form [22–24] (see Scheme 16.9).

displayed in Scheme 16.8 and well support the formation stackings. These results explain the swelling of the membrane as a function of the amount of charge.

Convincing arguments were made about the thin layer obtained by means of potentiostatic electrolyses (Schemes 16.9 and 16.11 as well as Figures 16.3 and 16.4). Thus, the average surfacial concentration of triphenylenes was established ($\Gamma_T \approx 3.5 \times 10^{-8}$ mol cm^{-2}), the volumic concentration of triphenylenes ($C = 3 \times 10^{-3}$ mol cm^{-3}), as well as the thickness of the polymeric film is estimated to be 0.1 μm, which would imply the stacking of about 300 mono-layers.

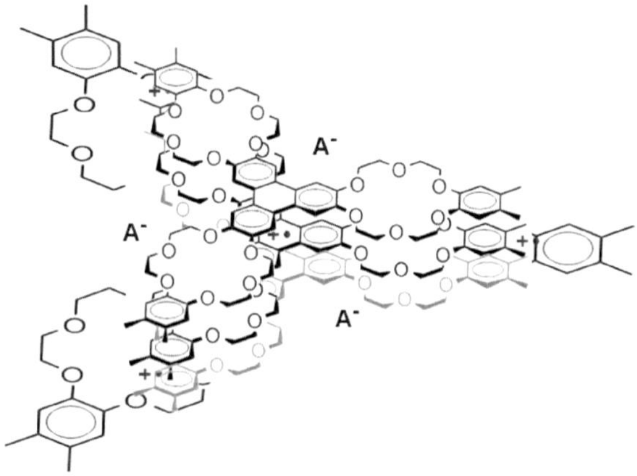

Scheme 16.9. A spatial representation of the polymeric form organized in stacking: columnar both for triphenylenes and for crown ethers that permits both the concomitant motion of electrons and cations. In the doped structure (oxidized form, shown here), anions are dispersed around stackings [32].

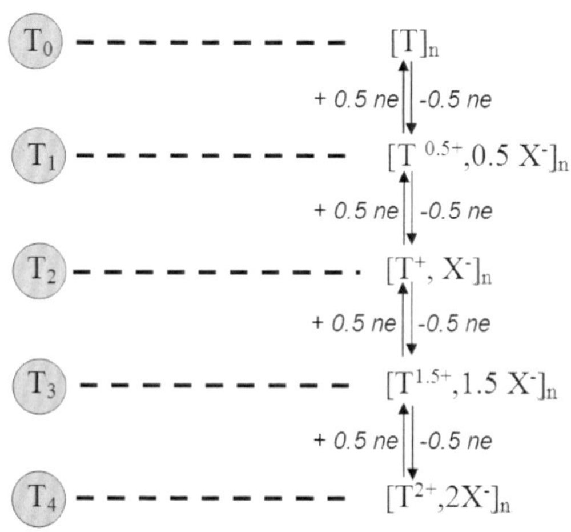

Scheme 16.10. Results of preparative electrolyses on dibenzocrown ethers considered in the scheme as systems essentially organized as triphenylene stackings $[T]_n$ with different oxidation states (from T_0 to T_4) all indentified by mean of voltammetric and microcoulometric studies. The per-oxidized system T_4 corresponds to triphenylene dications, in principle, without any correlation between each other (absence of stacking). The material under any oxidation state is stabilized by insertion of cations, in principle, inside crown cavities. The swelling of forms T_4 (coulombic repulsion) is reversible, under certain experimental conditions (artificial muscle induced by electrical impacts?) [13].

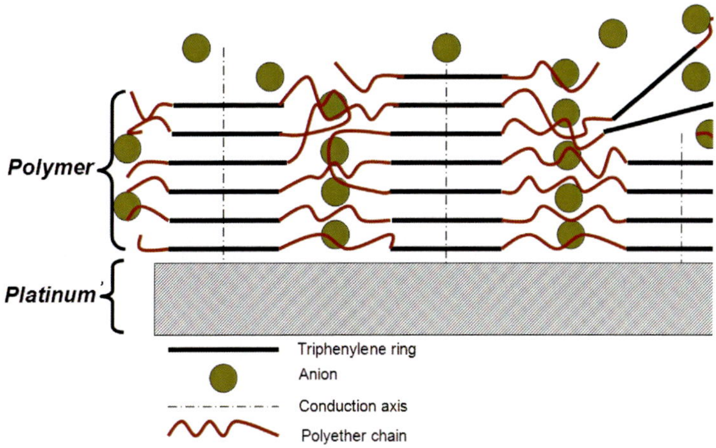

Scheme 16.11. A proposed structure of the polymer deposited onto platinum [32]. The presentation corresponds to the state [T_1], meaning that all triphenylenes are doped to the 0.5 electron degree. Layers are obtained by electrolysis in the presence of TAAX salts (anions: BF_4^- in green). Triphenylenes are expected to be organized in stacking providing a conductivity through the deposit. It is expected that anions are localized besides stackings but without connection with crown-ethers. One may imagine several conducting stackings (oxidized triphenylenes) closely connected with poly-ether channels (in red). As said further, anions can been replaced by alkaline cations taking place in crown-ethers columns when polymer deposits are (totally) reduced (chemically or electrochemically): see Section 16.4.

16.3.2. *About complexing properties of poly-dibenzo crown ethers*

Experimentally, it appeared that the best level of extraction of alkaline cations is obtained with partially undoped resins using an organic solvent for dissolving alkaline salts. It is important to notice that the complexing process of inorganic cations lead to an almost total reduction of resins toward an entire undoping procedure (Table 16.1) and then an apparent disappearance of stacking.

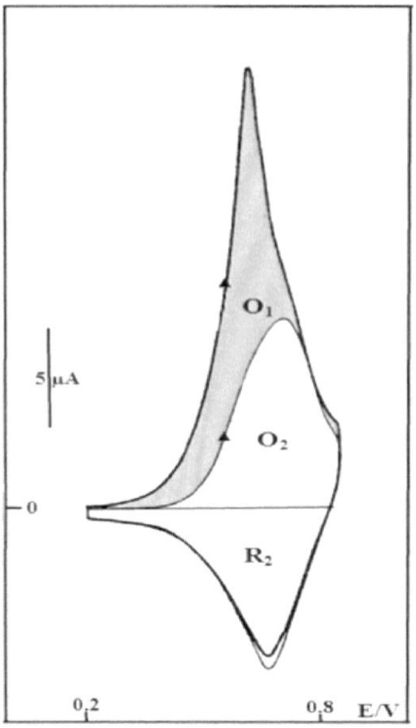

Figure 16.3. Voltammetric response (scan rate: 0.1 V s^{-1}) of a platinum anode (I mm^2) modified by an anodic deposit (applied potential: +1 V, amount of electricity: 2×10^{-4} C) from DB18C6 (concentration 2×10^{-3} M in a mixture CH_2Cl_2–CH_3–CN 9:1). The reversible couple O_2–R_2 corresponds to the rough electrolysis product [T_1] while the total step $O_1 + O_2$ that illustrates the oxidation of the resin initially *totally* reduced [T_0] at −2.5 V.

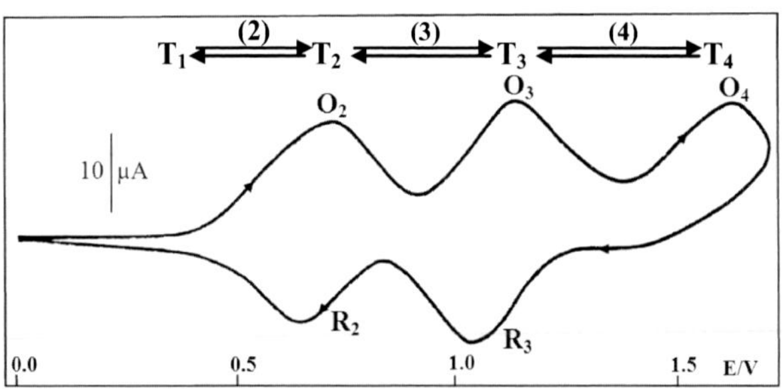

Figure 16.4. Opposite to the preceding scheme, the redox behavior of the resin after formation at +1 V and remaining doped at room conditions at the state [T_1], which would match with stacking of triphenylenes oxidized at different steps of +0.5 e each. One may note that the last oxidation step (leading to the form triphenylene di-cation) is stable despite the strong electrophilicity of the di-cation in the absence of trifluoroacetic acid in the solvent [13].

Table 16.1. Complexing results of alkaline cations by polyDB18C6(BF_4) in acetonitrile.

Cation	X_i	X_f	N_G	N_A
K^+	~50	0.1	0.31	*0.24*
K^+	27	5.6	0.49	*0.46*
Na^+	27	0	0.24	*0.29*
Li^+	*27*	*7*	*0.21*	*0.08*

Notes: X_i: initial doping level, X_f: doping level at the end of extraction, N: average number of cations per crown-ether obtained either gravimetric method (N_G) or by analytical way (N_A) [13].

Additionally, morphologies of layers (external face) were found to be dramatically changed by extraction of alkaline cations first dissolved in acetonitrile (see Figure 16.5). The extraction of radioactive strontium ion as well as that of Ag^+ and Au^{3+} by neutral membranes (T_0) were successfully achieved. Upon time, it appears that the material is progressively reduced and that gives to the membrane a spectacular metallic aspect (especially with gold).

16.4. Generation of Poly-Electrides

The world of "electrides" was quite recently described by Dye [36, 37]. It corresponds to the possibility of making reacting crown-ethers with alkaline metals as solutions of free electrons. For example, complexes of the forms {Cs^+(18-crown-6)$_2$ e^-} or {Cs^+(15 crown-5)$_2$ e^-} could be identified. From this concept, any π-acceptor connected to an tailored crown ether could react with a metal (room temperature) and then create an electride, the electron being stabilized on the LUMO of the acceptor and resulting alkaline cation naturally trapped and localized in an adjacent crown ether. This principle was applied to poly-dibenzocrown ethers and adjuncts and the usable potential window (> −1.8 V). Additionally, examples of grafting and/or insertion at or in gold metal appear much less documented [4].

Owing to the inherent difficulties bound to the synthesis of ready-to-use ligands (or organic donors) well adapted to grafting processes, the immobilization of an extensive palette of available functions has been certainly slowed down. We attempt to present here a general method that permits the development of pyridine electrodes at materials like carbons and gold providing highly adjustable material surfaces. However, pyridine is not cathodically electroactive in its neutral form [5], except at lead cathode [6]. Recent works based on nitrogen ligands and their electro-catalytic capacities were reported [7]. In the field of surface modification, a specific immobilization of pyridine itself could be proposed by employing a diazonium intermediate connected to a rigid phenylethynyl link [8, 9].

In the present chapter, a new and certainly valuable method for the generation of pyridine electrodes is proposed, owing to the unexpected electro-chemical reactivity (in non-aqueous polar solvents in the presence of tetraalkylammonium (TAAX) salts) of electrophilic reagents belonging to the family of 4-ω-haloalkylpyridines. Quite amazingly, these derivatives are quite readily electroreducible at gold surfaces (Scheme 17.1), especially iodides, despite the fact that neither pyridine nor chloro and bromoalkanes show reduction waves at gold or glassy carbon at potentials > −2 V. Also, the method could be extended to bipyridines (BiPy): unexpectedly but successfully, for the first time, the creation of a {BiPy}–gold

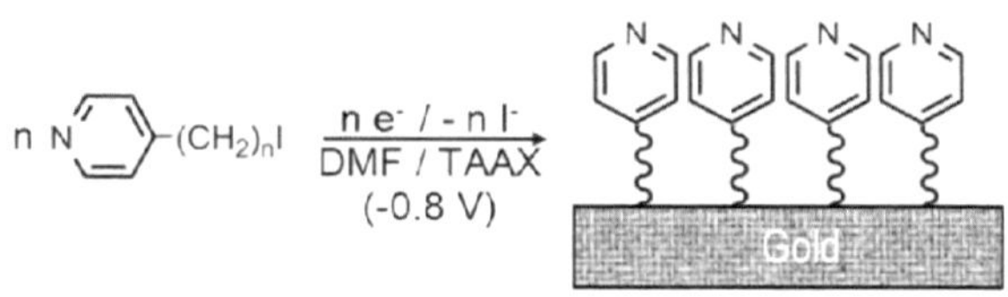

Scheme 17.1. Principle for a cathodic attachment of pyridines onto solid surfaces (carbon and gold) and additionally used for the immobilization of bipyridines for leading to {BiPy}–gold electrodes [14].

electrode was obtained and tested positively in the presence of transition metals.

17.2. Experimental Conditions

In this presentation, 0.1 M solutions of tetra-*n*-butylammonium and tetramethylammonium tetrafluoroborates ($TBABF_4$ and $TMABF_4$) in dimethylformamide (DMF) are conveniently used. It is worth mentioning that procedures given hereafter do not require extremely dry solutions (water content <1000 ppm). All electro-chemical experiments were performed under an inert atmosphere by bubbling dry argon through the solution.

The reported potentials refer to aqueous $Ag/AgCl/KCl_{(sat)}$ electrode. The electro-chemical instrumentation has been previously described [4].

All electrodes used in voltammetry had apparent surface area of 0.8 mm^2. Before use, gold and glassy carbon electrode surfaces were carefully polished with silicon carbide paper (Struers 500 and 1200) or with Norton polishing paper (types 02 and 03) and rinsed with water, then alcohol, and at last with acetone. After recurrent scans or fixed potential electrolyses, electrodes were sonicated for at least 2 min. Finally, before analysis, electrodes were dried with a hot air flow (60°C) for about 30 s. Coulometry measurements and electrolyses of pyridines substituted by alkyl halides reported in this chapter were carried out using three-electrode cells with a total catholyte volume of about 5 mL. A fritted glass separated the two compartments.

The preliminary work presented here essentially focuses on the grafting using pyridines of general formula $p[X{-}(CH_2)_n - C_5H_4N]$ (**1**) $X = I$, $n = 7$, (**2**) $X = Br$, $n = 8$ and (**3**) $X = Cl$, $n = 7$. The mechanism leading to the establishment of pyridine layers, presumably involving radical transients and pyridinium salts, at least with iodides, and their reactivity and applications in the field of surface catalysis are discussed in the course of this contribution.

Importantly, haloalkyl pyridines and bipyridines used in this work were synthesized from 4-methylpyridine (Aldrich) using

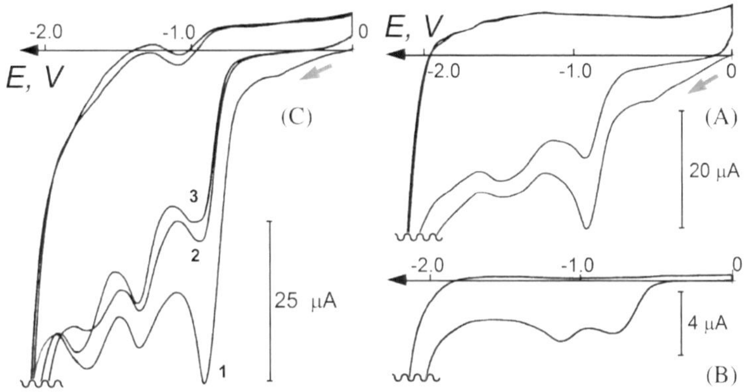

Figure 17.3. Voltammetric responses of a solution of $BiPy(C_7Cl)_2$ in DMF/0.1 M $TBABF_4$. Concentration: 2.6 mmol L^{-1}. v = 50 mV s^{-1}. (A) GC electrode; area: 7 mm^2; first three scans. (B) Smooth gold electrode; area 0.7 mm^2. (C) Response after addition to the initial solution of Py–C_8–Br (concentration 8 mmol L^{-1}) at a GC electrode [14].

solution of 6-iodohexylferrocene leads (after ultrasonic rinsing) to obtaining a very sharp fully reversible signal assigned to the immobilization of ferrocene via an alkyl pyridinium transient. Quite presumably, this reduction peak can be assigned to the indirect reduction of an iodoalkane via the transient pyridinium ion formed with pyridine. Because of ferrocene marking, the current integration leads to a superficial concentration $\Gamma_{Py\text{–}Fc} = 9\times10^{-10}$ mol cm^{-2} that supports pyridinium moieties are mainly integrated in a monolayer. Moreover, the surfacial concentration of ferrocene was separately estimated through (curve D) its reversible anodic voltammetry ($\Gamma_{Fc} = 3\times10^{-9}$ mol cm^{-2}). This value could be regarded as larger than expected, but the exact setting of immobilized ferrocene (onto gold or onto pyridine) remains for the moment uncertain.

17.4. Reactivity of Bromo- and Chloroalkyl Pyridines as Solid Surfaces

Quite astonishingly, the reduction of ω-haloalkylpyridines at gold electrodes leads to very similar results: as a matter of fact, ω-bromooctylpyridiene and ω-chlorooctylpyridine exhibit identical two-step patterns with peaks at −0.78 and −1.04 V. These two steps could simply be assigned to two successive one-electron transfers, the first one leading to a free alkyl radical readily grafted to the surface. This behavior is rather classical at Pd or Au electrodes [13]. The absence of a sharp growing peak (as already observed with iodides, (see Section 17.3)) could be due, in the case of chlorides and bromides, to a quite slow formation of the intermediate pyridinium salt. The modified electrode after several scans with Br–R–Py affords an irreversible step (E_p = −1.05 V), assigned to the reduction of a pyridinium cation. In order to support this assumption, pyridine deposition onto GC and Au was checked by treating the produced surfaces with 0.1 N HCl or $HClO_4$ solutions: after rinsing and drying, FTIR spectra of these electrodes show (Figure 17.4) vibrations that could be assigned to immobilized pyridinium salts and, additionally, a corresponding voltammetric peak appears within the potential range of pyridinium reduction. Inhibition of the produced surfaces by alkylpyridine deposits could also be characterized by EIS experiments [14].

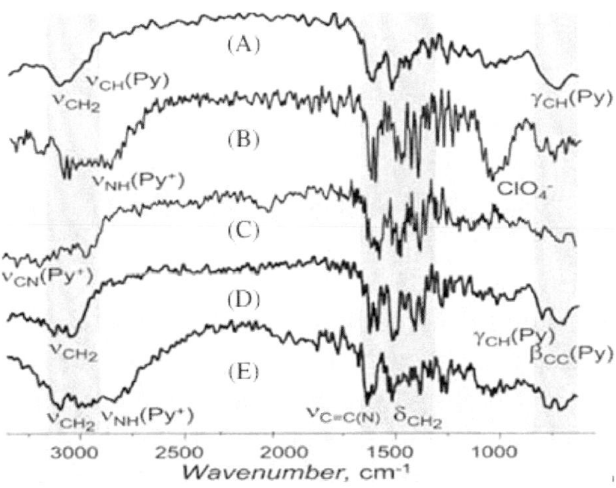

Figure 17.4. FTIR responses of the pyridine layers at carbon surfaces: (A) GC modified using $Py(CH_2)_8Br$. (B) Same electrode after dipping in 0.1 M $HClO_4$. (C) GC–$(CH_2)_8Py$ electrode reacted with $C_5H_{11}I$. (D) Au electrode modified using $Py(CH_2)_7Cl$. (E) Same electrode after contact with 0.1 M HCl [14].

17.5. Introduction to a General Mode of Bipyridine Graftings

The reduction of complexes *in situ* formed by 2,2'-BiPy or by other "non-innocent" N-heterocyclic ligands with halosilanes [15] and halogermanes [16] appears as an example of electro-catalysis, allowing to direct the dissociative two-electron reduction of these haloderivatives to one-electron cleavage of the M–X (M = Si, Ge; X = F, Cl, Br, I) bonds. This is due to the lucky overlapping of two circumstances: (i) the ability of the elements of the group 14 for hypercoordination (most nucleophilic reactions of these compounds are nucleophile-assisted) and (ii) a remarkable (often exceeding 1–1.3 V) decrease of the reduction potential of the electroactive ligands by their coordination with the Lewis acid halo compounds. Silicon- and germanium-centered radicals, generated from such reductions, can be observed by spin-trapping; otherwise they dimerize or react with appropriate radicalophilic reagents. Additionally, results gathered above with haloalkylmono-pyridines in Section 17.3 encourage the ongoing grafting of different BiPy (such as (**4**)), for which the structure is detailed below:

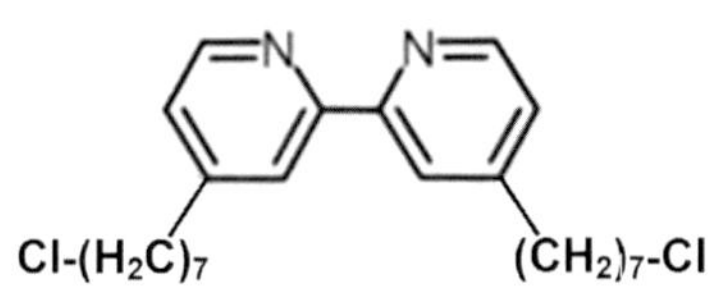

BiPy (4)

From Figure 17.3, it is observed that BiPy **4** (despite the two chlorine substituents) could be attached [14] to the gold interface (astonishing but convenient catalytic reaction at gold with R–Cl moieties), while, in contrast, a similar process exhibited in Figure 17.3 (curve A) does not permit any attachment of BiPy onto the GC surface. In curve C is shown a tentative mixed electrolysis. The project was to build a mixed layer of pyridines and BiPy **4** on the same electrode, which actually efficiently works at Au, as seen in Scheme 17.4. The possibility for attached BiPy to afford the formation of corresponding bipyridinium salts was verified (Figure 17.5(D) as well as Schemes 17.4. and 17.5).

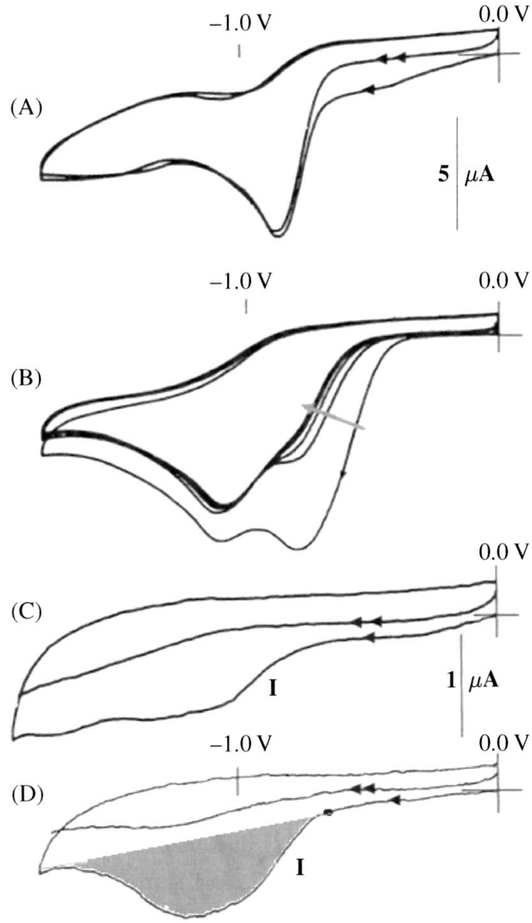

Figure 17.5. Voltammetries of BiPy **4** (concentration: 4.4 mmol L^{-1}) in DMF + 0.1 M $TBABF_4$ [14]. (A) At a GC electrode (area: 0.8 mm^2), scan rate: 50 mV s^{-1}. (B) At a gold electrode (same surface area), scan rate: 100 mV s^{-1}. After grafting under the condition of curve shown in (B), the grafted BiPy could be reacted with 4-methylbenzylbromide (in C) and with 1-iodopentane (in D). The integration (in dark grey) step (I) — first scan — exhibits a strong level of coverage (15×10^{-9} mol cm^{-2}), which could justify the formation of a multi-BiPy-layer induced (at the level of (B)) by catalytic generation of radicals from the reduction of C–Cl bonds.

disclosing an unexpected field in the cathodic reactivity of metals such as (at least) platinum, palladium, and also rhodium. This list is probably not exhaustive. It was discovered by Simonet [23–37, 40, 41] that platinum may react using electron transfer with common electrolytes (alkaline or TAAX salts). The cathodic behavior of platinum is totally changed by using super-dry solutions with common electrolytes, specifically in aprotic organic polar solvents. That started with the obtaining of reduction steps with TAAX salts and alkaline cations at fairly negative potentials (typically displayed in Figure 18.1). Why can tetramethylammonium tetrafluoroborate in DMF when "reduced" at platinum or palladium exhibit reversible

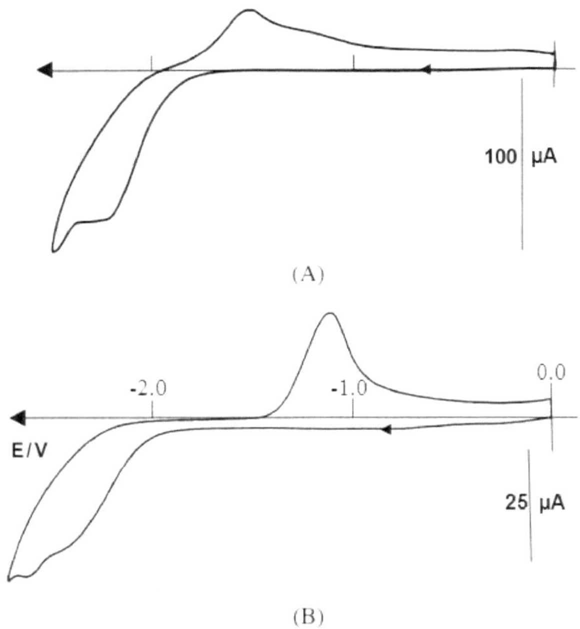

Figure 18.1. Voltammetries displaying the reversible charge of platinum and palladium electrodes in super-dry conditions. Solvent: DMF. Salt: 0.1 M.L^{-1} Me_4NBF_4 (*tetra-n*-butylammonium tetrafluoroborate ($TMABF_4$)). Presence *in situ* of activated alumina. Scan rate: 100 mV s^{-1}. Surface areas of electrodes: 0.8 mm^2. (A) Au–Pt electrode obtained by galvanostatic deposit of Pt onto smooth gold from an acidic solution of H_2PtCl_6. Charge: 10^{-4} C mm^{-2}. (B) smooth Pd electrode. Reference electrode: Ag/AgCl/KCl_{Sat} electrode [45].

charge processes? The elements undergoing cathodic reactions of precious metals and their applications in analysis and catalysis are discussed in the following.

18.2. Experimental Conditions Required for a Cathodic Charge of Platinum and Palladium

In most of the experiments described in this chapter, the solvent is mainly DMF. It is essential to use almost perfect aprotic conditions in order to avoid hydrogen evolution, which is the competing reaction to the cathodic charge of the substrate. The method used for getting dry solvents consists in adding *in situ* activated neutral alumina in the electro-chemical cell. In this way, the maximum level of water in the electrolytic solution may currently reach 50 ppm. For this, alumina was activated by heating at 320°C during 4 h under vacuum. Also, DMF was almost currently analyzed (by the Karl Fischer method) to ascertain the water content. Additionally, a permanent storage of the salt–solvent over alumina ensured best results. In most of the experiments described in the present chapter, the electrolyte concentration was 0.1 M. Alternatively, alkaline and earth-alkaline metal iodides (potassium, sodium, lithium, caesium iodides) were used (because of their larger solubility). TAAX salts (puriss grade) were of >99.7% purity. They were dried by conventional methods before use.

Cyclic voltammetry investigations were carried out in a standard three-electrode cell. For analytical studies, the working electrode was a smooth disk of metal (here essentially Pt and Pd) having a surface area of 0.8 mm^2. The counter-electrode was a GC rod separated by a glass frit from the catholyte. Most of potentials are referred to the aqueous SCE. However, it appeared of importance to use also the Ag/AgI/0.1 M $INBu_4$ system (in DMF) to avoid possible water diffusion in the cell. The potential shift is –0.52 V vs. SCE.

Prior to experiments, platinum and palladium electrodes were carefully polished with silicon carbide paper of successively smaller sizes (18–5 μm) and then by diamond powder (6 and 3 μm). Finally, electrodes were rinsed with ethanol and acetone and then dried with a hot air flow. Lastly, for macroelectrolysis investigations, Pt sheets (99.99% purity, surface area of about 1 cm^2, thickness 0.05 mm^2) were employed.

Coulometric and EQCM experiments were carried out on thin metallic layers prepared by deposition of metals from solutions of 10 g L^{-1} H_2PtCl_6 in 0.1 M HCl onto polished gold disks (2×10^{-3} cm^2). The plating was achieved in a galvanostatic mode (current 10^{-2} A cm^{-2}). All the experiments were performed with gold substrates because this substrate was found to be very weakly reactive toward salts within the potential ranges explored. Gold substrate could be a gold microelectrode polished before each deposition.

For achieving electro-chemical quartz microbalance experiments (EQCM), much larger electrode consisting in gold-coated quartz crystal were used. Mass balance experiments were carried out with an oscillator module quartz crystal analyzer connected to a potentiostat. The instrumentation has been described elsewhere [25, 40]. Platings were also achieved by using galvanostatic procedure. The deposited mass was additionally confirmed by EQCM. The EQCM measurements were performed in a Teflon cell equipped with a GC counter-electrode, a reference electrode, and the metalized quartz crystal working electrode. The apparent area of the quartz crystal was about 0.2 cm^2.

Micrograviметric data, reported in terms of mass change, were calculated using the Sauerbrey equation, which links the resonant frequency and mass change. The EQCM was found to be a valuable method to quantify, under the given experimental conditions, the mass cathodic change of concerned metals (platinum and palladium) in the presence of a large variety of salts.

18.3. Platinum Charge in Super-Dry Conditions

18.3.1. *Behavior in the contact of alkali metal salts*

First of all, the charge of platinum was initially observed in the presence of alkaline cations (Li^+, Na^+, K^+ and Cs^+) always associated to iodide anion. Thus, alkali metal iodides display, at a Pt microcathode, a quasi-reversible peak (as displayed in Figure 18.2). It could be

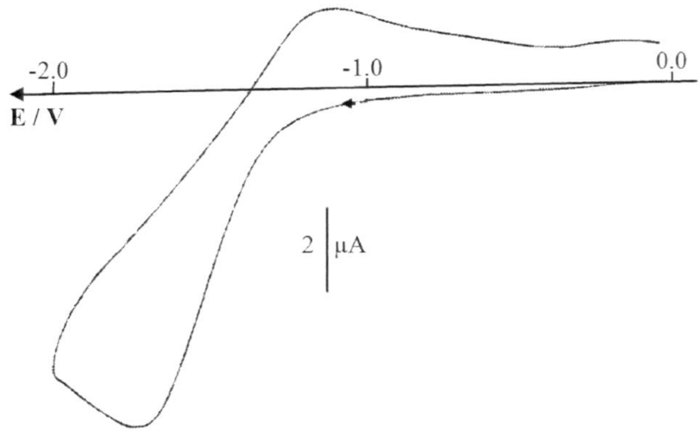

Figure 18.2. Voltammetric response of a platinum disk (surface area: 8×10^{-3} cm^2) in a solution of NaI 0.1 mol L^{-1} in DMF. Super-dry conditions. Scan rate: 200 mV^{-1}. Ag/AgI/I$^-$ electrode [27].

Table 18.1. Stationary platinum cathode, area 8×10^{-3} cm^2: Voltammetric data relative to charge–discharge processes for some alkaline iodides in super-dry DMF.

Electrolyte (0.1 M)	E_{pc} (V)	I_{pc} (μA)	E_{pa} (V)	$E_{pc} - E_{pa}$ (V)
LiI	−2.84	−5.0	−2.42	0.42
NaI	−2.07	−10	−1.72	0.35
KI	−2.17	−6.4	−1.62	0.55
CsI	−2.18	−5.4	−1.77	0.41

Notes: E_{pc} and E_{pa}: Peak potentials and I_{pc}: Peak currents (cathodic **c** and anodic **a**). Scan rate: 200 mV s^{-1}. Potentials are referred to the SCE.

demonstrated that the anodic step corresponds to the oxidation of the cathode material clearly generated at the level of the cathodic step. Charges noticed with alkali metals ions required generally rather negative potentials ($E < -2$ V vs. SCE) as displayed in Table 18.1. All cathodic steps described under those experimental conditions are diffusion-controlled.

One may notice that quite small current densities for cathodic steps are related to a "reduction" of platinum. It is expected (and shown hereafter) that the limiting diffusion reveals a slow insertion of both cations and salt expected to be mostly reacting under ion-pairs into the metallic bulk. Also, these cathodic steps are not rendered smaller when activated alumina is progressively added to the electrolytic solution, which allows to exclude almost totally the reduction of residual water. On the other hand, the reduction of alkali metal cations for leading to the corresponding metal at the platinum surface is unlikely. Moreover, limit currents were established to depend on both the concentration of the salt and on the nature of the cation. Specifically, in the presence of alkali iodides, currents observed at platinum follow the order:

$$Cs^+ < K^+ < Na^+ \ll Li^+$$

In order to get more information about these charge–discharge processes, employing the coulometric technique appeared helpful. The idea was that the amount of electricity stored in the material (or at its surface) during the cathodic process should be able to be anodically restored. However, it is unlikely that the charge system can be found fully reversible owing to traces of residual water. Thus, charge values found during the cathodic reduction, in the course of rather long fixed potential electrolyses, were believed to be excessive obviously disturbed by residual moisture reduction.

The actual explanation bound to the existence of a cathodic charge stored in material and the access to its evaluation lies in the discovery of a reversible redox process for platinum process (by measuring the recovery of the amount of electricity Q_a by anodic oxidation (see Figure 18.3)). Thus, values Q_a become invariant with long-enough charge processes and can be assigned to the total reduction of platinum. For this, the amount of platinum consumed by the redox process has to be very small. Experimentally, they correspond to thin deposits of Pt at gold electrode surfaces (the choice of a gold substrate is supported by the fact that this metal does not give appreciable insertion of salts [44]). Consequently, all experiments described hereafter will take into account those maximum charge values.

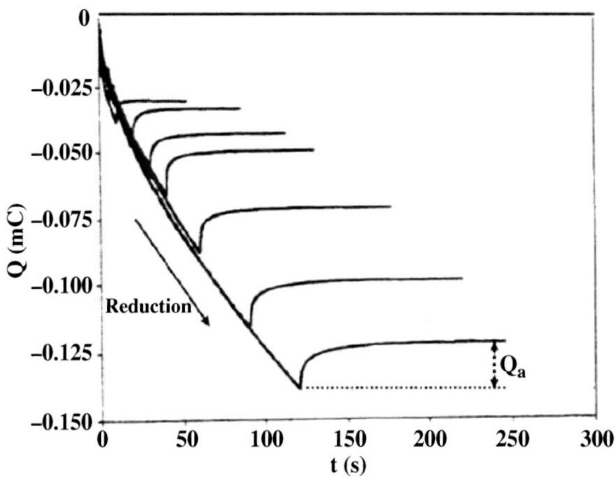

Figure 18.3. Successive charge–discharge processes of a platinized smooth gold disk (area: 0.8 mm^2) at potentials of −1.65 V (charge) and 0 V (discharge). Electrolyte: 0.1 M NaI in super-dry DMF. Amount of deposited platinum: 60 μg. Variation of electricity amounts Q_a (recovered electricity during the re-oxidation process) upon charge times. Q_a values upon amounts of charge gets a limit that is assigned to the external platinized layer saturation [27].

In order to quantify exactly the proportion of metal involved in the insertion process, very thin layers of platinum deposited under galvanostatic conditions on a gold or GC surfaces were achieved and used as working electrodes. The aim was to adjust the amount of deposited platinum. When gold (which wouldn't afford any appreciable insertion of electrolyte) was used as the substrate, the accuracy of the amount of Pt was also determined by the EQCM technique. Experimental conditions are those given in Figure 18.3: the length of the electrolysis time (cathodic) has to be long enough to ensure the ***total*** charge of the deposited platinum layer. Thus, as displayed in the example displayed in Figure 18.3, the reduction of Pt in contact of a solution ICs/DMF permits to obtain the Q_a limit and it is remarkable that different Q_a values appear almost perfectly proportional to various electrodeposited Pt amounts. The slope of lines $Q_a = f$ (Pt amount) was found to be 2 with an accuracy of 2%. The assumption relative to a reaction in mass is therefore verified and then greatly supports the hypothesis of an insertion of alkali metal cations M^+ inside platinum layer. Thus, in the case of CsI, the iono-metallic formula:

$$[\mathbf{Pt_2^-, M^+, X}]$$

could be proposed. However, it is now necessary to ascertain the insertion of which species (X being quite certainly neutral, possibly the solvent or the salt) is concomitantly involved in the process. Consequently, the control of the Pt sample is required to be achieved in the course of the charge. Experiments performed by using the EQCM technique help to define the true formula of iono-metallic layers (Figure 18.4).

Therefore, it ***definitively*** appears that the cathodic charge operated at the level of the redox step observed for Pt metal, in the presence of different alkali metal iodides, generates a mass increase. Interestingly, the maximum mass increase is proportional to the initial mass of deposited metal and does not depend apparently on

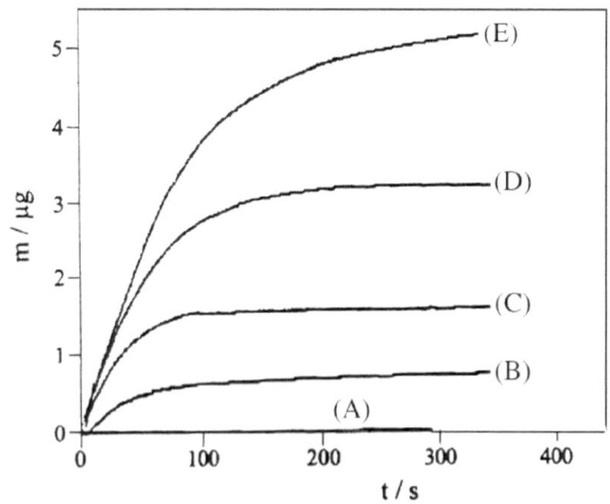

Figure 18.4. EQCM experiments: charge of platinum upon time in the presence of 0.1 M CsI in super-dry DMF. Applied potential: −2.3 V vs. SCE. (A) Unmodified golden quartz. (B, C, D and E) the same substrate covered with different amounts of platinum. Deposited masses of Pt are 0.9, 1.5, 3.5 and 6 μg, respectively [27].

Table 18.2. Global determination of stoichiometries of iono-metallic phases $[Pt_2^-, M^+, (MX)_y]$ from coulometric measurements and use of the ECQM technique (values *y*) obtained with platinum in super-dry DMF in the presence of alkaline iodides [27].

Electrolyte ($0.1\ mol \cdot L^{-1}$)	Platinum
LiI	$[Pt_2Li(LiI)_{1,06\pm0,06}]$
NaI	$[Pt_2Na(NaI)_{0,98\pm0,08}]$
KI	$[Pt_2K(KI)_{0,97\pm0,08}]$
CsI	$[Pt_2Cs(CsI)_{0,88\pm0,05}]$

the salt concentration in the liquid electrolyte bulk. As a matter of fact, in the special case of cesium iodide, the linear mass increase until a total saturation (for a time > 200–300 s) enables proposing a general formula ***at saturation*** of the platinum phase. When taking into account several salts, coherent stoichiometries for iono-platinum phases were carried out, which are displayed in Table 18.2. This approach establishes the remarkable uptake of one electron by two atoms of platinum, concomitantly with the insertion of one cation ***together*** with one equivalent of the electrolyte salt (or an ion pair). Accordingly, a realistic stoichiometry of the fully modified phase is coherent with:

$$[\mathbf{Pt_2^-, M^+, MI}]$$

The coherent accuracy (Table 18.2) of EQCM experiments permits to leave out the insertion of the solvent in a large amount. That should also exclude any additional minor parent phases. It is expected that the mass increase of the Pt interface might be very large and the insertion of bulky ions (such as TAAX cations) leads to noticeable swellings, which is perfectly verified. Consequently, as discussed below, the electronic conductivity of the external layer dramatically drops down and globally the ion insertion leads to the material passivating. With massive pieces of platinum, owing to the rapid lack of conductivity, the insertion remains limited within a thin layer and does not exceed a few microns. The reversibility of the charge process allows expecting that such reduced phases are efficient reducing species and this assumption is perfectly verified by experiments. Consequently, the stability of these phases in the presence of air is generally limited and mainly generates superoxide.

18.3.2. *Charge process in the presence of TAAX salts*

The noticeable difference existing between ammonium salts and alkaline iodides toward platinum appears to be a certain lack of the reversible character of the charge with bulky ammonium cations (when > NEt_4^+). Also, fairly negative potentials are then necessary to get at least rather slow charge processes. Discharges could be obtained, but the potential difference ΔE between these charge–discharge processes is in most of the cases significant and may actually reach >1 V. As previously described above with alkaline iodides, coulometry and EQCM technique were efficiently used to determine stoichiometries of platinum phases with a large palette of TAAX salts. Rather generally, voltammetries exhibit quite slow charge processes (transfer coefficient values $\alpha < 0.2$). Especially for smooth massive platinum, the electro-chemical rate (in contact with a large palette of TAAX) appears strongly dependent on the size of both the anion and the cation. In a certain way, the ease of the discharge process appears to be correlated to the rate of the charge process.

In all the cases, results relative to saturated forms were coherent and formula, whatever the size of the considered TAA^+ cations (Table 18.3) and the structure of the associated anion, have the general form:

$$[Pt_2^-, R_4N^+, R_4NX]$$

The difference between charge and discharge in terms of energy may reach very large values. With bulky TAAX salts, the intensity of the cathodic step is significantly greater than the current observed for the discharge. It has been shown that the charge amounts are notably dependent on the potential boundary. Thus, in the case of

Table 18.3. x and y values corresponding to stoichiometric experiments relative to phases [Pt_x^-, R_4N^+, $(R_4NX)_y$] obtained at platinum deposits with several TAAX salts.

Electrolyte (0.1 mol·L^{-1})	Applied potential (V)	x value	y value
Me_4NI	−1.90	2.04 ± 0.12	0.97 ± 0.04
Et_4NI	−2.20	2.08 ± 0.20	0.99 ± 0.06
Et_4NBr	−2.20	1.91 ± 0.11	1.05 ± 0.17
Bu_4NI	−2.30	2.06 ± 0.09	0.95 ± 0.06
Hex_4NBr	−2.35	2.05 ± 0.15	0.92 ± 0.06
Oct_4NBr	−2.35	2.07 ± 0.07	1.02 ± 0.05

Note: Potentials are referred to Ag/AgI/(0.1M)I^- electrode [27].

$n\text{Bu}_4\text{N}^+$, if the boundary of the forward scan does not exceed −2.3 V, the mass increase reaches a maximum and the backward scan exhibits a progressive but effective discharge all along the scan. On the contrary, when more negative potentials are reached, the gained mass suddenly collapses and suggests an abrupt instability of the implied cations in the insertion process. Therefore, strong decays of currents in the course of the charge at very negative potential are expected to accompany the electro-chemical degradation of the electrolyte and that, indirectly, suddenly stops the charge process (Figure 18.5). One then may obtain the electro-chemical reduction of the salt leading to electro-generated bases that trigger via a Hofmann degradation type (the degradation of NR_4^+ (when R ≠ Me) via a base catalysis). This hampers insertion and forbids continuing charge process (Scheme 18.1).

18.3.3. *Evidencing smooth platinum surface changes by SEM*

It was underlined that charge processes swelling could result in insertion, especially when rather bulky cations such as Cs^+ or Bu_4N^+ and anions like PF_6^- are involved. However, when samples are used according to procedures described before, it is worth recalling that those materials were afterward unavoidably oxidized by air when

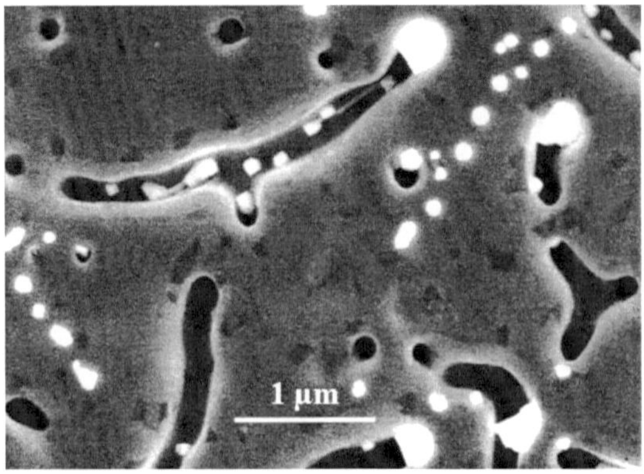

Figure 18.7. Electro-chemical corrosion of platinum by reduction (–2. V vs. ECS) by contact with CsI (0.1 mol L^{-1}). Super-dry conditions. Amount of charge: 10 C cm^{-2} [27].

insertion. Still in the presence of CsI, but when injecting much lower amounts of electricity, platinum surface is unexpectedly changed into holes and channels (Figures 18.6(A) and 18.7). Alternatively, the platinum charge in the presence of TAAX salts produces amazing swellings with layers of spheres, which could be unaltered by undoping (loss of TAA^+, producing porous Pt layers of high specific surface and spectacular growing of spheroids as shown in Figure 18.6(B)). This appreciable enlargement of the active Pt surface was simply verified by covering of the interface by a PtO monolayer and its coulometric determination by reduction [35]). Additional views of Pt surfaces after reductions are displayed in Figures 18.8 and 18.9.

18.3.4. *Following of the platinum charge–discharge in real time: Use of in situ EC–AFM*

A decisive support for a dynamic knowledge of the platinum charge process operating concomitantly with the ion insertion in the material was brought by Hapiot in the Rennes Electro-chemistry Laboratory given his indisputable experience in the field of the

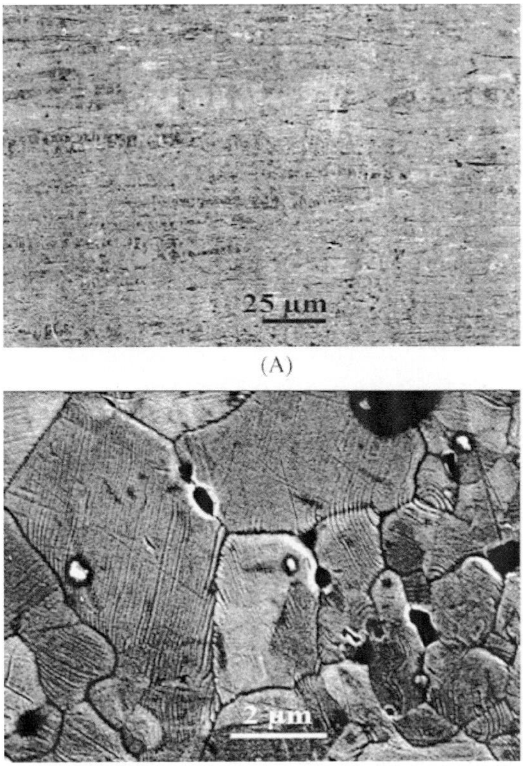

(B)

Figure 18.8. Surface changes of smooth platinum under electro-chemical reduction (super-dry conditions) in a solution: DMF + 0.1 M mol L^{-1} TBAI. (A) Pt surface prior to reduction. (B) After reduction at −2.5 V vs. SCE. Amount of electricity: 10 C cm^{-2}. Note changes at the grain surface that turns being profoundly striated and pierced by cavities [27].

EC–AFM imaging technique. For this, an experimental cell that allowed obtaining *in situ* charge and discharge of platinum in polar (and somewhat corrosive) inert atmosphere has been conceived [37–39].

As previously expected, platinum's reduced phase [Pt_n^-, M^+, MX] could be electro-chemically re-oxidized. In order to check the real reversibility of platinum's cathodic modification at the micrometer scale, and follow the structure changes of the Pt surface in

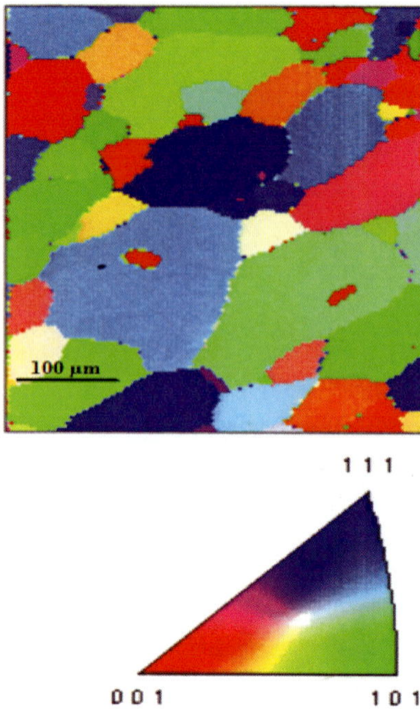

Figure 18.9. Revelation (backscattering MEB) of grain orientation and boundaries by means of cathodic polarization under electro-chemical treatment of platinum under super-dry conditions. Experimental conditions are of the same as in Figure 18.7 [45].

real time, the charge–discharge process was strictly achieved *in situ* by the EC–AFM imaging technique. Thus, the electro-chemical transformations of platinum surfaces and the AFM imaging were, at the same time, completed. The great novelty of such experiments was to thoroughly follow the progressive change of the interface morphology upon the amount of the injected charge in real time as a function of the nature of the supporting electrolyte. By means of the *in situ* EC–AFM technique, decisive observations help to corroborate data already obtained by SEM. In particular, this technique permits to follow the progressive swelling of platinum during the charging process and to verify the structural reversibility of the overall process. Initial experiments were performed with alkaline iodides (solution: DMF + 0.1 mol L^{-1} of the salt). The chosen cathode material was Pt/MgO and its chemical stability toward the liquid electrolyte was verified by AFM by an immersive contact for 1 h. Experiments then proceeded through charge injections, which were performed by repetitive scans. After each scan, the sample was disconnected but maintained in the DMF solution under argon. The injected charge was quantified and the image recorded. The stability of surface changes during different observation times was checked and any significant modifications could be noticed for time durations not longer than 10 min.

Typical AFM topography images obtained upon the applied charge density are displayed in Figure 18.10. One may observe structure changes of Pt samples upon applied potential before and during the scans. Thus, prior to the charge, bright and dark areas (shown by image **a** correspond respectively to the platinum plates) are obtained in the course of the epitaxial growth of the Pt deposits under precise deposition conditions. Now, in the course of charge processes, limited at a boundary potential E_r less negative than –0.8 V, AFM images did not show any significant change. In contrast, image b, beyond boundary potential E_r (–1.4 V vs. SCE) that corresponds to the threshold of platinum corrosion process and small grains start to grow on the Pt platelets, while the global platinum structure roughly remains. When the potential boundary is moved up to E_r = –1.7 V/SCE, which favors a much larger insertion (in the given example, injection of a charge density of around 30 × 10^{-3} C cm^{-2}), huge changes appear in the metal surface (see image (c)). Then, bulky spheroids appeared for strongly perturbing the initial structure. Thus, progressively, the surface roughness increases until a value of 70–100 nm is reached. Such spectacular morphological changes obtained by surface swelling, clearly depending on the injected charge ***definitively*** support the reactivity of platinum toward the electrolytes. It is then much likely that swellings attest to the formation of iono-metallic phase, implying the electrolyte.

Crystallographic data are not available so far for structural forms such as [Pt_2^-, M^+, MI], for which the corresponding crystallographic parameters are obviously very different from those of

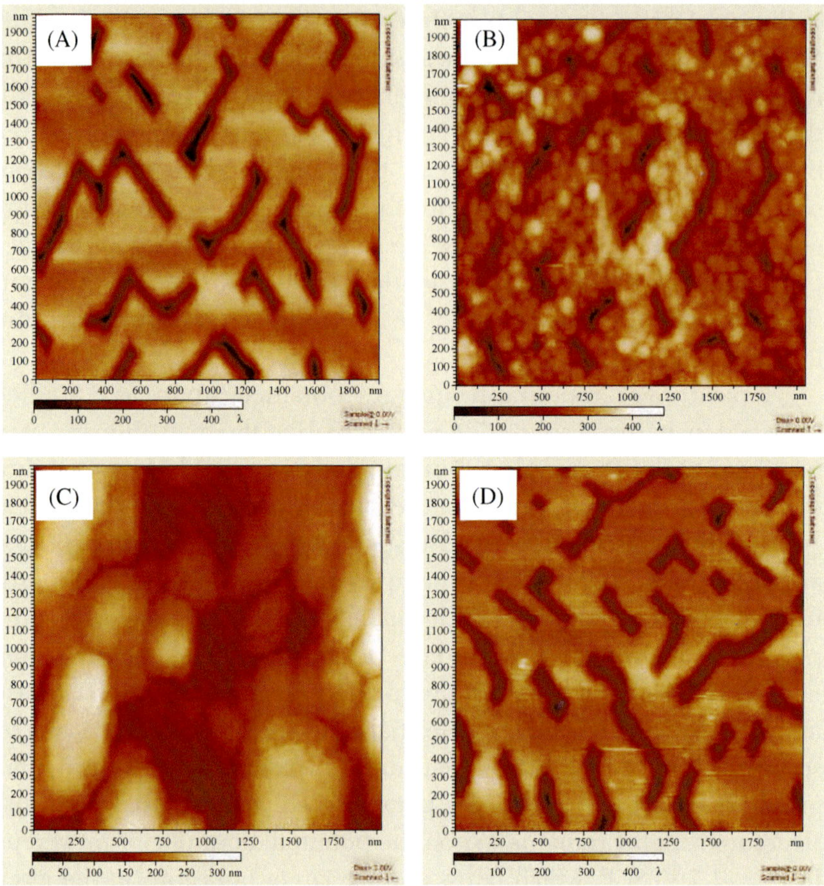

Figure 18.10. EC–AFM images for the *in situ* modification of platinum in a solution of CsI (0.1 mol L^{-1}) in DMF. (A) Pt substrate in solution before electrochemical reduction. (B) After injection to the surface of a charge density of 5.3 mC cm^{-2}. (C) After passing a charge of 43 mC·cm^{-2}. (D) Structure of the surface sample after total electro-chemical re-oxidation. Images displayed here concern surface areas of 2×2 μm^2 [43].

(100)Pt. Any attempt to get information in that sense failed because the electro-generated phase is not compatible with the MgO substrate and causes an irreversible layer desegregation and dispersion in solution.

Finally, if a (0.0 V) potential is now applied to the reduced platinum layer sample [Pt_2^-, Cs^+, CsI], one then may note the almost total re-oxidization of the charged product. If the initial injected reductive charge is made sufficiently low to avoid layer desegregation, the surface morphology of the sample as displayed in Figure 18.10(D) is almost identical to the initial pattern of the virgin sample [38, 39, 42, 43].

Similar EC–AFM experiments were performed with several salts such as CsI and KI. The remarkable structural reversibility of the process and the stability of the layer after ***several*** redox cycles between 0.0 and −1.6 V was demonstrated and confirms the existence of charge–discharge processes. Especially with all alkaline iodides salts, recovering of the initial structure of the Pt sample after modification was verified either after electro-chemical re-oxidation or after contact of the reduced sample, for few hours, to air. Additionally, TAAX salts were similarly tested by AFM–EC for totally confirming charge reversibility as displayed above with alkaline iodides. In particular, clear results were obtained with tetramethylammonium iodide. It is worth noting that bulky TAA+ ions, less stable at very negative potentials as depicted by Figure 18.5, led to less meaningful results.

18.3.5. *Modified platinum phases as reducing reagents*

The stability of different reduced platinum phases (generated both in the presence of alkaline and TAAX salts) has been already discussed above. It was established that any platinum phase reacts fast with dioxygen (presence of air) as written below:

$$[Pt_2^-M^+, MX] + O_2 \longrightarrow O_2^{\bullet-} + M^+ + 2\,Pt + MX$$

$$[Pt_2^-M^+, MX] + O_2^{\bullet-} \longrightarrow O_2^{2-} + M^+ + 2\,Pt + MX$$

However, until now, there is no precise indication about the exact redox potential of the reduced phases that correspond to alkaline and TAAX salts. Additionally, the intrinsic chemical reactivity of

Table 18.4. Reduction of some organic π-acceptors (concentrations: 5 mmol L^{-1}) by the Pt–CsI phase.

π-acceptor	Standard potential (V vs. SCE)	% of the phase $[Pt_2Cs(CsI)]$ having reacted
Chloranile	−0.27	100
Phenanthrenequinione	−0.58	95
p-Dinitrobenzene	−0.61	100
3, 5-Dinitrobenzyl chloride	−0.80	100
Acenaphthenequinone	−0.83	98
Nitrobenzene	−1.10	92
9-Fluorenone	−1.28	67
Benzophenone	−1.75	6

Note: Use of a platinum sheet (area: 3.5 cm^2) after a charge of 1 C, which corresponds to an excess of reactant to the acceptor. The right column displays the percentage of the organic substrate reduced through electron transfer after 90s. Ratios were evaluated by means of voltammetries at a micro Pt electrode [27].

some of electro-generated phases was tested toward a large palette of organic π-acceptors (displayed in Table 18.4) and experimentally visualized by Figure 18.11 with the intense color of as-produced anion radical at the Pt surface by means of electron transfer.

Striking experiments were achieved with different organic π-acceptors (nitro-aromatic and cyano-aromatic compounds, aromatic ketones such as fluorenone) placed in contact with reduced platinum phases: one may determine the reducing power of the platinum phase acting as an electron donor leading then to reduced forms detected either by specific deep color of their anion radical (or dianion) or by means of Electron Spin Resonance (ESR) analysis in the case of single electron transfer (in the case of aromatic ketones). However, poorly efficient acceptors such as aromatic hydrocarbons (anthracene, pyrene, naphthalene) react very slowly with such reducing agents.

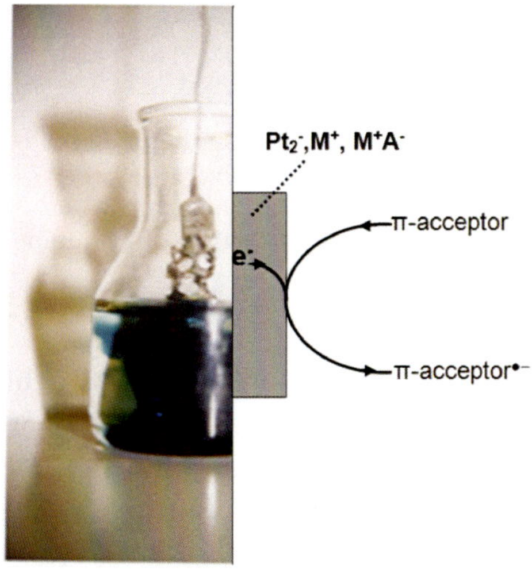

Figure 18.11. Reduction of 2,4-dinitro-toluene. Dark blue generation of its radical anion by contact with a platinum phase [Pt_2Bu_4N, Bu_4NI] potentiostatically produced separately in super-dry DMF. Electrode total area: 6 cm^2. Potential −1.8 V vs. SCE. Maximum charge 52 C. Dinitro-toluene (concentration 10 mmol L^{-1}) was dissolved in dry DMF maintained under insert atmosphere [45].

As revealed by Table 18.4, π-acceptors having a standard potential < −1.8 V do not react fast with electro-generated platinum phases.

Quite similarly, carbon–heteroatom bonds can be irreversibly cleaved by reducing platinum phases. As a matter of fact, alkyl and aryl iodides could be reduced by most of the [Pt_2M, MX] phases. Arylalkylsulfones ($ArSO_2$–R) behave similarly. For example, 2,5-dinitrobenzyl chloride is readily and quantitatively cleaved under these conditions. Many organic compounds RX possessing good leaving groups are cleaved according to Scheme 18.2 (general case of a two-electron reduction owing to the reducibility of the transient radical $R^{•}$).

Thus, phases like [Pt_2^-, Na^+, NaI] should react with molecules like aryldiazonium salts that display a reduction peak at around

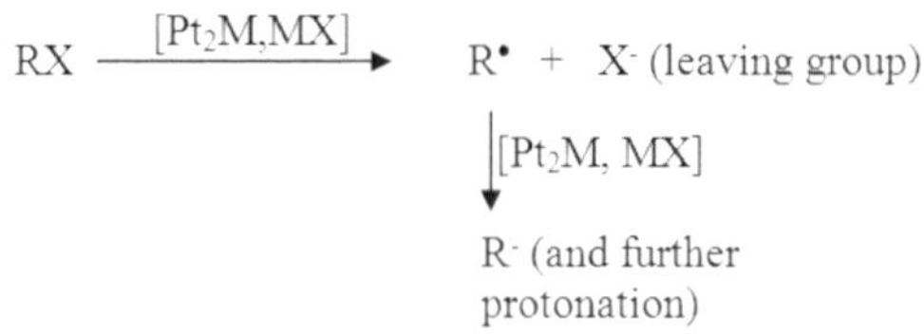

Scheme 18.2. Chemical reactivity of platinum phases.

0.0 V/SCE. The procedure consists first in achievement of the electro-generation of [Pt_2^-, Na^+, NaI] at –1.7 V/SCE during 100 s, then immediately after, to plunge the charged platinum sample into a deoxygenated DMF solution containing 0.1 mol L^{-1} of 4-nitrophenyl diazonium tetrafluoroborate [43].

18.4. Cathodic Charge of Palladium in Super-Dry Media

Similar studies were performed with smooth palladium under conditions strictly identical to those given with platinum (see Section 18.3): super-dry conditions, solvents: DMF or ACN, salts: alkali earth iodides and TAAX salts. Therefore, as a first example, a palladium electrode can be reversibly charged in the presence of CsI (Figure 18.12).

Under these conditions, charge and discharge processes could be also achieved with palladium. The whole electro-chemical process appears totally irreversible possibly limited by the diffusion rate of ions in the metal. However, one can note that the amount of charge electricity is equivalent to the discharge amount. Currents (cathodic and anodic whatever the chosen salt) at Pd electrodes are much larger than the ones already mentioned (see above in Section 18.3) with platinum electrodes under similar conditions.

With alkaline iodides, cathodic peak current was found to be diffusion-controlled and reduction potentials in the case of alkaline iodides are rather close to those already observed with platinum. With TAAX salts, TMAI displays a quasi-reversible system. On the contrary, more bulky TAA^+ cations considerably contribute to

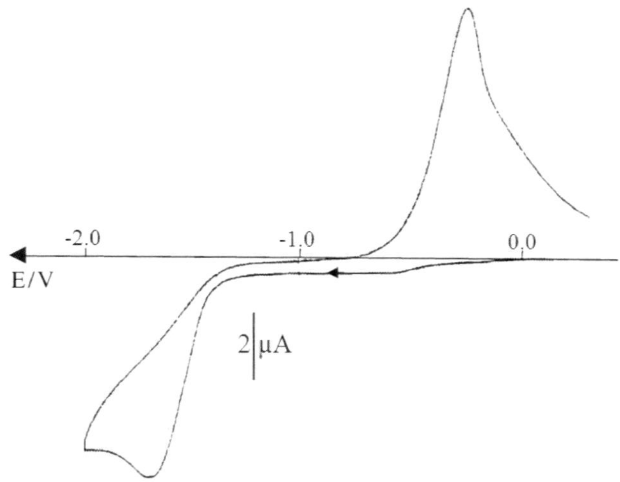

Figure 18.12. Voltammetry of a smooth palladium electrode (area: 0.8 mm^2) in a solution of CsI (0.1 mol L^{-1}) in super-dry DMF. Scan rate: 200 mV s^{-1} [27].

enlarge the irreversibility of the charge–discharge process. Thus, the potential difference (ΔE) between the cathodic and anodic peaks is about 0.7 V with Me_4NI, while it reaches almost 1.7 V with Hex_4NBr.

Reduced forms of palladium could be established, like with platinum, from limited charges in DMF with a large array of salts. Concerning alkaline metals, it has been found that the recovered charge Q_a (toward the mass of palladium deposited onto an inert substrate such as gold) corresponding to the number of electrons effectively stocked is proportional to the number of the involved palladium atoms. Then, Li^+, Na^+, K^+, and Cs^+ exhibit a stoichiometry that corresponds to one charge element and a two-palladium atom pair. Additionally, EQCM experiments were performed. Thus, the concomitant insertion of the salt MI with M^+ is then confirmed, and stoichiometries of the obtained limit phases correspond to:

$$[\mathbf{Pd_2^-, M^+, MI}]$$

Table 18.5. Values of x and y corresponding to experimental stoichiometries $[Pd_xNR_4(NR_4)_y]$ obtained from coulometric and gravimetric measurements on palladium sheets in the presence of different electrolytes at controlled potential (solvent: DMF under super-dry conditions) [27].

Electrolyte (0.1 mol L^{-1})	Applied potential (V)	x values for $(R_4N)Pd_x$	y value for $[Pd_xNR_4(NR_4X)_y]$
Me_4NI	−1.90	2.13 ± 0.09	0.92 ± 0.02
Me_4NCl	−1.90	4.02 ± 0.10	1.01 ± 0.03
Et_4NI	−2.20	2.01 ± 0.08	0.95 ± 0.05
Et_4NBr	−2.20	4.01 ± 0.16	0.98 ± 0.03
Bu_4NI	−2.30	2.10 ± 0.15	0.92 ± 0.04
Bu_4NBr	−2.30	4.07 ± 0.16	0.98 ± 0.04
Bu_4NCl	−2.30	4.05 ± 0.13	0.95 ± 0.04
Hex_4NBr	−2.35	4.09 ± 0.18	0.99 ± 0.03
Hex_4NCl	−2.35	4.04 ± 0.04	1.02 ± 0.08
Oct_4NBr	−2.35	4.04 ± 0.21	1.00 ± 0.08
Oct_4NCl	−2.35	4.18 ± 0.09	0.98 ± 0.09

The experimental accuracy is good and supports that obtained phases, at least at the level of charge saturation, do not have added solvents. These results obtained with palladium perfectly corroborate with alkaline iodide stoichiometries already obtained with platinum.

When charge processes are achieved in solutions containing TAAX salts, important discrepancies should be highlighted with palladium–R_4NX reduced phase stoichiometries. Table 18.5 exhibits phase formulas with a large number of TAAX salts and it appears that the nature of the anion may alter the stoichiometries of the obtained phases. More specifically, harder is the salt anion (strong donor effect), weaker is the energy storage inside the palladium phase, which could explicate the obtaining of phases of the form $\{Pd_4^-$, TAAX$\}$. These data are remarkable since the level of insertion does depend both on the bulkiness of the TAA^+ cation and the donor efficiency of the anion (Figure 18.13).

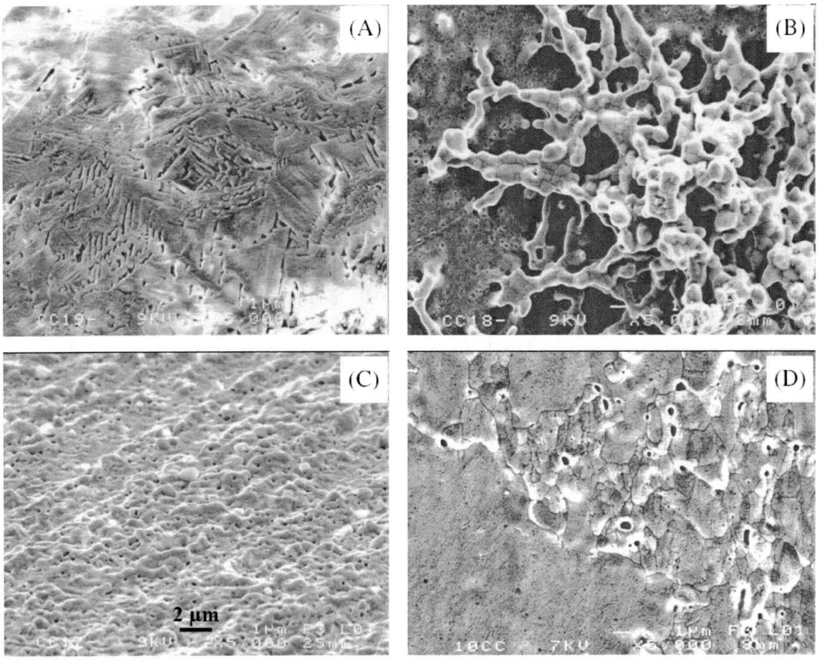

Figure 18.13. Morphology changes by insertion of different electrolytes in smooth palladium. The magnification is the same in the four pictures. All surface views correspond to palladium modifications in solution containing 0.1 mol·L^{-1} of salts. No sonication. Rinsing with water and acetone after reduction. (A) Pd polarized at −1.6 V in NaI, total charge: 24 C cm^{-2}, (B) in CsI at −1.8 V, charge: 25 C cm^{-2}, (C) in Me_4NI at −1.8 V, charge: 2 C cm^{-2}, and (D) in Bu_4NI at −2.3 V. Charge 1.2 C·cm^{-2}. The generation of extremely open materials suitable for the insertion of π-acceptor reduced forms or electro-generated organic anions (e.g., obtained from the *in situ* cleavage of alkyl iodides) by mean of exchanges in the Pd matrix between organic anions and salt anions ([27] and Chapter 19).

18.5. Concluding Perspectives

Platinum and palladium belonging to the column VIII of Mendeleev's classification may produce, under well targeted experimental conditions, unexpected insertion of ions (both cations and anions) by the way of reversible cathodic charges. These charged layers all exhibit precise stoichiometies that quite certainly present a structure of iono-metallic complexes. To a certain extent, they could look like hetero-polyanions, implying both Pt and Pd with anions derived from metalloids. Since they are very sensitive to air, their exact structure could not be, until presently, established by X-ray analysis.

Importantly, structures of the reduced forms could be viewed as a new family of reducing species by electron transfer (in fact quite analogous to amalgams) as well as catalysts dispersed in thick and conducting layers, opportunely pierced with numerous channels and holes estimated suitable for the migration of charges and ions when potentials are applied. Such highly porous structures (obtained in the course of charge processes) are printed in the metal bulk and do not totally disappear in the course of re-oxidation operations. This apparently permits obtaining fluid and open materials through which bulky TAAX cations and solvated alkaline iodides can easily migrate.

When the insertion is done for reaching a stoichiometry of two metal atoms per stored electron, it leads to the salt itself (the cation, for stabilizing the coulombic neutrality inside the bulk matrix) and the cation/anion pair was found to be concomitantly involved and allows impressive swelling of the interface. The large insertion of ions compared to the weak amount of platinum or palladium was expected to lead progressively, in the course of the charge, to the formation of a non-conducting layer, which contributed to stopping the insertion process. Consequently, the thickness of the transformed metal, at the electrolyte interface is weak, certainly not more than a few microns.

In this first approach that concerns only Pt and Pd, electrochemistry now appears to be a suitable tool in the generation, the growing as well as fine studies for application of these new iono-metallic phases. It is quite certain that phases that imply TAAX salts cannot be produced without polarization of the metal. Reactions with bulky ions and possible exchanges (or incorporations) of a large array of π-acceptors or their reduced forms at the metal surface or within a reaction layer close to the interface is certainly rich with possibilities for decorating or functionalizing metals. This is more precisely developed in Chapter 19.

References

[1] J. Vocke, *Chem. Listy* **62** (1968) 497.

[2] J. Vocke, A.M. Kardos, *Coll. Czch. Comm. Commun.* **33** (1968) 2560.

[3] A.J. Bard, L.R. Faulkner, *Electro-chemical Methods. Fundamentals and Applications,* John Wiley & Sons, Inc., New York, Second Edition, (2001).

[3] O.R. Brown, in *Physical Chemistry of Organic Solvent Systems,* ed. A.K. Covington, Plenum Press, New York, p. 747 (1973).

[4] S.D. Ross, M. Finkelstein, R.C. Petersen, *J. Am. Chem. Soc.* **92** (1970) 6003.

[5] G. Braauer, G. Düsing, *Z. Aorg. Allg. Chem.* **328** (1964) 154.

[6] J.D. Littlehailes, B. Woodhall, *J. Chem. Soc. Chem. Comm.* (1967) 665.

[7] L. Horner, H. Neumann, *Chem. Ber.* **98** (1965) 1715.

[8] V. Svetlicic, P.B. Lawin, E. Kariv-Miller, *J. Electroanal. Chem.* **284** (1990) 185.

[9] E. Kariv-Miller, P.D. Christian, V. Svetlicic, *Langmuir* **11** (1995)1817.

[10] E. Kariv-Miller, V. Stetlicic, *J. Chem. Soc., Faraday Trans.* **83** (1987) 1169.

[11] E. Kariv-Miller, P.B. Lawin, *J. Electroanal. Chem.* **247** (1988) 345.

[12] J. Simonet, H. Lund, *J. Electroanal. Chem.* **75** (1977) 719.

[13] J.O. Besenhard, H.P. Fritz, *J. Electroanal. Chem.* **53** (1974) 329.

[14] J.O. Besenhard, E. Theodoridou, H. Moehwald, J.J. Nickl, *Synth. Met.* **4** (1982) 211.

[15] J. Berthelot, M. Jubault, J. Simonet, *J. Chem. Soc. Chem. Commun.* (1982) 759.

[16] E. Zintl, J. Goubeau, W.Z. Dullenkopf, *Z. Phys. Chem. Abst. A* **154** (1931) 1.

[17] D.G. Adolphson, J.D. Corbett, D.J. Merryman, *J. Am. Chem. Soc.* **98** (1976) 7234.

[18] P.A. Edwards, J.D. Corbett, *Inorg. Chem.* **16** (1977) 903.

[19] A. Cisar, J.D. Corbett, D.J. Merryman, *J. Am. Chem. Soc.* **16** (1977) 2482.

[20] JD. Corbett, D.G. Adolphson, P.A. Edwards, *J. Am. Chem. Soc.* **97** (1975) 6267.

[21] O. Hammerich, V.D. Parker, *Electrochimica Acta,* **18** (1973) 537.

[22] J. Heinze, *Angew. Chem.* **96** (1984) 823.

[23] J. Simonet, Y. Astier, C. Dano, *J. Electroanal. Chem.* **451** (1998) 5.
[24] J. Simonet, E. Labaume, J. Rault Berthelot, *Electrochem. Commun.* **1** (1999) 252.
[25] C. Cougnon, J. Simonet, *Electrochem. Commun.* **3** (2001) 2509.
[26] J. Simonet, *Electrochem. Commun.* **402** (2002) 5.
[27] C. Cougnon, Ph.D. thesis Université de Rennes (2002).
[28] C. Dano, Ph.D. thesis, Université de Rennes (1998).
[29] C. Cougnon, J. Simonet, *J. Electroanal. Chem.* **507** (2001) 226.
[30] J. Simonet, *Platin. Met. Rev.* **46** (2002) 180 and unpublished works.
[31] J. Simonet, *Electrochem. Commun.* **5** (2003) 439.
[32] C. Cougnon, J. Simonet, *J. Electroanal. Chem.* **531** (2002)179.
[33] C. Cougnon, J. Simonet, *Electrochem. Commun.* **4** (2002) 266.
[34] C. Cougnon, J. Simonet, *Platin. Met. Rev.* **6** (2002) 9.
[35] J. Simonet, *J. Electroanal. Chem.* **578** (2005) 79.
[36] P. Hapiot, J. Simonet, *Electroanal. Chem.* **23** (2010) 105.
[37] J. Simonet, *Electrochem. Commun.* **53** (2015) 15.
[38] J. Guilane, C. Lagrost, M. Guilloux-Viry, J. Simonet, M. Delamar, C. Mangeney, P. Hapiot, *J. Phys. Chem.* **111** (2007) 5701.
[39] J. Guilane, M. Guilloux-Viry, C. Lagrost, P. Hapiot, J. Simonet, *J. Phys. Chem.* **109** (2005) 14925.
[40] J. Simonet, *Platin. Met. Rev.* **50** (2006) 180.
[41] J.Simonet, *J. Electroanal. Chem.* **93** (2006) 3.
[42] J. Ghilane, M. Guilloux-Viry, C. Lagrost, J. Simonet, P. Hapiot, *J. Am. Chem. Soc.* **129** (2007) 6654.
[43] J. Guilane, Ph.D. thesis, Université de Rennes (2005).
[44] J. Simonet, *J. Electroanal. Chem.* **578** (2005) 79.
[45] J. Simonet, Unpublished observations.

Chapter 19

The Electro-Chemical Insertion of π-Acceptors in Platinum and Palladium: Reactive Organo-Metallic Phases

19.1. Introductory Concept

Most organic chemists interested in electro-chemistry and organo-metallic chemistry (irrespective of the overall conditions concerning solvents and salts) consider metals such as platinum and palladium (and many other precious metals) as totally unreactive toward electrons and ions. Therefore, these metals can play the role of ideal electrodes, free from side-reactions with the surrounding environment. Contrary to this concept, Chapter 18 was principally devoted to the cathodic reactivity of platinum (applied potentials < -2 V vs. Ag/AgCl electrode), which enables the cathodic insertion of several electrolytes in the presence of super-dry polar aprotic solvents.

One of the most important and intriguing observations made in Chapter 18 is actually the formation of weakly conducting iono-metallic layers, covering the metal until a quasi-total inhibition of electron transfer occurs. With platinum, the general formula for the formation of such layers was determined [1–6] to be

$$\{\mathbf{Pt_2^-\ M^+,\ MX}\}$$

with M^+ being TAA^+ or alkaline metal cations (principally Na^+ and Cs^+). Many anions have been employed so far (e.g., BF_4^-, PF_6^-, ClO_4^-, Cl^-, Br^-, etc.). As demonstrated in this reaction, the genesis and the stability of iono-metallic structures depend on the essential participation of an inorganic anion. However, as shown previously, ***inorganic anions can be exchanged with organic anions***, principally those electro-generated at the surface and within the metallic bulk in the course of additional electrolyses. Therefore, the present focus is on the insertion of electro-generated anion radicals and dianions (from organic acceptors) formed at a potential less negative than the threshold of the cathodic step assigned to the formation of the iono-metallic layer, which then leads to the insertion of any salt listed above in order to favor the exchange between an inorganic anion and an organic anion or dianion. Examples given in Rennes [1] were obtained in the course of the mixed charge of platinum in the presence of π-acceptors under super-dry conditions (involving an electrolyte such as $TBABF_4$ or CsI in dimethylformamide (DMF)). As shown in Figure 19.1, the insertion of

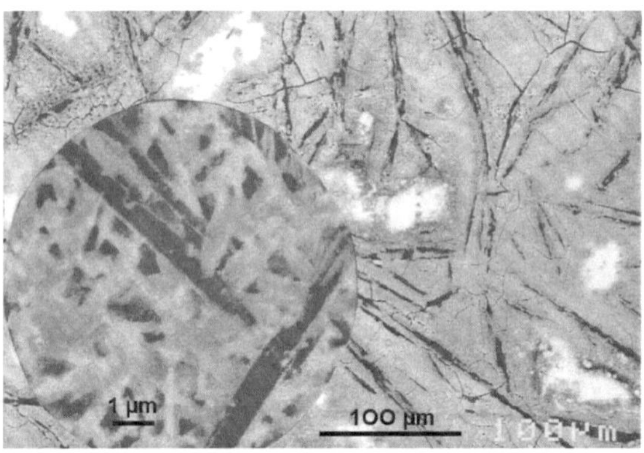

Figure 19.1. Cathodic reduction (and insertion) of acenaphtenoquinone APQ (concentration: 10 mM L^{-1}) in DMF/CsI at −2.0 V (under super-dry conditions), initial current: 1.2 mA cm^{-2}, total amount of electricity: 27 C cm^{-2}. In the inset, another magnification of the phase is shown. The EDX analysis establishes that white crystals are made of pure platinum while black needles point to the presence of carbon and oxygen atoms and are assigned to APQ crystals (oxidation of the transient APQ anion radicals by dioxygen during the work-up) [11].

organic species in their reduced form(s) (all from electro-active entities soluble or in the state of anionic nanoparticles) could be achieved under very simple conditions. A potential of about –2 V is applied to the platinum cathode, and the organic compound (such as a π-acceptor) is chosen to be electro-active within this potential range. Additionally, one may imagine that the irreversible formation of mono-anions — for example, carbanions derived from the cathodic cleavage of carbon–heteroatom bonds — could also lead to such insertion as well as the cathodic doping of organic conducting polymers (for example, electro-generated poly-fluorenes owing to the relative acidity of the fluorene C–H function in the 9-position). Basic forms of such polymeric substances could be obtained simply at the interface of electro-generated bases (EGBs), obtained possibly by the reduction of electroactive impurities that act as base catalysts.

First of all, in the case of platinum, all this appears as a new world within the concept of conducting organometallic surfaces, layers, and/or materials, which can be possibly generated in this way. Figure 19.1 shows that a large diversity of organic compounds become suitable candidates to be embedded in the metal mass (depending on the current density and the total amount of electricity). After electrolysis, the resulting complexes {platinum, ions, reduced forms of an organic acceptor} are oxidized by air in the course of the work-up. The chemical-modified metal surface, now being used as an electrode, may lead to a new process of redox catalysis where the catalyst is not immobilized by conventional methods (e.g., via radical grafting), but by an efficient and finally an easy embedment process. The nature of the whole exchange is summarized in Scheme 19.1.

Alternatively, it is now important to introduce the cathodic capacity of palladium by which reduced acceptors can be inserted by electron transfer, organic anions could be embedded in a certain way, and polymeric anionic forms can also be inserted when they are electro-generated (possibly under the action of EGBs), as depicted in Scheme 19.2.

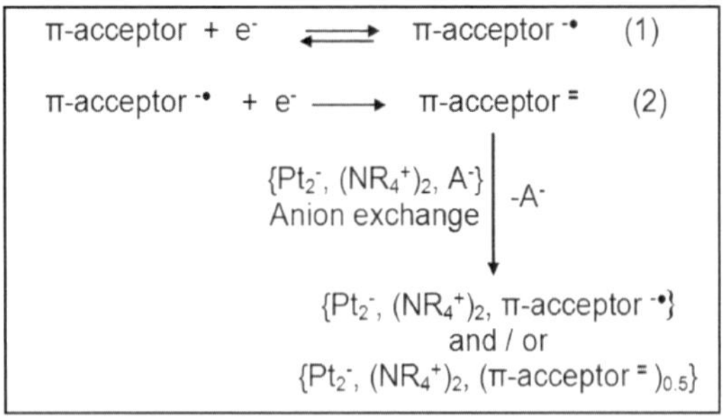

Scheme 19.1. Anion exchange permitting the insertion of π-acceptor reduced forms into the platinum metal mass.

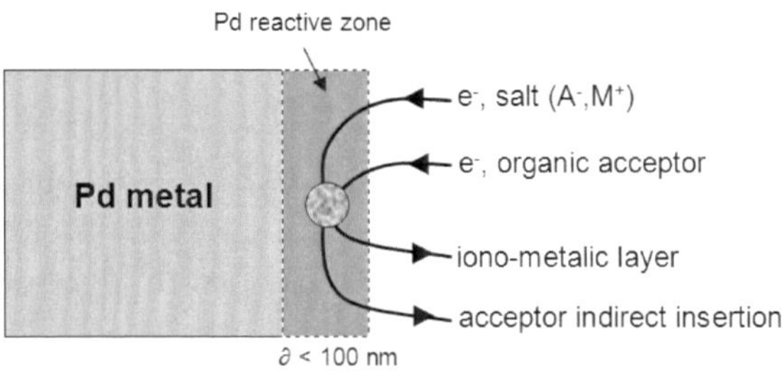

Scheme 19.2. Concomitant charge of palladium (super-dry conditions, electrolyte A^-, M^+) together with the reduction of an organic acceptor [11].

19.2. Reversible Electrical Charge of Palladium in the Presence of Different Salts in Super-Dry Electrolytes

In Chapter 18, devoted to the reduction of platinum and palladium in super-dry conditions, the reversible reduction of Pd (with a total recovery of charges) was highlighted: actually, the whole process involves the electro-chemical charge from its threshold at about −2 V, followed, in the course of the anodic scan, by a sharp typical discharge step under normal conditions, mainly involving DMF or acetonitrile (ACN) as solvents containing a $TAA^+BF_4^-$ salt in super-dry conditions [7–10]. Thus, the amount of electricity injected into the metal was **almost totally** recovered in the course of the discharge. The amount of current (as also the amount of electricity involved)

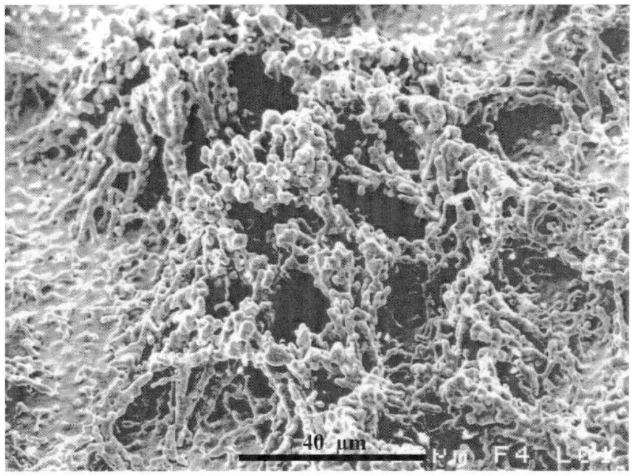

Figure 19.2. Smooth palladium surface modified by a cathodic treatment in DMF + CsI ($E = -2.2$ V, amount of electricity: 25 C cm^{-2}) for achieving a highly rough interface [6].

was observed to be much higher than that observed with platinum under similar experimental conditions. This observation may justify the choice of palladium as an ideal material to construct new matrices based on the insertion of organic anionic forms.

In the present survey that focuses on the insertion of organic molecules of high molecular weight into the metal, the porosity of the metal is certainly an important factor that deserves to be taken into account. Especially with palladium, the pre-treatment of the metal (for instance, reduction in the presence of CsI dissolved in DMF) was found to deeply transform the working surface into a spongy layer as depicted in Figure 19.2. More practically, galvanostatic deposits of palladium onto gold substrates (from a solution of $PdCl_2$ in 0.1 N HCl) also lead to the generation of porous surfaces. Such alternative is advisable since gold does not generate any noticeable cathodic charge in the presence of standard electrolytes (see Chapter 18).

A typical experiment that may demonstrate the insertion of organic anions (in principle, generated *in situ* concomitantly with the cathodic charge of palladium) is depicted in Figure 19.3. It concerns the reversible cathodic charge of Au–Pd interface achieved

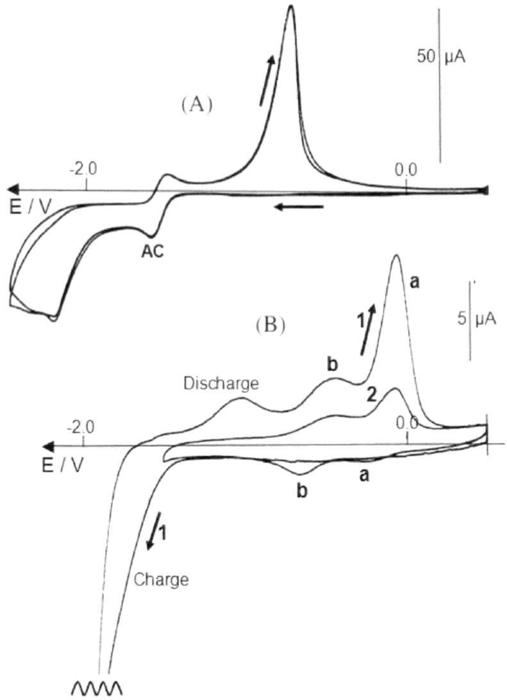

Figure 19.3. Doping of palladium metal deposited onto gold in the presence of a π-acceptor such as acenaphthylene **AC**: concentration 8×10^{-3} mol L^{-1}. Deposition of Pd onto gold by means of a galvanostatic electrolysis. Amount of electricity: 7×10^{-3} C cm^{-2}. Solvent: DMF + $TMeABF_4$. Scan rate: 50 mV s^{-1}. (A) Charge process in the presence of **AC**; (B) after rinsing of the modified surface, evidence of the presence of two reversible steps between +0.7 and −1.2 V (in a solution free of AC) that are assigned to the presence of acenaphthylene or its reduced forms (anion radical and dianion) presumably inserted into the iono-metallic complex generated during the cathodic charge [11].

in the presence of acenaphthylene in DMF (electrolyte $TMeABF_4$). In the course of scans displayed in Figure 19.3A, one may distinguish the specific charge of the palladium deposit ($E < -2$ V) and its discharge evident through a sharp anodic peak (≈ −0.8 V). At this electrode material, the reversible first reduction peak of acenaphthylene (−1.6 V, marked AC) is also observed. Scans until a positive potential range (until +0.8 V) do not generate any oxidation step.

A potentiostatic electrolysis of the solution (involving both Pd + AC) is then achieved at −2.6 V (until an amount of 7×10^{-3} C of electricity passed through the cell). It has to be stressed that a highly reducing potential (−2.6 V) is necessary to involve the AC dianion in the insertion process. Under these conditions, after rinsing with water and then acetone and drying using a hot air stream, the electrode is tested in a dry DMF–$TMeABF_4$ solution. The starting potential is fixed at −0.7 V as depicted in curve B. Two reversible steps **a** and **b**, located at quite positive potential values, are obtained and then they disappear in the course of recurrent scans: one additionally notices the presence of the partial charge and discharge of the electrode. After several scans, the total extinction of steps **a** and **b** was observed while a new doping of the electrode beyond −2 V leads to a regeneration of reversible steps **a** and **b**. It is estimated that during the progressive reduction in the course of scans of iono-metallic forms, the associated reduced forms of AC are progressively obtained along the cathodic scans. Importantly, a new doping phase of palladium enables regeneration of steps **a** and **b**, which fully supports the doping of the organic acceptor in the Pd metallic bulk (Figure 19.3).

Such a feature is obtained with many π-acceptors (for example, triphenylene, dinitrobenzene, anthraquinone (AQ), chloranile, bibenzyle, diacethylbenzene, etc.). Astonishingly, peak potentials assigned to the presence of acceptor-reduced forms (considered now as donors toward iono-metallic palladium phase) are all significantly different and shifted to a more positive potential range (+0.2 to +1.0 V). It is expected that the first step **a** is assigned to the anodic response of the organic radical anion fully inserted as a donor species into the metallic phase while step **b** corresponds to the presence of the dianion in the Pd bulk. The impressive potential shifts would result in an excess of donor species in the palladium complex. In the course of scans toward negative potentials, peak **b** is found to be larger, while in the course of the reverse scan, peak **a** (associated with the presence of the anion radical) turns out to be fairly huge, which supports a prior dianion oxidation. Significantly, the increase of steps **a** and **b** when redoping along a short scan until the Pd charge threshold is reached may easily regenerate reducing centers

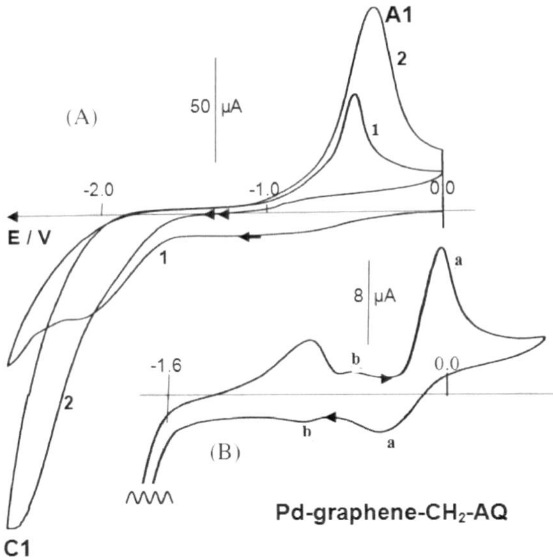

Figure 19.9. Complimentary to Figure 19.8, graphene is now decorated with AQ preferably fixed onto graphene flakes by radical addition of AQ–CH_2–Br in the presence of a catalyst (Chapter 10). Of note is the weak cathodic current at $E < -1$ V in the course of the first scan (reduction of AQ) and small current at the C1 step compared that at the second scan. After a multi-scan voltammetry (starting at −1.8 V), signals of AQ reduced forms appear in B as displayed in Scheme 19.1 and which are compared with the signal shown in Figure 19.6 with AQ alone [11].

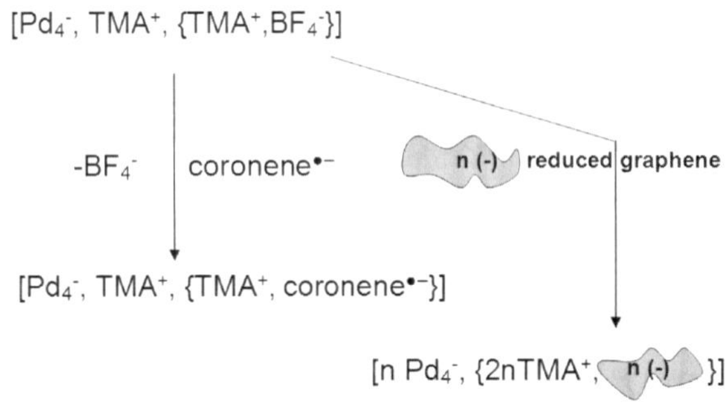

Scheme 19.3. The reactivity of electro-generated anionic species from coronene (case of the anion radical) and that of deposited graphene flakes (implication of poly-anion obtained by a multi-electron transfer). The apparent involvement of those organic anions in the palladium iono-metallic complexes is quite reasonable.

19.4. Super-Oxide as a Donor toward Charged Palladium and Platinum

It is also worth mentioning the *in situ* reactivity of electro-generated super-oxide with charged palladium and platinum. Briefly, as shown in Figure 19.10, the reduction of dioxygen in DMF mixed with $TMeA^+$ salts leads to a fast inhibition of the electrode surface, which forbids the charge of the metal and partially the reduction of O_2, which turns out to become totally irreversible. An anodic treatment of the electrode up to +1.0 V allows regeneration of the electrode with an evolvement of dioxygen (Scheme 19.4) [11].

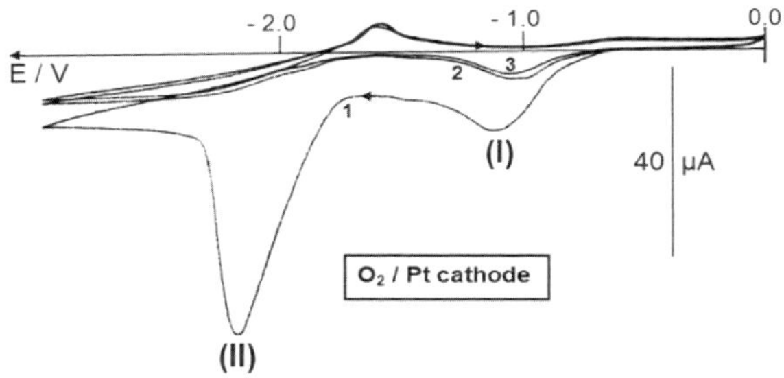

Figure 19.10. Voltammetries of a saturated solution of di-oxygen in DMF/$TMeABF_4$ at a smooth Pt electrode. Surface area: 0.8 mm^2. Scan rate: 100 mV s^{-1}. In the course of the first scan, O_2 is reduced according the irreversible step (I), while platinum is reduced (possibly partially modified by electro-generated super-oxide) according the large peak (II). This reduction step disappears from the second scan and the dioxygen step almost totally vanishes. An anodic scan (not shown) within the anodic domain (until + 1.0 V) allows oxidation of a supposed Pt/O_2 complex for totally restoring the scan 1 [11].

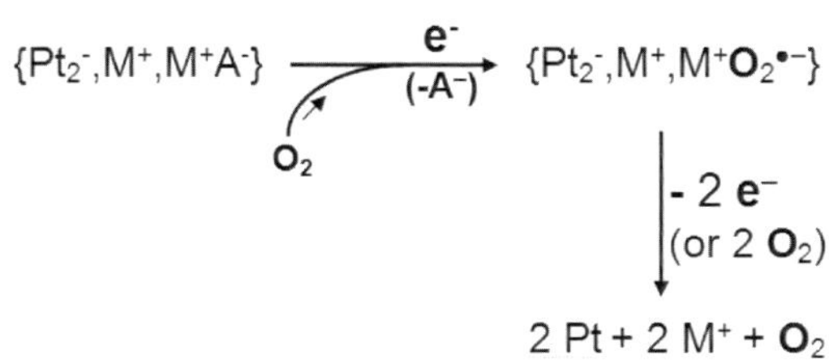

Scheme 19.4. Reaction of the platinum iono-metallic phase with electro-generated super-oxide. The possible reaction with dioxygen dianion is not mentioned [11].

19.5. Additional Comments and Concluding Remarks

Finally, this short presentation shows an unexpected mode of insertion of reduced and/or polyanionic forms, possibly electro-generated in metals such as palladium and platinum matrices. The method is, for the moment, exclusively based on the exchange of anion of the electrolyte involved during the cathodic doping process of the concerned metal. Importantly, the insertion is conditioned to a preliminary treatment that implies the n-doping of the metal under specific unusual conditions (exclusive use of super-dry electrolytes). Preliminary experiments emphasize that the insertion of inorganic and organic anions is feasible as well. A schematic form of the insertion of an external layer in palladium is shown in Scheme 19.5. Organic anions can be anion radicals, dianions, and poly-anions generated from the reduction of very large π-acceptors or they could be generated from the doped organic structures derived from the n-doping of organic conducting polymers. Therefore, using such specific recipes, generating new polyvalent materials may be possible.

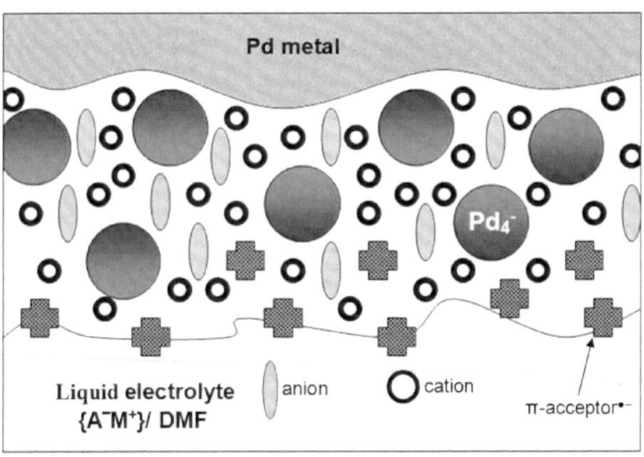

Scheme 19.5. Schematic view of the modified Pd layer {Pd_4^-, $2M^+$, acceptor$\bullet^-$} after a partial insertion of a reduced form of a π-acceptor in the original iono-metallic coating [11].

Until now, ion exchange has remained the mode of insertion, which is principally achieved via mixed reduction processes (e.g., doping of the metal concomitantly with the reduction of the acceptor). Generally, an intermediate phase is generated at quite negative potentials and it is immediately oxidized by air during the work-up. For example, this is the case displayed in Figure 19.1 with acenaphthenoquinone inserted in platinum. Quite unexpectedly, the intermediate phase is not obtained when the same acceptor is reduced at palladium (Figure 19.4), in which case a deep blue paramagnetic zone (apparently connected to the acceptor anion radical) appears, with E° of the order of 0 V vs. Ag/AgCl, which is highly different from that of the starting acceptor, which was definitively much more negative. It is important to underline that the behavior of the immobilized species is strongly influenced by the surrounding species, which forms the iono-metallic layer.

Lastly, in this introductory series, only exchanges implying anions were considered. One could quite easily conceive ion permutations that would imply cations of the electrolyte (e.g., Cs^+, Na^+, K^+, or $TMeA^+$) with electro-chemically generated organic cation radicals or simply cations (e.g., generated by an *in situ* non-Kolbe reaction) with the condition that the anodic range (that leads to the discharge) is essentially taken into account.

References

[1] J. Simonet, *Electrochem. Commun.* **53** (2015) 15.
[2] J. Simonet, Y. Astier, C. Dano, *J. Electroanal. Chem.* **451** (1998) 5.
[3] J. Simonet, E. Labaume, J. Rault Berthelot, *Electrochem. Commun.* **1** (1999) 252.
[4] C. Cougnon, J. Simonet, *Electrochem. Commun.* **3** (2001) 2509.
[5] J. Simonet, *Electrochem. Commun.* **402** (2002) 5.
[6] C. Cougnon, Ph.D. thesis, Université de Rennes 1 (2002).
[7] J. Simonet, *Electrochem. Commun.* **5** (2003) 439.
[8] J. Simonet, *Platin. Met. Rev.* **46** (2002) 180 and unpublished works.
[9] C. Cougnon, J. Simonet, *J. Electroanal. Chem.* **531** (2002) 179.

[10] P. Hapiot, J. Simonet, *Electroanal. Chem.* **23** (2010) 105.
[11] J. Simonet, Unpublished Results. To be noticed that the cathodic behavior of dioxygene looks very similar to that of CO_2 as developed in Chapter 20. The idea could be the development of a 3D superoxide platinum or palladium electrode.
[12] J. Simonet, *Electrochem. Commun.* (2017) in press.

Chapter 20

Cathodic Carboxylation of Metals in Thick $\{M–CO_2\}_n$ Layers: Electro-Chemical Sequestration of CO_2

20.1. Introduction

The electro-chemical behavior of carbon dioxide and its reactivity in electro-chemistry (valorization via certain reduction processes, modes of sequestration, environmental considerations, modification of surfaces, catalysis) has emerged as a significant challenge. Carbon dioxide is an almost unavoidable side-effect of human activity. In this book, the use of CO_2 was essentially focused on its electrophilicity to modify, under cathodic activation, several allotropic forms of carbons [1] like graphite, glassy carbon (GC) and graphene (see Chapter 11). Its valorization via electro-chemical processes at different catalysts under protic solutions has been extensively documented [2–13], and was therefore not a primary objective here.

Numerous contributions based on the reduction of CO_2 into methanol, carbon monoxide, methane and salts such as oxalates, etc. could be a partial solution to decrease waste, which is unfortunately still increasing in quality. Despite this, most process costs concerning degradation of CO_2 appear unable to solve this problem even partially.

Presented hereafter, through a very elementary electrochemical process, is a new development which would permit the reduction and/or the reversible sequestration of CO_2 into current transition metals and concomitantly creating carboxylated layers at the surface of these metals (ease of functionalization, anti-corrosion gain). Although these studies still remain under investigation today [14], many metals appear to be affected by such carboxylation processes. This chapter, essentially a preview, is focused on a few metals such as gold, palladium, platinum and copper. Other materials such as iron, rhodium, silver, titanium, cobalt or nickel appear to show promising results, but these are not investigated in-depth in this preliminary presentation. Additionally, since cathodic gas insertion corresponds to a chemically reversible process, it could be that the present approach generates interest for the development of new accumulators using a low-cost raw material which, when dissolved in an electrolyte is not consumed by the process.

The catalytic conversion of carbon dioxide into renewable fuel chemicals (CO_2 saturated at room temperature in aqueous solutions or in organic polar solvents mixed with proton donors) was achieved generally at smooth transition metal electrodes and with wide variety of catalysts which unfortunately did not have high selectivity [2, 3]. A totally different approach concerns experiments that must specifically be conducted in ***dry polar aprotic solvents*** and containing ***exclusively*** tetramethylammonium $TMeA^+$ salts (essentially associated with tetrafluoroborate and perchlorate). Iodides, bromides, chlorides and triflate could be used as well, but a part of the present study concerns the behavior of the electrode within the anodic range, and it appears, for the moment, judicious to leave them aside. Moreover, the necessary use of $TMeA^+$ salts in the present study could be highly justified, since they remain among the only agents that do not trigger the Hofmann degradation (source of proton in basic media) and consequently do not allow, in the course of the charge process, the protonation of the CO_2 anion radical, which could be considered the principal intermediate in the electrochemical process in aprotic conditions.

This short presentation, firmly limited in volume, does not include a long list of transition metals, which are potential candidates for the insertion of CO_2 by cathodic means under the conditions briefly underlined above. This preliminary presentation focuses on the evidence of insertion of carbon dioxide essentially into precious metals and copper.

20.2. Reactivity of Carbon Dioxide at a Negatively Polarized Gold Electrode

The catalytic capacities of gold are quite developed in organic chemistry, [15] but much less so in electro-chemistry, and it remains less attractive than platinum as the current electrode material. It has been essentially used as a tool for fixing thiolates [16]. Let us quote also the reduction of aryl-diazonium salts on gold, which results in a primary immobilization layer further functionalized via click chemistry [18]. A catalytic electro-chemical reduction of iodoalkanes onto gold via a preliminary adsorption was recently disclosed and represents a powerful way to form stable monolayers on this metal [19].

The first evidence of the sequestration of CO_2 in gold [20] was obtained simply at GC electrodes galvanostatically capped with a thin layer of gold. Using the following explicit conditions, very sharp reduction steps were observed at about −1.5 V vs. Ag/AgCl reference electrode with saturated carbon dioxide solution in dimethylformamide (DMF) (0.2 mol L^{-1} at room temperature [21]). Additionally, with small alkaline metal cations (Li^+, Na^+ or K^+) as well as with a $TMeA^+$ cation, an unexpected progressive and fast electrode inhibition occurs upon several scans or after a short fixed-potential electrolysis at $E < -1.7$ V. This phenomenon could be assigned to the insertion of CO_2 into gold, allowing the formation of a thick iono-metallic multi-strata layer (less conducting than pure metal), which grows in the course of the electrode charge. This carboxylated interface is presumably inherent to the three following factors: (i) transient formation and stabilized CO_2 anion radical, (ii) the metal particle (of large interface, and possibly acting as catalyst) and (iii) the small sized of the involved cations ($M^+ = Na^+$ or $TMeA^+$) considered more or less electrophilic. Those three entities associated in a nanometric volume are expected to produce a transient form that can be written as $\{Au–CO_2^-,M^+\}$, which should be reactive toward di-oxygen, moisture and some organic π-acceptors (reducing properties). Using now a smooth gold fold, multi-scans up to −2.2 V, progressively generate the formation of a weakly conducting layer that may reach values superior to 10^{-7} mol cm^{-2} based on the bulkiness of the $TMeA^+$ cation inserted within the required volume. Such hypothetical multi-layers undergo a decomposition in the anodic domain at about +1.7 V for releasing CO_2 trapped earlier in Au. Coulometric analyses demonstrated that insertion (cathodic) and release (anodic) steps are almost perfectly equivalent, which allows us to consider the sequestration process of carbon dioxide as (almost totally) chemically reversible.

Now, more practically, the cathodic reduction of CO_2 at smooth gold can be depicted in Figure 20.1. Thus, the first scan exhibits a very sharp cathodic peak that rapidly decays upon successive scans. The peak does not totally disappear irrespective of the number of scans, suggesting then that the material modified by CO_2

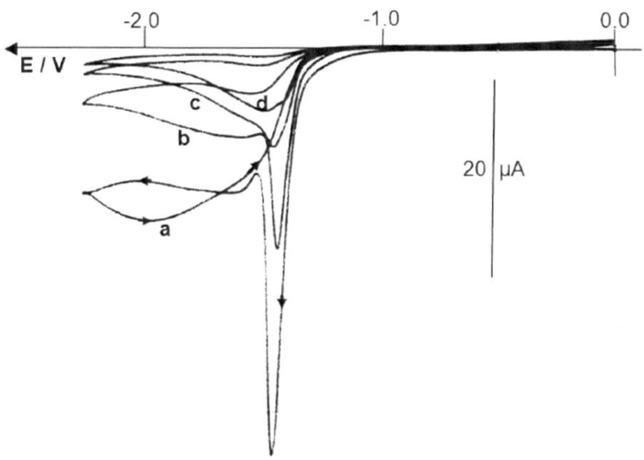

Figure 20.1. Repetitive voltammetries at smooth gold electrode, of a saturated solution (room temperature) of CO_2 in DMF containing 0.1 M $TMeAClO_4$. Scan rate: 100 mV s^{-1} [20].

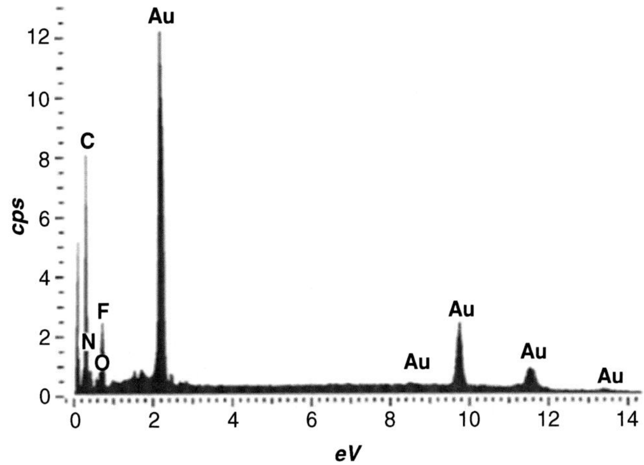

Figure 20.2. EDX analysis of the deposit created by the reduction of CO_2 at a gold pristine in the presence of DMF/$TMeABF_4$ as the electrolyte.

(concomitantly by the salt cation) remains weakly conductive after the cathodic reaction insertion. The EDX spectra (Figure 20.2) of modified samples confirms the presence of carbon and oxygen at the interface. Fourier transform infrared spectroscopy (FTIR) spectra also strongly support the sequestration of carbon dioxide and the carboxylation of the surface (Figure 20.3). Lastly, SEMs relative to the morphologic modification of the surface after polarization of the gold sample in the presence of CO_2 underline the formation of large flakes and/or swelling zones (Figure 20.4), which fully support an insertion process.

As a matter of fact, experimental results obtained during the reduction of CO_2 at gold surfaces appear remarkable. It was already known that other TAA^+ salts used as an electrolyte under the same conditions do not allow us to reproduce results displayed with the $TMeA^+$ cation. Spectacularly sharp reduction peaks followed by strongly inhibiting processes suggest the formation of a poorly conducting organo-metallic species first at the interface and progressively growing inside the metal bulk. More specifically, as featured in Scheme 20.1, the electro-chemical technique may provide a very easy way to form an intermediate in a reaction of Au and CO_2 (tentatively

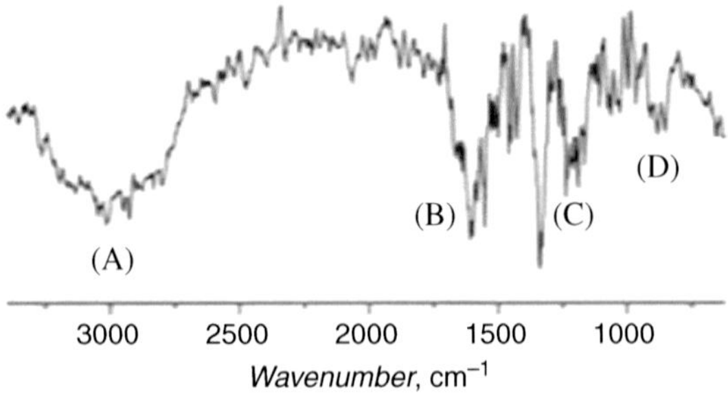

Figure 20.3. FTIR of the deposit onto the pristine gold. (A) ν_{OH}, ν_{CH}, 3200–2600 cm^{-1}; (B) $\nu_{C=O}$, 1700–1600 cm^{-1}; (C) δ_{C-O}, δ_{CH}, 1440–1230 cm^{-1}; (D) δ_{OH}, 930 cm^{-1} [20].

Figure 20.4. A SEM view for a gold pristine sample after cathodic carboxylation (−2.0 V, DMF, TMeAClO$_4$) and then reoxidation by air [14].

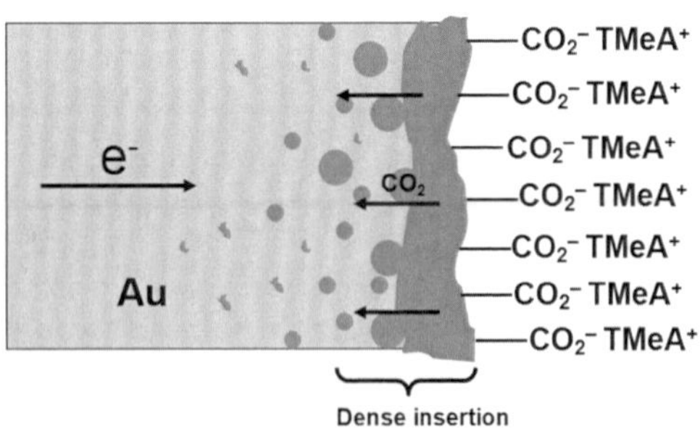

Scheme 20.1. Process of cathodic carboxylation of gold (superficial and internal) in the presence of tetramethylammonium cations (TMeA$^+$) [20].

written as {Au–CO_2–M} or {Au–CO_2^-, M^+}, with M^+ representing the cation of the electrolyte).

20.2.1. *Comments on electro-chemically driven insertion/desorption of CO_2*

At a smooth gold electrode, the solution of CO_2/DMF/TMeA$^+$ does not show any oxidation stages up to +2.0 V. On the contrary, a gold electrode, preliminarily carboxylated between −1.1 and −1.9 V under conditions shown in Figure 20.1 (e.g., three consecutive scans up to −2.2 V) triggers a reactive insertion of CO_2 up to a superficial concentration (estimated to be 4.5×10^{-7} mol cm^{-2}) on the basis of an one-electron process (catalytic formation of CO_2 anion radical), displaying a quite different response shown in Figure 20.5 (anodic branch of the voltammogram) with the presence of a large irreversible step (**II**) at E_p = +1.76 V. A neat pre-peak (**III**) is tentatively assigned to the oxidation of the superficial carboxylated layer. Current integration of peak (**II**), attributable to the emission of CO_2 through the anodic scission of the Au–CO_2 bond, looks like a "metallic" non-Kolbe reaction still being, presumably, one-electron bond scission, suggests that the total amount of CO_2 present in the multi-layer reaches 4.0×10^{-7} mol cm^{-2}. This value is astonishingly close to that obtained during total a cathodic charge (that permits an almost complete passivating of the electrode) at the level of step (**I**) and should confirm that the chemical reduction of the gas does not occur.

Several experiments relative to gold electrodes devoted to CO_2 charge–discharge processes do confirm a rather good parity between cathodic and anodic coulometric values. Furthermore, anodic discharge permits us to check the emission of CO_2 by means of a gold electrode used as an analytical tool. Thus, the discharge of a carboxylated gold plate (≈1 cm^2) near a smooth gold electrode permits us to partly collect and then demonstrate the local evolvement of carbon dioxide under the conditions of the formation of step (**I**) described above.

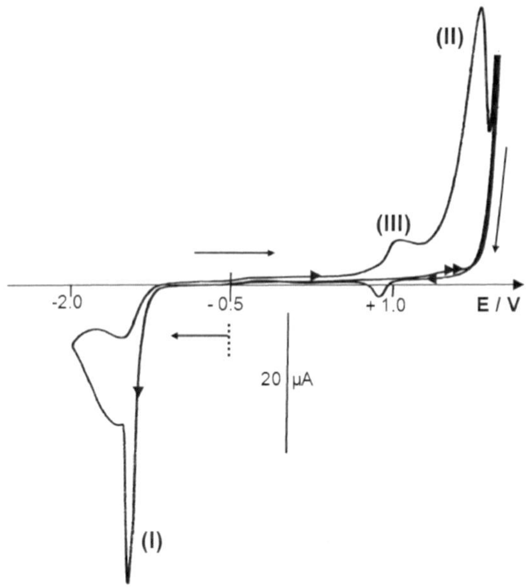

Figure 20.5. Complete view of a charge-carboxylation (step **I**)/discharge-decarboxylation (step **II**) process of a smooth Au electrode (area: 0.8 mm^2) in the presence of a saturated solution of CO_2 in DMF/$TMeAClO_4$. Scan rate: 50 mV s^{-1}. The first scan starts from −0.5 V toward cathodic potential (−2.0 V) to be followed by an anodic exploration. Note that a second anodic scan is drawn and underlines the total absence of oxidation steps until solvent oxidation. Reference electrode: Ag/AgCl/KCl_{sat} [20].

20.2.2. *Additional remarks on the reversible sequestration of CO_2 by gold*

SEM images of Au interfaces after reduction (Step (**I**) support a superficial and profound change of the pristine gold sample (Figure 20.4) with apparent swelling especially at the electrolyte surface where the pressure of CO_2 is the largest. In a way, the morphology of the surface looks compatible with insertion of CO_2 (presumably via the formation of {Au–CO_2^-,M^+} phase) followed by desorption (in the course of rinsing and the oxidation by air) and displays — at least with Au — wavy deposits totally different from the original pristine morphology. Moreover, EDX and FTIR analyses achieved on sonicated samples well support the existence of carboxylation. Lastly, Nyquist plots obtained from such samples indicate a large increase of the interface impedance [20] even after sample reoxidation by air, which could support a remanence possibly due to a partly irreversible insertion of $TMeA^+$ cations.

Some conclusive remarks should insist on the possibility of charging gold electro-chemically by means of CO_2. Carboxylation processes appear to take place both at the interface of the metal–electrolyte and in mass (Scheme 20.1) with the formation of a thick layer of {Au–CO_2-cation} covering the external face of the metal sample. It is expected that the process is supposed to rest on the specific formation and the reactivity of a carbon dioxide anion radical leading to a weakly conducting salt, which is slowly stocked in the immediate vicinity of the metal interface. Importantly, the reaction is found to be chemically reversible and the iono-metallic carboxylate is anodically reactive to permit the gas release in a large anodic step.

Hereafter, the insertion reaction of carbon dioxide is extended to some other current transition metals and it was found that this process is almost general under experimental conditions comparable to those described above with gold.

20.3. Carboxylation of Silver

Silver is considered as a fascinating metal for its catalytic properties, especially in the domain of the scission of carbon–halogen bonds [22, 23]. Despite its capacities to act as a reducing entity, attempts to modify surfaces or insert ions (as developed in Chapter 18 with platinum and palladium) inside Ag matrixes, were not achieved so far.

Research into silver for possible capacity to insert small electroactive molecules that can act as vectors to diffuse into the metal under their reduced form(s), imply the paired insertion of the electrolyte. Consequently, insertion of carbon dioxide varies because its reduction potential depends on the electrode material and the electrolyte. Moreover, as described in this chapter, the reactivity of CO_2 is high both as electrophilic species or, at the stage of the anion radical, may depend on the applied potential for carrying out any reaction at the electrode. Thus, as already mentioned in this chapter, it exhibits a strong capability to react with cathode materials like GC, graphite and graphene at potentials where its own reduction does not interfere.

Very recently [24], it was found that the paired reduction/insertion of CO_2 at smooth silver surfaces is obtained specifically in the presence of certain tetraalkylammonium (TAAX) salts in polar aprotic solvents. The great novelty of this approach is that a reduced form of CO_2, presumably its anion radical, occurs at the metal at −1.6 V vs. Ag/AgCl. Thus, similar to gold, the reduction of CO_2 undergoes a catalytic process depending on the metal together with a large inhibition process observed ***only*** in the presence of $TMeA^+$ cations. This fact is clearly evidenced by Figure 20.6, which shows a strong catalytic effect followed by a sudden inhibition process induced presumably by formation (or insertion) of a fairly stable organometallic intermediate, which can be written as {Ag–CO_2–TMeA}. Other more bulky TAA^+ cations (tetra-*n*-butyl and tetra-*n*-hexylammonium cations) do not provoke a similar inhibition process (still in Figure 20.6) at

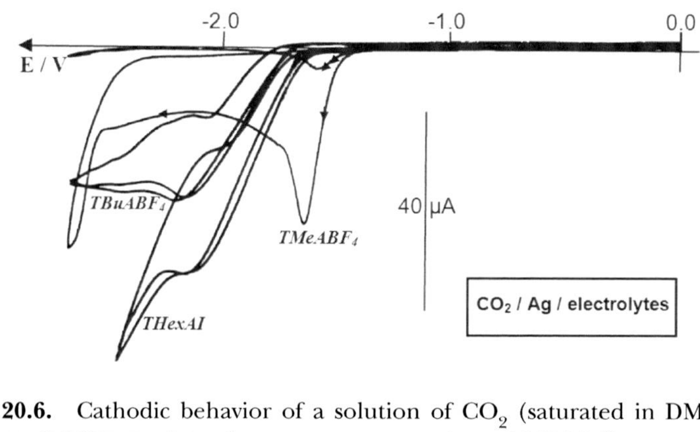

Figure 20.6. Cathodic behavior of a solution of CO_2 (saturated in DMF in the presence of different electrolytes at a concentration of 0.1M L^{-1}) at a smooth Ag electrode (surface area: 0.8 mm^2). Scan rate: 50 mV s^{-1}. Through all the processes used electrolytes, two very highly different processes are displayed. With $TMeA^+$, a sharp peak is obtained at the first scan, while a quasi-total extinction (in the course of the second scan) of the reduction process (almost no current up to −2.8 V) appears, presumably provoked by a sudden lack of the conductivity o wing to the paired insertion of CO_2 and $TMeA^+$ into silver. In contrast, the use of $TBuA^+$ and $THexA^+$ constantly displays a slow reduction process, which does not change within successive potentials sweeps [14].

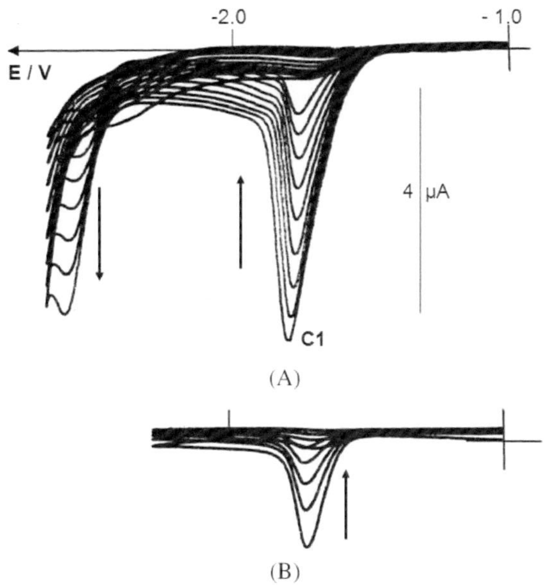

Figure 20.7. Voltammmetric reduction of a solution (in A) of CO_2 (saturated in DMF, super-dry conditions, electrolyte $TMeABF_4$) at a smooth Ag electrode (area: 0.8 mm^2). Scan rate: 50 mV s^{-1}. The first cathodic peak **C1** gradually decreases upon scans followed by the merging of a new step at extremely negative potential ($E < -2.7$ V) assigned to the reduction of the inhibitor $\{Ag\text{–}CO_2^- TMeA^+\}$ generated at the level of **C1**. On the contrary, while DMF solution is made free of CO_2 by an efficient bubbling of argon (10 min at 0 V), it appears that the electrode bulk is still filled by a remnant presence of CO_2, presumably formed by the slow reduction of $Ag\text{–}CO_2\text{–}TMeA^+$. Note that the reduction of carbon dioxide is progressive and probably limited by the diffusion of the gas (or/and that of the ammonium cation) contained in the metal or within a thick layer at the surface. The total current integration from all curves in (B) reaches 10^{-7} mol cm^{-2}. It can be shown that the internal reduction of CO_2 allows recreating the organometallic species inside Ag. Additionally, curves (B) are almost adsorption-like quite differently from those displayed in (A), suggesting that CO_2 at least remains present in the metal [14].

all and generate reduction steps, which could be assigned to the reduction of CO_2 in the presence of those cations.

Voltammetries, shown in Figure 20.7(A), point to the exclusive presence of $TMeA^+$, and the progressive current decay of the main reduction step in the course of recurrent voltammetries suggests a growing inhibition process, possibly provoked in time by a thick layer of {Ag–CO_2–M} of weak conductivity. The capped electrode — after reduction — maintained at 0 V under a flow of argon permits us to completely remove the CO_2 atmosphere inside the cell. However, the silver electrode reacts with the remnant CO_2 still inside the metal (Figure 20.7(B)), which strongly supports the Ag matrix containing quite large amounts of unmoving carbon dioxide, but totally absent from the liquid electrolyte.

Lastly, it is of importance to underline the organo-metallic Ag–CO_2^- intermediate is anodically destroyed and liberates oxidized forms of silver, which have so far not been identified [24]. SEM investigations also demonstrate the large perturbation at the surface of silver (Figure 20.8 and Scheme 20.2) where amorphous or crystalline deposits are seen.

Figure 20.8. Surface modification of silver after cathodic carboxylation in DMF/ TMeABF$_4$. Applied potential: −1.7 V vs. Ag/AgCl. Charge: 1.1 C cm^{-2}. Use of a saturated solution of CO_2 in DMF [14].

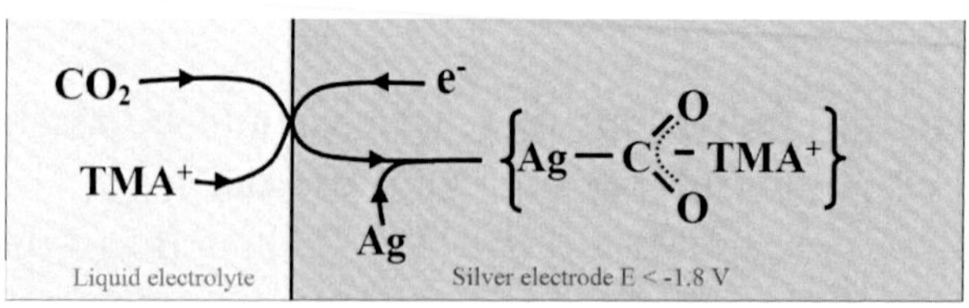

Scheme 20.2. Carboxylation of silver by cathodic means [24].

20.4. Carboxylation of Palladium, Platinum and Rhodium

20.4.1. *Palladium*

The reactivity of Pd, Pt and Rh toward carbon dioxide appears a little different from the results given above with gold and silver. The fundamental discrepancy would lie in the reduction step assigned to the sequestration of CO_2 (step **C1**), which is a totally separate part of the charge process (in **C2**, as fully described in Chapter 18) involving these metals. A typical charge–discharge process is displayed in Figure 20.9 at the palladium surface.

The carboxylation of the palladium surface may lead to large differences in the matter of boundary conditions between a smooth palladium electrode cyclized in DMF + $TBuABF_4$ (absence of CO_2) and the modified surface with a gap > 0.35 V (Figure 20.10).

20.4.2. *Platinum*

It is clear that the behavior of platinum toward CO_2 is very close to that of palladium. Recall that both metals (named M) may be charged in the presence of salts C^+, A^- (Chapter 18) to lead to iono-metallic structures of general stoichiometry:

$$[M_n^- C^+, \{C^+ A^-\}] \text{ with } n - 2 \text{ or } 4 \text{ in the case of Pt and Pd, respectively.}$$

The reduction of CO_2 is clearly different from the metal charge: the insertion of carbon dioxide may obey a type of exchange between the iono-metallic form given above, in which the reducing power was demonstrated in Chapter 18. A mechanism based on the exchange of the inorganic anion, according to the process displayed below in the specific example of platinum (Scheme 20.3), is observed.

Thus, in the presence of a π-acceptor, an electron transfer from the iono-organic form (now carboxylated) could occur and leads in de-aerated conditions (like argon atmosphere) to the π-acceptor anion radical (with $C^+ = TMA^+$) as associated cation (Figure 20.11).

20.5. Carboxylation of Copper

Copper plates dipped first in diluted sulphuric acid and then rinsed with water and dried at 60°C may be easily carboxylated. Large amounts of CO_2 are probably inserted into the metal, but the easy oxidation of Cu at $E \geq 0.5$ V prevents any reliable anodic quantification.

Under these conditions, the copper surface carboxylation could be successfully achieved (Figure 20.14). Quite probably, the amount of CO_2 inserted in Cu metal is roughly in the same proportion as that already obtained with other metals (see above with gold or platinum). Surface carboxylation coverage indirectly obtained with amino-acceptors (like 1-aminoanthraquinone) or with 2-bromomethylanthraquinone shows the capping of a rather thick active layer of $3 \pm 2 \times 10^{-9}$ mol cm^{-2}. Additionally, carboxylation procedures lead to strongly hydrophilic copper surfaces (Figure 20.15).

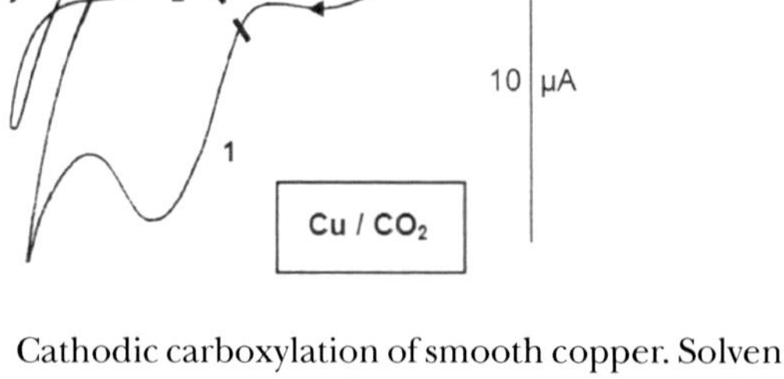

Figure 20.14. Cathodic carboxylation of smooth copper. Solvent: DMF + $TMeABF_4$. Cu electrode surface area: 1.8 mm^2. Scan rate: 100 mV s^{-1}. Visualization of the six first cathodic sweeps that provoke an almost total current vanishing up to −2.2 V. On the right part of the figure, is drawn the beginning of the first anodic branch (after cathodic reduction) that displays a strong inhibition of Cu oxidation [14].

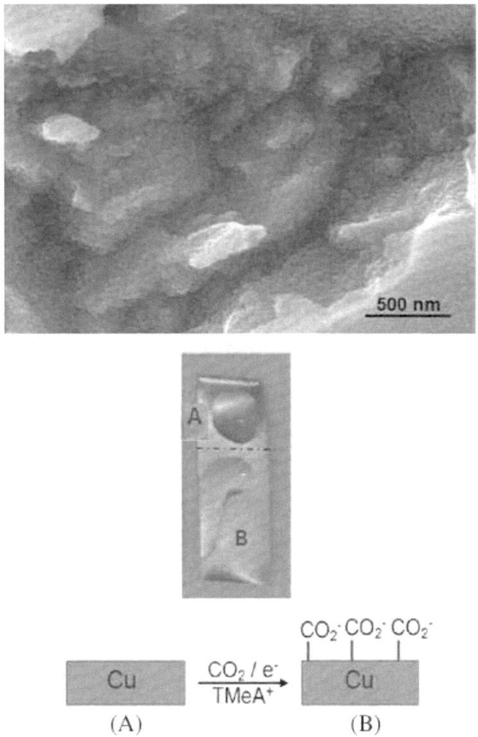

Figure 20.15. Chemical modification of copper pristine surfaces under cathodic charge in the presence of CO_2. First, the huge digging of the surface revealed by SEM that stresses the efficiency of CO_2 to insert and burst the material. Moreover, the water ability of the copper surface is underlined below: (A) Water droplets deposited on smooth copper. (B) Bottom part: evidence for a large wetting ability by opposition with the water drop in contact with the untreated Cu zone [14, 17].

20.6. Conclusive Remarks About the Reductive Insertion of Small Electro-active Molecules into Transition Metals

In this chapter, a few examples devoted to the insertion of CO_2 are described with metals that are, in principle, among the most used in current electro-chemistry as electrode materials. Thus, iron and copper exhibit spectacular anodic corrosion inhibitions. Quite attractive results were obtained with titanium [14]. However, zinc or cadmium, cobalt, nickel, etc. have not so far led to any investigation in this field.

We have not developed the insertion process of di-oxygen, particularly in Pt, Pd or Au in this book [14]. We predict that a long series of small electro-active molecules are candidates for behaving quite similarly to CO_2 under the conditions presented above. In particular, SO_2, already reported to behave as electrophilic species with reduced forms of acceptors, should be tested. Also NO_2, NO, N_2O and even CO (for a carbonylation of several metals) could be envisaged. Moreover, small organic electroactive molecules (formaldehyde, acetone) are presently tested at platinum and palladium electrodes [14].

Lastly, CO_2, seen as a "natural" resource, cheap, and in large excess, could be used as a new class of metal–CO_2 gas rechargeable batteries with large thin electrodes made of a low-cost metal. The huge potential difference between charge and discharge (almost 3 V) could direct research in this direction. Let us also recall that alkaline metal salts (Li^+, Na^+, Cs^+ or K^+, as sulphates, tetrafluoroborates or hexafluorophosphates) can be used as well.

References

[1] J. Simonet, *Electrochem. Commun.* **21** (2012) 22.

[2] D. Pletcher, *Electrochem. Commun.* **61** (2015) 97.

[3] I. Ganesh, *Renewable and Sustainable Energies Reviews* **59** (2016) 1269.

[3] J.H.Q. Lee, S.J.L. Lauw, R.D. Webster, *Electrochem. Commun.* **64** (2016) 69.

[4] G.O. Larrazabal, A. J. Martin, S. Mitchell, R. Hauert, J.P. Ramiorez, *J. Cat.* (2016), in press.

[5] W. Lv, R. Zhang, P. Gao, L. Lei, *J. Pow. Sources* **253** (2014) 276.

[6] W. Lv, J. Zhou, F. Kong, H. Fang, W. Wang, *Int. J. Hydrogen Energy* **41** (2015) 1585.

[7] J. Yuan, P. Wang, C. Hao, G. Yu, *Electrochim. Acta* **193** (2016) 1.

[8] F. Zhou, S. Liu, B. Yang, P. Wang, A.S. Alshammari, Y. Deng, *Electrochem. Commun.* **46** (2014) 103.

[9] F. Zhou, S. Liu, B. Yang, P. Wang, A. S. Alshammari, Y. Deng, *Elecrochem. Commun.* **46** (2014) 103.
[10] Q. Wang, H. Dong, H. Yu, H. Yu, *J. Pow. Sources* **279** (2015) 1.
[11] O. Scialdone, A. Galia, G.L. Nero, F. Proietto, *Electrochim. Acta* **199** (2016) 332.
[12] M. Alvarrez-Guerra, A. Del Castillo, A. Irabien, *Chem. Eng. Res.* **92** (2014) 622.
[13] C.L Zhao, J. Wang, J.B. Goodenough, *Electrochem. Commun.* **65** (2016) 9.
[14] J. Simonet, unpublished observations.
[15] G.C. Bond, C. Louis, D.T. Tompson, *Catalysis by Gold* (Imperial College Press London, 2006).
[16] M. Frasconi, A. D'Annibale, C. Favero, F. Mazzei, A.M. Santucci, *J. Am. Chem. Soc.* **112** (1990) 4301.
[17] J. Simonet, *Electrochem. Commun.* (2017), in press.
[18] J. Lher, B.E. Williamson, B.S. Flavel, A. Downard, *Langmuir* **25** (2009) 13503.
[19] V. Jouikov, J. Simonet, *Electrochem. Commun.* **56** (2015) 20.
[20] V. Jouikov, J. Simonet, *Electrochem. Commun.* **59** (2015) 40.
[21] M. Jitaru, *J. Univ. Chem. Tecnol. Metall.* **42** (2007) 333.
[22] A.A. Isse, B. Giacomo, L. Falciola, R. Rossi, P.R. Mussini, *J. Appl. Electrochem.* **39** (2009) 2217.
[23] A.A. Isse, A. De Guisti, A. Gennaro, L. Falcola, P.R. Mussini, *Electrochim. Acta* **51** (2006) 4956.
[24] J. Simonet, *Electrochem. Commun.* **58** (2015) 11.
[25] S. Wawzonek, R. Berkey, E.W. Blaha, E.M. Runner, *J. Electrochem. Soc.* **102** (1955) 235.
[26] S. Wawzonek, D. Wearring, *J. Am. Chem. Soc.* **81** (1959) 2067.

alloys, 2, 18
allylic, 103
allyl-trimethylsilane specific reactivity, 250
amalgams, 11
ammonium salts, 183
anchoring of redox polymers, 151
anhydrous solvents, 7
anion radical, 3, 5
anodic coupling of anisole, 308
anodic deposition of allylic groups at carbon, 254
anodic oxidation of carbon polycarboxylate, 204
anodic polymerization of dibenzocrown ethers, 315
anodic trimerization of benzo 15-crown 5 at a redox cell, 312
Anodic Trimerization of Orthodimethoxy Benzenes, 309
appropriate organic electrophiles, 135
aprotic polar solvent, 4
Ar–Ar linkage generated by electro-chemistry, 123
Ar–Ar linkages, 125
Ar-CH2-Br, 191
arenization of glassy carbon, 6
aryl bromides, 127, 130
aryl halides, 3
aryl iodides, 126
aryl radicals, 70, 190
as a free radical trap, 151
attached graphene–alkyl link to, 151
attachment of acethylenic moieties at gold, 262
attachment of (CH_2)nCOOH groups, 240
attachment of Graphene at GC Surfaces, 149
attachment of graphene flakes using CH_2I anchors, 260
attachment of pyridines, 327
attachment processes of graphene layers, 174
Au–Ag NP, 99
Au–graphene interface, 297
Au–graphene interface as a spin-trap, 302
Au–graphene working electrodes, 296
average amount of the inserted TAA+ cations, 189
average level of grafting, 223

B

1,1'-binaphthyl, 128
1-bromo-6-chlorohexane, 236
1-bromobutane, 42, 44
1-bromodecane, 60
1-bromooctane, 42–44
9-bromomethyl anthracene, 191
9-bromomethylpyrene, 191
9-bromophenanthrene, 131
Bauld, 11
Bechgaard, 20
benzylic, 103
benzyl radicals onto graphene, 295
benzyltrimethylsilane (BTMS): oxidation at carbon anodes, 255
Bernier, 16
bielectrophilic species, 135

bi-electrophilic species like ICnFcCnI, 148
Bipy solid electrodes, 19
bis-(ω-iodoalkyl)ferrocenes, 136
bromo-cyclopentane, 235
building versatile surfaces, 260
bulk polymer/ferrocene/carbon/solvent/salt, 148
Bunnett, 11

C

(copper–DB3) complex, 169
C60, 15
capacities of ArXs to form dimers, 125
cap GC with very thin deposits of graphene, 176
capping, 2
capping carbon surfaces by Cu–Pd layers, 62
carbon activation by silver, 272
carbon as chargeable material, 230
carbon electrode functional capacities, 200
carbonaceous materials, 16
carbonaceous surfaces, 2
carbon carboxylation, 198
carbon–halogen bonds, 29
carbons, 2
carboxylation, 18
carboxylation in surfaces, 25
carboxylation of carbons, 197–198
carboxylation of copper, 404
carboxylation of deposited graphene, 217
carboxylation of graphene, 177
carboxylation of palladium, platinum and rhodium, 399
carboxylation of silver, 395
carboxylation of silver by cathodic means, 398
catalytic cleavage of 9-bromofluorene, 117
catalytic conversion by iodide ions, 132
catalytic cyclodimerization, 13
catalytic properties, 156
catalytic reactions
at solid conductor interfaces, 89
catalytic reactivity, 35
catalytic scission of carbon–halogen linkages, 51
catalytic scission of chemical bonds, 156
catalytic uses, 151
cathode material, 29
cathode materials (copper, nickel, silver, palladium, 125
cathodic attachment of pyridines onto solid surfaces, 328
cathodic carboxylation of graphenes, 215
cathodic carboxylation of Metals, 387
cathodic charge by insertion, 182
cathodic charge of platinum, 22
cathodic charge of platinum and palladium, 345
cathodic charge of varieties of graphite, 175
cathodic charge process, 186
cathodic corrosion induced by free alkyl radicals, 267

cathodic generation of Ar–Ar bonds, 126
cathodic reactivity of carbons, 199
cathodic reactivity of functionalized electrophiles F–RXs, 237
cathodic reduction of CO_2 at smooth gold, 390
cations migration through graphene, 185
C–Br bond cleavage, 65
Charge at Ag–graphene electrode, 292
charge-carboxylation/discharge-decarboxylation processes, 394
charged fine carbon particles, 136
charged graphene layer as an efficient polynucleophile, 187
charge–discharge of palladium–graphene layers, 380
charge-discharge processes of graphite, 174
charge of GC and swelling, 228
charge process in the presence of TAAX salts, 352
charging graphene with different cations TAA+, 183
chemical carboxylation of GC, 202
chemical doping of silver by palladium, 31
chemically grafted redox polymer at carbon, 136
chemical modification of fine silver powder by palladium, 91
chronopotentiometry of an HOPG sample, 207
cleavage of the C–I bond, 126
cleavages of carbon–halogen bonds at Cu–Pd electrodes, 56
CO_2, 2
CO_2 reactivity at polarized Au, 389
CO2 reduction at gold, 391
cobalt, 72
columnar, 21
commercial silver fine powder, 101
complexing properties by multicrown-ethers, 319
composite, 191
composite cathode materials, 81
composite GC-graphene material, 177
concomitant insertion of Ag–Pd particles and RX, 279
concomitant insertion of electrons and salts, 19
conducting organic polymers, 4
conducting (or redox) polymer, 136
conducting polymers, 177
conducting solid surfaces, 2
constant-potential reduction, 175
convenient electrode material, 175
conventional glassy carbon electrode, 57
copper, 2, 18, 72, 98
copper: a remarkable cathode material, 164
copper deposits onto electrified surfaces, 169–170
copper grids, 54
copper–palladium alloys, 10, 52, 65
coronene, 8, 23

coronene charge at palladium, 380
corrosion of Cu micro-crystallites, 168
corrosion through electro-generated radicals, 277
coupling reaction, 5
coverage levels, 228
coverage of GC by secondary alkyl groups, 235
coverage of palladium onto platinum, 161
coverages of metal NPs onto carbon, 161
creating graphene-overlaid GC interfaces, 194
creation of new composite and reactive materials, 174
creation of polynucleophilic centers, 176
cross-coupling reactions, 18
crown ethers, 21
crystallized nanoparticles, 55
Cu^0 NPs deposits, 167
CuPd alloys, 64
Cu/Pd interface, 52
cyclodimer, 12

D

1,3-dibromopropane, 155, 157
2,2′-dipyridine, 19
α,ω-dibromoalkanes, 141, 167
α,ω-di-electrophilic, 138
α,ω-dihaloalkanes (I–R–I), 135, 142, 284
α,ω-diiodoalkanes, 138, 140
DB18C6, 20
decorating unsaturated solid materials, 175
decoration of carbon surfaces, 238
dendritic-like growth, 55
deposited galvanostatically onto GC, 126
deposited ionophoric redox materials, 307
deposited onto inactive materials, 155
deposition of charged graphene, 189
deposition of graphene layers, 173
deposition of NPs at carbon electrodes, 159
deposition of NPs of metals like Au, Pt, Pd, Rh, and Cu, 156
deposition of palladium and gold NPs onto graphite and GC, 163
deposit of silver, 101
deprotection processes, 52
deuteriation reactions, 14
Diels–Alder reactions, 11–12
different oxidation states (from T_0 to T_4), 318
diiodoalkanes, 142
dimer yields, 47
dimethylformamide (DMF), 5
direct exfoliation of natural graphite, 194
disaggregation of graphene layers, 177
disordered copper–palladium alloy, 51
disproportionation, 4, 9
dissolved metal, 11
DMSO, 12
doped by transition metals, 79

doped graphene, 175
double cleavage/immobilization with a TMS–R–X compound, 263
dry medium, 7

E

3D electrode, 91
easy deposition of nanoparticles NPs, 155
effective reactivity of graphite with CO_2, 209
effective reactivity of reduced p-acceptors, 284
efficient alkyl-ferrocenyl capping of polarized GC, 240
efficient methods of grafting various functional groups, 70
efficient redox probes, 136
electrically charged micro or nanozones, 242
electrified solid substrates, 177
electro-activity of graphite, 205
electro-catalysis by transition metals, 77
electro-catalytical properties, 155
electro-catalytic process, 48
electro-catalytic properties, 70
electro-catalytic reduction of alkyl bromides, 155
electro-chemical alkylation, 3
electro-chemical catalysis, 1, 90
electro-chemical deposition of $Fc–C_{12}^{\bullet}$, 78
electro-chemical functionalization, 177
electro-chemical GC surfaces digging, 278
electro-chemical grafting, 3
electro-chemical insertion of π-acceptors in platinum and palladium, 371
electro-chemically prepared "graphite salts", 175
electro-chemical modification of carbons, 197
electro-chemical oxidation of carbons, 89
electro-chemical reactivity of 3-dibromopropane at copper, 1, 165
electro-chemical reactivity of DB3 toward metals, 170
electro-chemical sequestration of CO_2, 387
electro-chemical splitting of graphitic materials, 177
electro-chemical tailoring of many solid conductors, 178
electro-chemistry of organo-trimethylsilanes (R–TMS), 264
electrode materials, 140, 197
electro-deposition of a large series of metals, 156
electro-deposition of functionalized polymers, 89
electrodes metal-Ag NP, 99
electrodes of two-dimensional and three-dimensional configurations, 90
electro-generated bases, 12
electro-generated Ni0 catalyst, 19
electro-generated nucleophiles, 4
electro-generated reactive carbanions (nucleophilic species), 69

electro-generated species, 4
electro-generated super-oxide, 21
electron diffraction pattern, 56
electron transfers, 2
electrophilic properties, 4
electrophilic reagents, 175
electrophilic species, 137
EPR experiments, 118
EPR spectra from the reduction of FluHBr in PC/THF, 119
exfoliation, 17, 173, 177
exfoliation of natural graphite at a gold substrate, 296
exfoliation of natural graphite by ammonium salts, 175
exfoliation procedure, 178
exfoliation processes, 182
exfoliation process for a graphite deposit onto gold achieved, 298
exfoliation process for a thick graphite deposit onto a gold disk, 299

F

9-Fluorenyl radical, 115
9-fluorenyl units on the GC-graphene, 191
Fc–C6–I, 74
ferrocene, 228
ferrocene distribution into matrixes, 146
ferrocene layers, 136
ferrocene surfacial concentrations, 136
ferrocene-terminated self-assembled SAM mono-layers, 146
ferrocene units, 136
fine Ag–Pd powder, 94
fine carbon (GC, graphite, or graphene) powder, 148
fine suspension of silver–palladium alloy grains, 91
fixed-potential electrolyses, 174
fluidized bed electrodes, 90
fluidized bed electrolysis, 101
fluidized bed processes, 177
fluorenation of solid surfaces, 116
fluorene grafting onto metals, 120
fluorenyl bromides, 103
formation of –(FluH2)n–chains, 120
formation of nucleophilic material, 200
free alkyl radicals, 104
free radicals, 3
free radicals addition to graphene, 293
from 4-(7-iodoheptyl), 333
fry, 3
functionalization, 173
functionalization of GC, 236

G

galvanostatic deposit, 90
galvanostatic deposit of Pd, 73
GC and platinum surfaces can be readily copperized, 170
GC–C6–ferrocene, 85
GC doped with palladium, 72
GC electrode modified with graphene, 215
GC–graphene electrodes, 82
GC-graphene materials, 182
GC–graphite electrodes, 79
GC-graphite interfaces, 184

GC modified through TMS–$(CH_2)_6$–I, 258
GC network, 151
GC/Pd electrode, 73
GC polycarboxylates, 202
GC, silver, silver–palladium, and copper, 140
GC using Ni, Cu, Rh, and Ag as catalysts, 77
generate a palladized copper alloy, 53
generate branched redox polymers at interfaces, 136
generating propargyl radical ($Pg^{\bullet}$), 105
generation of benzyl radicals, 113
generation of bipyridine layers, 338
generation of metal NPs at a polarized surface, 159
generation of poly-electrides, 321
generation of propargyl radical, 109
generation of pyridine electrodes, 330
generation of radicals, 200
glassy carbon, 4
gold, 17–18, 71, 98
gold electrode after graphite exfoliation, 302
gold–graphene interface, 300
gold–graphene interface in TEA–DMF solution, 303
grafting, 2
grafting of dodecylferrocene to GC, 80
grafting of iodo-alkyl chains –(CH2)n–I, 8
grafting of some Fc–Cn–I, 76
grafting onto active electrode substrates, 104
grafting onto electrode surfaces, 69
grafting using pyridines, 329
graphene, 2, 151
graphene, 173
graphene layer is doped with Pd and Au nanoparticles, 150
graphene-like surface, 174
graphene modification by redox groups, 178
graphene nano-flakes, 17
graphene nanosheets, 173
graphene network, 192
graphene network associated with silver, 294
graphene oxide, 173, 177
graphene planes or flakes, 174
graphene reduction, 184
graphene species can be immobilized onto GC surfaces, 150
graphene suspensions, 177
graphite, 173
graphite-carbon interfaces, 177
graphite carboxylation, 191
graphite doped by gold, 81
graphite exfoliation, 174
graphite exfoliation in ionic liquids, 177
graphite foils, 173
graphite nanoinclusions in GC, 178
graphite nanoparticles, 176

H

helicene, graphene, 8
heterogeneous, 2
heterogeneous electron transfer, 6

heterogeneous primary electron transfer, 4
heterogeneous redox catalysis, 6
high complexity of GC, 175
high dielectric constant ε, 157
highly catalytic solid electrodes, 30
highly dissociating organic solvents, 157
highly oriented pyrolytic graphite, 14, 200
highly oriented pyrolytic graphite (HOPG), 175
highly versatile electrode coverages, 136
high polarization of C–Br bonds in PC, 158
homo-coupling, 9
homo-coupling reactions, 3, 18
homogeneous, 2
homogeneous redox catalysis, 4
how render graphene poly-nucleophilic, 222
H transfers, 9
hydrophobic polymeric layers at solid surfaces, 146
hydroxylation, 18

I

1-iodohexadecane, 39–40
1-iodonaphthalene, 126
1-iodopropylferrocene, 77
π,π-interactions, 174
ω-iodoalkylferrocenes, 142
ω-iodoalkylferrocenes grafting at GC, 77
ω-iodoalkylferrocenes used as a probe, 139
immobilization of activated polyaromatic molecules, 191
immobilization of alkanoid acid chains, 240
immobilization of allyl radical, 105
immobilization of a pure vinyl layer, 239
immobilization of benzyl groups at carbon anodes, 255
immobilization of FluH• radicals, 120
immobilization of free radicals at solid surfaces, 103
immobilization of functionalized alkyl chains, 227
immobilization of IC10FcC10I, 145
immobilization of organic bielectrophilic species, 135
immobilization of pyridine, 333
immobilization of RXs at carbon, 256
immobilization of tetrathiafulvalenes, 242
immobilized ferrocene, 139
incorporation of redox, 146
indirect reduction of C–X bonds, 141
industrial graphene, 14
in matrices of porous organic polymers, 101
insertion, 2
insertion of electrolytes into platinum and palladium, 341
insertion of organic or inorganic cations, 175
insertion of p-acceptors, 8
insertion of TAA+ cations, 174

insertion of TBA+ cations into graphitized micro-zones, 232
in situ electro-generated radicals, 177
in situ generation of ArIs, 132
in situ produced free radicals, 175
in-situ reactivity of electrophilic agents, 201
in suspension, 101
intercalation, 173
interfacial processes, 1
interfacial reactions, 1
intrinsically catalytic electrodes, 10
iodoalkyl moieties grafted onto carbon, 257
ion diffusion rate through the carbon sheets, 174
iono-metallic layer, 22
ionophoric aromatic polyethers, 313
ionophoric species, 20
iron, 18, 98

K

Kornblum, 11
Koutecky's theory, 3

L

large energetic gains, 70
large potential shifts, 52
large-scale carboxylation of graphene, 220
layer, 98
layer of radical ions, 7
links between the GC surface itself and a graphene layer, 150
localized radical ablation on graphite, 282
location of functionalization, 193
Lund, 3, 65

M

1-methyl-naphthalene charge at palladium, 379
macro Ag–Pd cathodes, 46
macro-electrolyses, 47
materials such as felts, GC, graphites, 197
mechanical dispersion of the Ag–Pd powder, 97
mechanism of the immobilization, 228
mediator (Ag–Pd grains), 101
mediators, 4
mercury, 17
metal oxidative insertion into C–Br, 118
metalizing procedures, 98
metallic particles, 2
metals (gold, silver, rhodium, copper, platinum, and palladium, 177
micro-crystals at a polarized anode, 309
micro-graphitized and fullerenized [9] parts, 200
mixed injection of electrons and cations, 174
mixed reductions of deposited graphite, 149
modification of GC by graphene, 150
modification of GC via fractals, 279
modifications, 89
modifications of GC and graphite surfaces, 138

modified solid 3D electrodes, 8
moieties, 146
molecular electrode, 4
mono-dentate ligands, 19
mono-electronic, 65
mono-electrophilic molecules, 135
motion of electrons and cations, 21
multi-columnar materials, 20
multi-layer reactive systems, 16

N

nano-inclusions of graphite, 180
nanometer resolution, 54
nanostructural morphology of a copper deposit, 168
naphthalene, 3
naphthalene-M+, 15
natural graphite, 174, 205
new electrode materials, 148
new type of graphene electrode, 151
nickel, 72
nitriles, 228
non-Kolbe reaction, 204
novel class of reducing agents, 176
novelty, 2
NP dispersion onto a polarized carbon electrode, 158
nucleophilicity of the electrode material, 72

O

–O–O– bond, 18
O_2, 2
obtaining modified solid interfaces, 177
of pure silver nanoparticles, 98
one-electron scission of carbon–halogen bonds, 47
one-electron scissions, 3
one-electron step, 44
organic carbonates, 123
organic electrophilic reagents, 174
organic or inorganic mediators, 4
organometallic intermediates, 155
organometallic redox polymer, 146
organometallic spheres in suspension, 158
oxidation of allyltrimethylsilane (ATMS) at carbon anodes, 251
oxidation of dibenzocrown ethers, 317
oxidative insertion of metals, 125
oxidative insertion of the metal M inside C–Br bonds, 155

P

p-acceptors, 2
palladiated silver surfaces, 44
palladium, 2, 17–18, 71, 399
palladium catalysis, 73
palladium electrodes, 10
palladium is deposited, 161
palladium–R4NX reduced phase stoichiometries, 366
palladization, 29
palladized copper interface, 54
palladized silver, 29
palladized silver interfaces, 35
palladized surfaces, 30
parker, 20
particles, 155
PC or PC–EC mixtures, 130
$Pd^{(0)}$ catalysts, 18

Pd(II)-based salts, 64
peroxidate groups, 18
photochemical induction, 11
platinum, 2, 17, 71, 399
platinum charge in super-dry conditions, 346
platinum NP deposition onto graphite foils, 162
polar aprotic organic solvents, 174–175
polarized graphite as a trapping material for alkyl radicals, 281
polarography, 17
poly-crown-ether, 22
poly(DB18C6), 21
poly-electride, 21–22
poly-graphene, 191
poly-graphenic (possibly organometallic) structure, 151
polymeric matrix, 120
polymerization of fluorene, 120
polynucleophilic entities, 136
poly-nucleophilic materials, 16
polynucleophilic reagent, 176
poly-nucleophilic species, 18
polynucleophilic zones, 175
possible applications, 2
powerfully activating organic RBr compounds, 103
precious metals, 2
primary alkyl iodides, 35, 39
procedure for a carboxylation of natural graphite, 210
process, 107
process of cathodic carboxylation of gold, 392
producing modified electrodes, 52
propargylate anion, 104
propargylic, 103
propargylic derivatives, 105
propargylic pillar, 264
propargyl radical, 104
propylene carbonate (PC), 155
propyne units grafting, 260
pyrene, 5
pyridine, 333
{pyridine-bipy-gold} interface generation, 331
pyridine electrode, 19

Q

quasi-reversible system observed with GC, 232

R

radical addition (AQ–$CH_2^\bullet$), occurring at GC-graphene, 193
radical addition to inactive organic species, 69
radical $ICH_2^\bullet$, 259
radical reactivity toward GC, 272
reaction of charged GC with bromobutene, 239
reaction with carbon dioxide, 197
reactivity of ArX, 125
reactivity of DB3 at copper surfaces, 165
reactivity of palladium polarized interfaces, 130
redox catalyses, 4
redox catalysis, 2–3
redox catalysis by the deposited graphene, 194

redox marked radicals, 69
redox probes, 136
reduced GC as a macro-nucleophile, 201
reducing and catalytic materials, 156
reducing or oxidizing reagents, 4
reduction of aryl diazonium salts, 190
reduction of aryl halides, 123
reduction of carbon dioxide, 397
reduction of electrophilic layers, 8
reduction of marked alkyl iodides, 82
reduction of onium salts such as diazonium, 200
reduction of the adsorbed RI onto Au, 82
reductive attachment, 136
reductive de-bromination, 254
reductive homo-coupling, 40
reference electrode: SCE, 5
remnant suspension of Ag–Pd in the electrolyte, 97
reticulated by attached fluorenyl moieties, 120
rhodium, 2, 18, 71
R• radicals, 3
Russel, 11
RX immobilization via trimethylsilyl group, 256

S

SAM formation on graphene plans, 84
Savéant, 3, 11–12
schematic representation of the exfoliation process, 179
scission of carbon–heteroatom bonds, 52
scission of C–Br bonds, 59
selectivity, 1
self-elaborated 3D interfaces, 83
sequestration of carbon dioxide, 24
sequestration of CO_2, 399
sequestration of CO_2 in gold, 389
sequestration of organic acceptors, 24
several metallic supports (gold, platinum, palladium, iron, copper, nickel, 91
short linker, 3-diiodopropane, 1, 150
silver, 18
silver electrodes, 10, 95
silver–graphene and gold–graphene electrodes, 287
silvering different solid conductors, 98
silvering procedure, 98
silvering procedure of solid conductors, 99
silver nanoparticles, 89
silver–palladium, 10
silver–palladium alloy, 47
silver–palladium at onto solid conductors, 44
simple galvanostatic electrolyses, 12
small molecule reductive insertion, 405
small size nanoparticles, 18
smooth Cu or Ag plates, 128

smooth gold electrode, 82
smooth palladium, 30
smooth silver, 41
SN2 displacement occurring via the charged graphene, 303
SN2 reaction, 5
SNR1 type reaction, 12
SO2, 2
solid conductor grains, 90
solid conductors (glassy carbon (GC), silver, gold), 174
solid electrodes, 1
solid electronic conductors, 90
soluble radicalophilic molecules, 8
solvated electron, 11
solvents of high permittivity, 155
source of nanoparticles, 155
sources of free radicals, 70
species, 138
specific deposit of a catalyst, 52
specific palladium alloys, 3
spectrum of single-layer graphene (1LG), 300
Stability of the ionic deposition, 186
stable modified composite solid, 174
stacking, 21
stacks of a few graphene layers, 184
standard potentials, 44
standard potential values, 3
step, 107
strong activation process, 57
strong electrode catalytic effects, 127
strongly adsorbed C–Pd–X linkage, 65
strongly polarized carbon-heteroatom bond, 176
subnanomolar metal layers, 69
super-dry solvents, 52
superficial carboxylation, 204
superficial formation of a polyanionic graphene, 189
super-oxide donor toward palladium, 383
surface chemical activation, 189
surface functionalizations, 135
surface images after electrolysis, 280
surfaces or suspensions, 157
swelling of the carbon interface, 228
synthetic graphite, 191

T

TAA-amalgams, 176
TAAX amalgams, 201
tailoring of composite materials, 156
tailoring of surfaces, 52
tertiary alkyl iodides, 38
tetrathiafulvalenes, 228
texture and structure analyses, 54
texture for copper-palladium interfaces, 54
thermodynamic data, 3
TMS–CH_2–I as versatile reactive, 262
thick layer of {Au-CO_2-cation}, 395
thin layers of graphene, 177
three-dimensional (3D) electrodes, 80
three-layer graphene (3LG), 300
titanium, 18

TMS+ as a leaving group, 256
TMS grafting at graphite, 251
to solid electrode surfaces, 89
total conversion of an ArBr in ArI, 131
total material degradation via its exfoliation, 83
transient radical FluH•, 118
transition metals, 2–3, 17
transition metals as catalysts, 174
tremendous level of functionalization, 191
tridimensional {graphene-GC} electrodes, 177
triggering of radical chains, 11
trimerization of 2-dimethoxybenzene, 1, 309
trimethylsilyl derivatives oxidation, 250
trimethylsilyl (TMS) moieties deposition, 247
triphenylene dications, 318
triphenylene stackings [T]n, 318
triphenylphosphine, 19
two-electron cathodic bond cleavages, 52
two-step grafting via (i) charging GC carbon, 233
typical activation of a GC cathode, 198
typical charge-discharge process, 399

U

Ullmann's reaction, 124, 128
unsaturated inclusions, 175
Utley, 13

V

various forms (rods, plates, microelectrodes), 175
versatile reagents such as TMS–CH2–I, 263
versatile surface modifications, 194
vinylation of carbon surfaces, 238
vinyl pyridines, 13
voltammetric response of natural graphite, 206
voltammetries, 4
volumic electrodes, 128, 149

W

well-organized {carbon-cations} materials, 175
wettability of the surface after carboxylation, 219

CATALYTIC SCIENCE SERIES

(Continued from page ii)

Vol. 6 *Catalysis by Gold*
edited by G. C. Bond, C. Louis and D. T. Thompson

Vol. 5 *Supported Metals in Catalysis*
edited by J. A. Anderson and M. F. García

Vol. 4 *Isotopes in Heterogeneous Catalysis*
edited by Justin S. J. Hargreaves, S. David Jackson and Geoff Webb

Vol. 3 *Zeolites for Cleaner Technologies*
edited by Michel Guisnet and Jean-Pierre Gilson

Vol. 2 *Catalysis by Ceria and Related Materials*
edited by Alessandro Trovarelli

Vol. 1 *Environmental Catalysis*
edited by F. J. J. G. Janssen and R. A. van Santen